THE PHYSICS OF THE COSMIC MICROWAVE BACKGROUND

Spectacular observational breakthroughs by recent experiments, and particularly the WMAP satellite, have heralded a new epoch of CMB science 40 years after its original discovery.

Taking a physical approach, the authors probe the problem of the 'darkness' of the Universe: the origin and evolution of dark energy and matter in the cosmos. Starting with the observational background of modern cosmology, they provide an up-to-date and accessible review of this fascinating yet complex subject. Topics discussed include the kinetics of the electromagnetic radiation in the Universe, the ionization history of cosmic plasmas, the origin of primordial perturbations in light of the inflation paradigm, and the formation of anisotropy and polarization of the CMB.

This timely and accessible review will be valuable to advanced students and researchers in cosmology. The text highlights the progress made by recent experiments, including the WMAP satellite, and looks ahead to future CMB experiments.

PAVEL NASELSKY is a research scientist and associate professor at the Niels Bohr Institute and at the Rostov State University, Russia. He has written over 100 papers on CMB physics and cosmology, and has taught an advanced course on 'Anisotropy and polarization of the CMB'. He is a member of the ESA technical working group of the PLANCK project.

DMITRY NOVIKOV is an astronomer and research associate at the Astrophysics Group of Imperial College London and also a research scientist at the Astro Space Center of the P. N. Lebedev Physics Institute, Moscow. His main research interests and publications are in cosmology and astrophysics.

IGOR NOVIKOV is a professor at Copenhagen University and was Director of the Theoretical Astrophysics Center prior to its transfer to the Niels Bohr Institute. He is also a research scientist at the Astro Space Center of the P. N. Lebedev Physics Institute, Moscow. His main research has been on gravitation, physics and astrophysics of black holes, cosmology and physics of the CMB. He has been actively involved in the theory of the anisotropy of the CMB and development of the theory with applications to the observations from space- and ground-based telescopes.

Cambridge Astrophysics Series

Series editors

Andrew King, Douglas Lin, Stephen Maran, Jim Pringle and Martin Ward

THE PHYSICS OF THE COSMIC MICROWAVE BACKGROUND

PAVEL D. NASELSKY

Niels Bohr Institute, Copenhagen and the Rostov State University

DMITRY I. NOVIKOV

Imperial College London and the P. N. Lebedev Physics Institute, Moscow

IGOR D. NOVIKOV

Niels Bohr Institute, Copenhagen and the P. N. Lebedev Physics Institute, Moscow

Translated by Nina Iskandarian and Vitaly Kisin

CAMBRIDGE UNIVERSITY PRESS
Cambridge, New York, Melbourne, Madrid, Cape Town,
Singapore, São Paulo, Delhi, Tokyo, Mexico City

Cambridge University Press
The Edinburgh Building, Cambridge CB2 8RU, UK

Published in the United States of America by Cambridge University Press, New York

www.cambridge.org
Information on this title: www.cambridge.org/9781107403123

First published 2006
First paperback edition 2011

A catalogue record for this publication is available from the British Library

ISBN 978-0-521-85550-1 Hardback
ISBN 978-1-107-40312-3 Paperback

The evolution of the Universe can be compared to a display of fireworks that has just ended: some few wisps, ashes and smoke. Standing on a well-chilled cinder, we see the slow fading of the suns, and try to recall the vanished brilliance of the origin of the worlds.

Abbé George-Henri Lemaître, the late 1920s

Contents

Preface to the Russian edition

We wrote this book in 2001–2002. These years saw the launch and start of operations of the American satellite WMAP (Wilkinson Microwave Anisotropy Probe), which began a new stage in the study of the primordial electromagnetic radiation in the Universe. This stage brought a qualitative change to the status of modern cosmology which, using a metaphor suggested by Malcolm Longair, entered the phase of 'precision cosmology' in which the level of progress in theory and experiment was so high that the interpretation of observational data became relatively less urgent than the problem of measuring the most important parameters that characterize the state of gravitation and matter as they were long before the current phase of the cosmological expansion.

Paradoxically, the entire period of explosive development of cosmology happened virtually within the last three decades of the twentieth century; however, it brought together thousands of years of mankind's attempts to comprehend the basic laws governing the structure and evolution of the Universe. Regarded formally, this period coincided – although realistically it was genetically connected – on one hand with the penetration into the mysteries of structure of matter at the microscopic level and on the other hand with the sending of humans into space and with progress in space technologies that revolutionized the experimental basis of the observational astrophysics. One of the authors of this book (Igor Novikov) was involved in the creation of the modern physical cosmology and remembers very well the hot discussions raging in the 'era of the 1960s and 1970s' about the nature of the primordial fluctuations that gave rise to galaxies and galaxy clusters, about the possible anisotropic 'start' of the expansion of the Universe and about the 'hidden mass' whose status was for a long time underestimated by most cosmologists. Another aspect that attracted huge interest was the problem of pregalactic chemical composition of matter which was most closely connected with the 'hot' past of the cosmological plasma and which highlighted for the first time the paramount role played by neutrinos and other hypothetical weakly interacting particles in the thermal history of the Universe; in a wider sense, though, it also connected with the problem of the birth of life in the cosmos. Finally, a brief list of 'hot spots' of astrophysics and cosmology since the late 1970s cannot avoid the eternal questions: How and why did the Universe 'explode'? What was the 'first push' that triggerred the expansion of matter? What was there (if anything) prior to this moment? And how will the expansion of the Universe continue to unfold?

We should add that working on answers to some questions has inevitably generated new ones – for instance, was space-time always four-dimensional? Is it possible that we actually face here manifestations of more complex topology of the space-time continuum and, among other things, the existence of the yet unknown remnants of the early Universe, for example primordial black holes or other mysterious particles? And so forth. These and a whole range of other problems were reflected in the pioneer studies by Peebles (1971), Weinberg (1972, 1977), Zeldovich and Novikov (1983), and in some later works (see, for example, Kolb and Turner (1989), Melchiori and Melchiori (1994), Padmanabhan (1996), Partridge (1995) and Smoot and Davidson (1993)). Some of these problems acquired new status and took their rightful places among the so-called 'eternal' problems of natural sciences that will excite subsequent generations of cosmologists and will await the arrival of new Newtons, Einsteins and Hubbles. As could be expected, some of the hypotheses failed the test of time and sunk into the realm of the history of science, leaving behind a sort of monument to mankind's thinking. But a smaller fraction of hypotheses were verified experimentally and ascended to the sanctum of science, having changed our comprehension of the Universe and of the properties of space-time and matter.

One spectacular example of this sort of achievement of modern cosmology is the problem of the origin of the primordial electromagnetic radiation, better known as the cosmic microwave background (CMB), which covers the aspects of its spectral distribution, anisotropy and polarization. This book is mostly devoted to discussing this range of problems; it was written immediately after the completion of a number of successful ground-based and balloon experiments closely connected with the satellite project COBE, which was successfully completed in the mid 1990s. This project was preceded by a Russian project, RELIKT, that was the first dedicated space mission for the investigation of the CMB anisotropy. The COBE mission became part of the history of cosmology not only as the first experiment that measured the CMB anisotropy with the maximum angular resolution achievable at the time (about 7 degrees of arc), but also as an experiment that put an end to numerous discussions on the possible non-equilibrium of the CMB spectrum and on its deviations from Planck's law of the blackbody frequency distribution of quanta predicted by the theory of the 'hot Universe'.[1]

Metaphorically speaking, the post-COBE cosmology entered a new phase in its development, switching from a search for, let us say, the most probable evolutionary 'treks' to a detailed clarification of the causes of why one reliably established (within a certain time span, of course) particular mode of cosmological evolution of matter had been realized.

The relay race to create a realistic picture of the evolution of the Universe by measuring the CMB anisotropy was continued after COBE by the next generation of experiments (CBI, DASI, BOOMERANG, MAXIMA-1, and quite a few others), all of which provided conclusive proof of the existence of the CMB anisotropy on small angular scales of about 10 minutes of arc. At first glance, the progress of the experiment towards smaller angular scales looks modest at best. Indeed, we still lack 1.5–2 orders of magnitude in order to gauge the typical sizes of galaxy clusters recalculated to the moment of hydrogen recombination at which the Universe became transparent to radiation ($\sim$300 000 years after the onset of the expansion of the Universe). The reality is that it was with the CMB anisotropy and polarization that we were connecting the possibility of 'peeking' into the remote past of the Universe and of 'discovering' the signs of the future clusters on what we now refer to as maps of distribution of the CMB temperature fluctuations on the celestial sphere. Unfortunately this problem was

---

[1] To be precise, the COBE data limit the degree of non-equilibrium of the primordial radiation at the level of 10^{-4}–10^{-5}, which is practically equivalent to a complete absence of distortions. Nevertheless, even this small but possible degree of non-equilibrium proves to be very informative in that it places constraints on energy releases in the early Universe, especially during the period of non-equilibrium ionization of hydrogen and helium. This aspect of the problem is analysed in more detail in several chapters of the book.

found to lie beyond the technical possibilities of radioastronomy, not so much because today's receivers of primordial radiation lack sensitivity, but rather owing to the disruptive effect of various types of noise connected with the activity primarily within our Galaxy, with hot gas in galaxy clusters, the emission from intergalactic dust, and a number of other factors that safely shield the CMB anisotropy from us. However, from the standpoint of CMB physics, this negative outcome is still an outstanding positive result for the adjacent fields of cosmology and astrophysics, which achieved excellent progress in studying the manifestations of activities of various structural forms of matter in the Universe. It was the symbiosis of the adjacent fields of astrophysics that made it possible at the very beginning of the twenty-first century to come very close to solving one of the key problems of cosmology: the determination of the most important parameters that characterize the evolution of the Universe in the past, present and future, namely the Hubble constant, H_0, the current density of the baryonic fraction of matter, the density of the invisible cold component (the so-called 'cold hidden mass'), the value of the cosmological constant, Λ, the type and characteristics of the spectrum of primordial fluctuations of density, velocity and gravitational potential of matter, and other important parameters that will be discussed in the book. As applied to CMB physics, this symbiosis made it possible not only to outline the contours, but also to start a practical implementation of the PLANCK satellite mission – an experiment unique in the extent of pre-launch analysis of the anticipated effects and noise, capable of mapping the CMB anisotropy and polarization with unique angular resolution (on the order of 6 minutes of arc) with a record low level of internal noise of the receiving electronics, less by approximately an order of magnitude than in all currently operational grand-based, balloon and satellite experiments.

It should be noted that the PLANCK project will launch in 2007–2008. Although the objectives, namely the mapping of the CMB anisotropy and polarization with maximum possible coverage of the celestial sphere, are shared by the two missions, the PLANCK project is meant to provide the maximum possible sensitivity of the receiver electronics and to achieve it with a unique selection of frequency ranges for the observation of the CMB anisotropy and polarization. Furthermore, the objectives of the project include compilation of a catalogue of radio and infrared pointlike sources that would cover the frequency range 30–857 GHz in 19 frequency channels, mapping of galaxy clusters, plus a number of other tasks whose solution became possible thanks to the unique theoretical and experimental studies of the CMB anisotropy and the noise of galactic and extragalactic origin that accompanies it.

The following legitimate questions may be asked. Is it justifiable to present the CMB physics now, before the completion of these two new space missions which may drastically change our ideas about the evolution of the Universe and about the formation of anisotropy and polarization of cosmic microwave background and, who knows, about the formation of its large-scale structure? Would it be advisable to wait perhaps seven or ten years until the situation concerning the distribution of anisotropy on the celestial sphere has been clarified and then summarize the era of studying the CMB with certainty, being supported by the data of literally 'the very last experiments'? Answers to the above questions seem to us surprisingly simple. First – and this point is perhaps the most important – we are absolutely sure that no subsequent experiments will act as 'foundation destroyers' for modern cosmology. The foundations of the theory are too solid for that, and its implications are very well developed and carefully checked against observations. Secondly, the preparation stage for the WMAP and PLANCK missions stimulated unprecedented progress in the theory that needs further digeston and systematization. Suffice it to say that compared with the situation at the beginning of the 1990s, the CMB physics has progressed greatly, coming very close to predicting effects with an accuracy of better than 5%, requiring for their simulation modern computer networks and the development of new mathematical techniques for data processing. Finally, placed third in sequence but not in significance, the future space experiments, the PLANCK mission among them, have one obvious peculiar feature: they have been mostly prepared under the guidance of the generation of 'veterans', whereas the results will mostly be used by the generation of 'pupils'. We think that in this relay race of generations it is extremely important not to lose sight of the subject, not to disrupt the connection between the days of 'Sturm und Drang' of the 1970s–1990s when the foundations of the CMB physics were laid and, let us say, the 'days of bliss' that we all anticipate to arrive roughly by the end of the first decade of this century when the WMAP and PLANCK projects will have been successfully completed. This is the reason why we attempted in the book to stand back from discussing the general aspects of cosmology and to focus mostly on specific theoretical problems of the formation of the CMB frequency spectrum, its anisotropy and polarization and their observational aspects; we assume the reader to have at least some general familiarity with the foundations of the theory of the 'hot Universe', physical cosmology, probability theory and mathematical statistics, the theory of random fields and atomic physics.

We have attempted to demonstrate in what way the modern apparatus of theoretical physics can be applied to studying the properties of cosmic plasma and how the limits of our knowledge of such fundamental natural phenomena as gravitation, relativity and relativism can be expanded owing to their symbiotic relationship with astrophysics.

We are grateful to all our colleagues in the Astrocosmic Centre of the P. N. Lebedev Physics Institute (FIAN, Moscow), Rostov State University, Copenhagen University, the Theoretical Astrophysics Centre (Copenhagen) and Oxford University for supporting our work and for numerous discussions.

We are especially grateful to E. V. Kotok for her enormous work preparing the manuscript of this book, and also for her participation in a number of research papers quoted in it.

Preface to the English edition

The English translation of our book appears three years after the first Russian edition, which was published in 2003. Cosmology, and specifically the cosmology of the cosmic microwave background (CMB), is the most rapidly evolving branch of science in our time, so there have been several important advances since the first edition of this book. Some extremely important developments – the publication of new observational results (particularly the observations of the Wilkinson Microwave Anisotropy Probe (WMAP) space mission), the discussion of these results in numerous papers, the formulation of new ideas on the physics of the CMB, and the creation of new mathematical and statistical methods for analysing CMB observations – have arisen since the completion of the Russian edition, originally entitled *Relic Radiation of the Universe*. The term 'cosmic microwave background' used in publications in the West (and now often in Russia) is rather clumsy. 'Relic radiation', introduced by the Russian astronomer I. S. Shklovskii, is an impressive name that appealed to many astrophysicists; however, since CMB is used in the specific literature in the field, we had to call the English version of our book *The Physics of the Cosmic Microwave Background*, and we continue using this term throughout the book.

In the original Russian edition, we tried to give a complete review of all the important topics in CMB physics. In preparing this edition, we tried hard to incorporate most of the new developments; however, we preserve the original spirit of the book in not striving to encompass the entire recent literature on the subject (especially as this now seems to be impossible, even in such an inflated volume). Nevertheless, we hope that the English edition presents the current situation in CMB physics.

This edition also includes a new eighth chapter, entitled 'The Wilkinson Microwave Anisotropy Probe (WMAP).' This chapter describes in detail the primary results of the most important CMB project of the last few years. In addition to the references recommended in the Preface to the Russian edition, we recommend the following books devoted to the subject: de Oliveira-Costa and Tegmark (1999), Freedman (2004), Lachiez-Rey and Gunzig (1999), Liddle (2003), Partridge (1995), Peacock (1999) and Peebles (1993).

We also used this opportunity to correct misprints and some imperfections detected when rereading the Russian edition. We are grateful to our translators, Nina Iskandarian and Vitaly Kisin, for their valuable help in preparing the English edition.

And last but not least, while working on the English edition we enjoyed unfailing support from the Niels Bohr Institute, Copenhagen, and Imperial College London. We wish to express our sincere thanks to these institutions and the wonderful people there who helped make this edition possible.

I

Observational foundations of modern cosmology

1.1 Introduction

In a way, the entire history of cosmology from Ptolemy and Aristotle to the present day can be divided into two stages: a period before and a period after the discovery of the cosmic microwave background (CMB). The first period was the subject of hundreds of volumes of literature; now it is not only an integral part of science, but also marks a step in the progress of mankind. The second stage started in 1965 when two American researchers, A. Penzias and R. Wilson published their famous article in the *Astrophysical Journal*, 'A measurement of excess antenna temperature at 4080 Mc/s' (Penzias and Wilson, 1965), in which they announced the discovery of a previously unknown background radio noise in the Universe. Another article, in the same issue of the *Astrophysical Journal*, preceded the one by Penzias and Wilson; this was by R. Dicke, P. J. E. Peebles, P. Roll and D. Wilkinson (Dicke *et al.*, 1965) and discussed the preparation of a similar experiment at a different wavelength, but also interpreted the Penzias–Wilson results as confirming the predictions of the 'hot universe' theory. The radiation with a temperature close to 3 K discovered by Penzias and Wilson was described as the remnant of the hot plasma that existed at the very onset of expansion which then cooled down as a result of expansion.

Formally, the new stage in the study of the Universe was catalysed by several pages in one volume of a journal and began in this non-dramatic and almost routine way. Note that the 'child' wasn't born all that unexpectedly for astrophysicists. In the mid 1940s George Gamow had already published a paper (Gamow, 1946) in which he proposed a model of what became known as the 'hot' starting phase of cosmological expansion; this work stimulated the work of R. Alpher and R. Herman (Alpher and Herman, 1953), offering an explanation of the chemical composition of pre-galactic matter (see a review and references in Novikov (2001)).

The starting point for motivating all these authors was an attempt to explain specific features of the abundances of chemical elements and isotopes in the Universe. It was assumed that these were all produced at the very first moments of expansion of the Universe. Tables of the abundances of different isotopes show that isotopes with an excess of neutrons typically dominate. It followed that free neutrons should have existed in the primordial matter for a sufficiently long time – something that is only possible at extremely high temperatures. This stimulated the idea of the hot initial phase of expansion of the Universe. The first publications of the theory of the hot Universe contained a number of inconsistencies on which we will not dwell here. The reader can find the details in Weinberg (1977) and Zeldovich and Novikov (1983).

According to our current understanding, in the first three minutes of expansion of the Universe only the lightest elements were 'cooked', whereas the heavier ones were produced much later by nuclear processes in stars; the heaviest elements were born when supernovas exploded. It is important to note that Gamow, Alpher and Herman's main idea about the need for high temperatures of the primordial matter proved to be correct. For details on the modern theory of nucleosynthesis in the early Universe, see, for example, Kolb and Turner (1989) and Zeldovich and Novikov (1983). There was, however, another altogether funnier reason why the authors of the theory of the 'hot Universe' considered it necessary 'to cook' (literally) all the chemical elements in the very first seconds of the cosmological expansion. Namely that, in the 1940s, the value of the Hubble constant, H_0, and, consequently, the age of the Universe, were evaluated incorrectly. The Hubble constant was thought to be several times larger than the value deduced from modern measurements, so that the age of the Universe was as low as $(1–4) \times 10^9$ years, as against the value of $(13.5–14) \times 10^9$ years accepted now. This duration would not be enough for the synthesis of chemical elements in stars; consequently, Gamow and his colleagues came to the conclusion that all chemical elements must have been 'cooked' from the primeval matter.

We now know, owing to the available cosmochronological data, that the age of the Universe is far greater than the age of the Earth (4×10^9 years), and that the Earth was formed from the protoplanetary material that had been enriched by products of thermonuclear synthesis deep inside stars. Therefore the need to find an explanation for the chemical composition of matter, including elements heavier than iron, within the limits of the 'hot Universe' model has simply gone up in smoke, but the principal idea of the founders of this theory – the idea of high initial temperature and high density of cosmic plasma – passed the test of time.

Let us return, however, to the history of the discovery of the cosmic microwave background. Using somewhat inconsistent estimates, Gamow and his colleagues concluded that, owing to the hot birth of the Universe, the space that exists during this epoch must be filled with equilibrium radiation at a temperature of several kelvin. It would seem likely to us now that once a major prediction had been formulated, it demanded immediate testing, and that radioastronomers would have tried to detect this radiation. This, however, failed to happen. An outstanding American scientist, winner of a Nobel prize for physics, Steven Weinberg, wrote in *The First Three Minutes: A Modern View of the Origin of the Universe* (Weinberg, 1977) 'This detection of the cosmic microwave background in 1965 was one of the most important scientific discoveries of the twentieth century. Why did it have to be made by accident? Or to put it another way, why there was no systematic search for this radiation, years before 1965?' We mentioned above that Gamow and his colleagues predicted the probable presence of electromagnetic radiation with a temperature of several kelvin more than 15 years before its detection. Perhaps special radiotelescopes were required, with sensitivity unattainable at the moment? Apparently not; the necessary receivers were available. The main reason, in our opinion, was probably of a psychological nature. There is convincing evidence to support this view, and we will discuss this later.

In fact, numerous examples can be found in the history of science when predictions of novel phenomena, and in particular ground-breaking discoveries, occurred long before experimental confirmations were obtained. Weinberg (1977) provides us with an excellent example: the prediction, made in 1930, of the existence of the antiproton. Immediately after this theoretical prediction, physicists could not even imagine what kind of physical experiment would be capable of confirming or, as often happens, disproving this fundamental inference of the theory. It only became possible almost 20 years later when a suitable particle accelerator was built in Berkeley that provided impeccable confirmation of the prediction of the theory.

However, as we shall see below, in the case of this particular prediction the suitable receivers necessary to start searching for the microwave background already existed. Alas, radioastronomers simply did not know what it was they should search for. There was no proper communication between theorists and observers, and theorists did not really trust the not yet perfect theory of the hot Universe. Ideas on how it would be possible to detect the electromagnetic 'echo of the Big Bang' started to appear only in the mid 1960s, and even then only accidentally. Another reason why radioastronomers did not attempt to discover the CMB, and perhaps the most important one, was formulated by Arno Penzias in his Nobel lecture of 1979 (Penzias, 1979). The fact was that none of the work published by Gamow and his colleagues pointed out that the microwave radiation that reaches us from the epoch of cosmological nucleosynthesis, having cooled down to several kelvin owing to the expansion of the Universe, could be detectable, even in principle. In fact, the general feeling was quite the opposite; Penzias, in his Nobel lecture, formulated the widespread impression: 'As for detection, they appear to have considered the radiation to manifest itself primarily as an increased energy density.[1] This contribution to the total energy flux incident upon the earth would be masked by cosmic rays and integrated starlight, both of which have comparable energy densities. The view that the effects of three components of approximately equal additive energies could not be separated may be found in a letter by Gamow written in 1948 to Alpher (unpublished, and kindly provided to me by R. A. Alpher from his files). "The space temperature of about 5 K is explained by the present radiation of stars (*C*-cycles). The only thing we can tell is that the residual temperature from the original heat of the Universe is not higher than 5 K." They do not seem to have recognized that the unique spectral characteristics of the relict radiation would set it apart from the other effects.'

This, however, was understood by A. Doroshkevich and I. Novikov, who, in 1964, published a paper in *The Academy of Sciences of the USSR Doklady* entitled 'Mean density of radiation in the metagalaxy and certain problems in relativistic cosmology' (Doroshkevich and Novikov, 1964). The basic idea formulated in this paper has not lost its relevance even 40 years later. We shall assume for the moment that we know how galaxies of different type emit electromagnetic radiation in different wavelength bands. Choosing certain assumptions concerning the evolution of galaxies in the past and taking into account the redshifting of the wavelength of light from distant galaxies because of the expansion of the Universe, it is possible to calculate the intensity of radiation from galaxies in today's Universe for each wavelength. What we need to consider is that stars are not the only sources of radiation: indeed, many galaxies are powerful emitters of radio waves on the metre and decimetre wavelengths. Gas and dust in the galaxies also radiate. The nontrivial aspect of this is that if the Universe had been 'hot' at some point, the primordial radiation background has to be added to the radiation spectrum one wishes to calculate, and this is what Doroshkevich and Novikov (1964) accomplished. The wavelength of this radiation should be on the order of centimetres and millimetres and should fall within that range of spectrum where the contribution of galaxies is practically zero. Therefore, the cosmic microwave background in this wavelength range should exceed the radiation of known sources of radio emission by a factor of tens of thousands, even millions. Hence, it should be observable! Here is how Arno Penzias formulated it in his Nobel lecture: 'The first published recognition of the relict radiation as a detectable microwave phenomenon appeared in a brief paper entitled "Mean density of

---

[1] Penzias referred here to work by Alpher and Herman dated 1949.

radiation in the metagalaxy and certain problems in relativistic cosmology", by A. G. Doroshkevich and I. D. Novikov (1964). Although the English translation appeared later the same year in the widely circulated *Soviet Physics–Doklady*, it appears to have escaped the notice of the other workers in this field. This remarkable paper not only points out the spectrum of the relict radiation as a blackbody microwave phenomenon, but also explicitly focuses upon the Bell Laboratories twenty-foot horn reflector at Crawford Hill as the best available instrument for its detection!'

Note that the cosmic microwave background was indeed discovered in 1965 using precisely this facility.

The paper by Doroshkevich and Novikov was not noticed by observer astronomers. Neither Penzias and Wilson, nor Dicke and his coworkers, were aware of it before their papers were published in 1965. We wish to mention a strange mistake involving the interpretation of one of the conclusions in Doroshkevich and Novikov (1964). Penzias (1979) wrote: 'Having found the appropriate reference [Ohm, 1961], they [Doroshkevich and Novikov] misread its result and concluded that radiation predicted by the "Gamov theory" was contradicted by the reported measurements.'

Also, in Thaddeus (1972) one can read: 'They [Doroshkevich and Novikov] mistakenly concluded that studies of atmospheric radiation with this telescope (Ohm, 1961) already ruled out isotropic background radiation of much more than 0.1 K.' Actually, Doroshkevich and Novikov's paper contains no conclusion stating that the observational data exclude the CMB with temperature predicted by the hot Universe model. In fact, it states: 'Measurements reported in Ohm (1961) at a frequency $\nu = 2.4 \times 10^9$ cycles s^{-1} give a temperature 2.3 ± 0.2 K, which coincides with theoretically computed atmospheric noise (2.4 K). Additional measurements in this region (preferably on an artificial earth satellite) will assist in obtaining a final solution of the problem of the correctness of the Gamow theory'. Thus, Doroshkevich and Novikov encouraged observers to perform the relevant measurements! They did not discuss in their paper the interpretation of the value 2.4 K obtained by Ohm (1961), who used a technique developed specifically for measuring the atmospheric temperature (see discussion in Penzias (1979)).

This is not the end, however, of the dramatic episodes in the history of the prediction and discovery of the cosmic microwave background. It is now clear that astronomers came across indirect manifestations of the CMB long before the 1960s. In 1941, a Canadian astronomer, Andrew McKellar, discovered cyanide molecules (HCN) in interstellar space. He used the following method of studying interstellar gases. If light travelling from a star to the Earth propagates through a cloud of interstellar gas, atoms and molecules in the gas absorb this light only at certain wavelengths. This creates the well known absorption lines that are successfully used not only for studying the properties of interstellar gas in our Galaxy, but also in other fields of astrophysics. The positions of absorption lines in the emission spectrum of radiation depend on what element or what molecule causes this absorption, and also on the state in which they were at the moment of absorption. As the object of research, McKellar chose absorption lines caused by cyanide molecules in the spectrum of the star 'ε' of Ophiuchus. He concluded that these lines could only be caused by absorption of light by rotating molecules. Relatively simple calculations allowed McKellar to conclude that the excitation of rotational degrees of freedom of cyanide molecules required the presence of external radiation with an effective temperature of 2.3 K. Neither McKellar himself, nor anyone else, suspected that he had stumbled on a manifestation of the cosmic microwave background. Note that this happened long before the ground-breaking work of Gamow and his colleagues! Only after the discovery of the CMB, in 1966 were the following three papers published in one year: Field and Hitchcock (1966), Shklovsky (1966) and Thaddeus and Clauser (1966); later, Thaddeus (1972) showed that the excitation of rotational degrees of freedom of cyanide was caused by CMB quanta. Thus, an indication, even if indirect, of the existence of a survivor from the 'hot' past of the Universe was available as early as 1941.

Even now we are not at the end of our story. We shall return to the question of whether the experimental radiophysics was ready to discover the microwave background long before the results of Penzias and Wilson. Weinberg (1977) wrote that 'It is difficult to be precise about this but my experimental colleagues tell me the observation could have been made long before 1965, probably in the mid 1950s and perhaps even in the mid 1940s.' Was this indeed possible?

In the autumn of 1983, one of authors of this volume (I. Novikov) received a call from T. Shmaonov, a researcher with The Institute of General Physics, with whom Novikov was not previously acquainted. Shmaonov explained that he would like to discuss some details concerning the discovery of the cosmic microwave background. When they met, Shmaonov described how, in the middle of the 1950s, working under the guidance of the well known radioastronomers S. E. Khaikin and N. L. Kaidanovsky, he conducted measurements of the intensity of radio emission from space at the wavelength of 3.2 cm using a horn antenna similar to the one that Penzias and Wilson worked with many years later. Shmaonov very carefully measured the inherent noise of his receiver electronics, which was certainly not as good as the future American equipment (do not forget the time factor, which in those years was decisive as far as the quality of receivers was concerned), and concluded that he had detected a useful signal. Shmaonov published his results in 1957 in *Pribory i Tekhnika Eksperimenta* and also included them in his Ph.D. thesis (Shmaonov, 1957). The conclusion drawn from these measurements was as follows: 'We find that the absolute effective temperature of the radioemission background . . . is 4 ± 3 K.' Moreover, measurements showed that radiation intensity was independent of either time or direction of observations. Even though temperature measurement errors were quite considerable, it is now clear that Shmaonov did observe the cosmic microwave background at a wevelength of 3.2 cm; alas, neither the author nor other radioastronomers with whom he discussed the results of his experiments have given this effect the attention it deserved. Furthermore, even after the work of Penzias and Wilson was published, Shmaonov failed to realize that the source of the signal was the same; in fact, at the time, Shmaonov was working in a very different branch of physics. Only 27 years after he published those measurements did Shmaonov make available a special report on his discovery (see the discussion in Kaidanovsky and Parijskij (1987)).

Even this is not the last piece of the jigsaw puzzle! More recently, we have learnt that at the very beginning of the 1950s Japanese physicists made attempts to measure the cosmic microwave background. Unfortunately we were unable to find reliable contemporary or more recent references to these studies.

It is obvious that the drama of ideas and 'random walks' of the 1940s to the 1950s in search of manifestations of the cosmic microwave background is still waiting for its historian, while the period from 1965 to the present day is a well planned and orchestrated attack on the secrets of cosmic radiation, not only at radio wavelengths, but also in the optical, infrared, ultraviolet, x-ray and gamma radiation ranges.

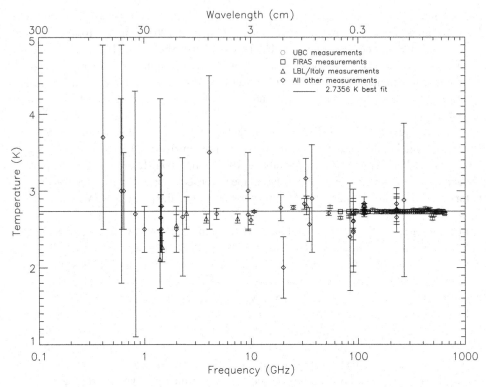

Figure 1.1 Thermodynamic temperature of the CMB as a function of radiation frequency and wavelength. Data from the FIRAS instrument are shown in the 100 to 600 GHz range. The horizontal line corresponds to $T_0 = 2.736$ K – the best approximation of the COBE data. For comparison, the data from other experiments are marked by squares and triangles. Adapted from Nordberg and Smoot (1998) and Scott (1999a).

1.2   Current status of knowledge about the spectrum of the CMB in the Universe

Only a year after the publication of the paper by Penzias and Wilson, their colleagues, F. Howell and J. Shakeshaft (Howell and Shakeshaft, 1966) measured the temperature of the cosmic microwave background at a wavelength of 20.7 cm and found it to be 2.8 ± 0.6 K. Similar values of temperature, but in the wavelength range 3.2 cm ($T = 3.0 \pm 0.5$ K), were reported in the same year by Roll and Wilkinson (1966) and by Field and Hitchcock (1966) ($T = 3.2 \pm 0.5$ K at a wavelength 0.264 cm), and by a number of other researchers in subsequent years.

Table 1.1 gives a complete list of published measurements of the CMB temperature from 408 MHz up to 300 GHz (Nordberg and Smoot, 1998). In spite of a large number of experiments ($\sim$60) that measured the CMB temperature, not all of them are equally informative. Quite often a high level of systematic errors led to considerable spreads of the average values of T_R. In this connection, Fig. 1.1 presents selective data for a number of experiments carried out over a period from the end of the 1980s to the beginning of the 1990s and manifesting an extremely low noise level (references to these experiments are given in Table 1.1).

Table 1.1. *Measurements of the CMB temperature*

Frequency (GHz)	Wavelength (cm)	Temperature (K)	Reference
0.408	73.5	3.7 ± 1.2	Howell and Shakeshaft (1967)
0.6	50	3.0 ± 1.2	Sironi *et al.* (1990)
0.610	49.1	3.7 ± 1.2	Howell and Shakeshaft (1967)
0.635	47.2	3.0 ± 0.5	Stankevich, Wielebinski and Wilson (1970)
0.820	36.6	2.7 ± 1.6	Sironi, Bonelli and Limon (1991)
1.4	21.3	2.11 ± 0.38	Levin *et al.* (1988)
1.42	21.2	3.2 ± 1.0	Penzias and Wilson (1967)
1.43	21	$2.65^{+0.33}_{-0.30}$	Staggs *et al.* (1996a,b)
1.45	20.7	2.8 ± 0.6	Howell and Shakeshaft (1966)
1.47	20.4	2.27 ± 0.19	Bensadoun *et al.* (1993)
2	15	2.55 ± 0.14	Bersanelli *et al.* (1994)
2.5	12	2.71 ± 0.21	Sironi *et al.* (1991)
3.8	7.9	2.64 ± 0.06	de Amici *et al.* (1991)
4.08	7.35	3.5 ± 1.0	Penzias and Wilson (1965)
4.75	6.3	$2.70 + 0.07$	Mandolesi *et al.* (1986)
7.5	4.0	2.60 ± 0.07	Kogut *et al.* (1990)
7.5	4.0	2.64 ± 0.06	Levin *et al.* (1992)
9.4	3.2	3.0 ± 0.5	Roll and Wilkinson (1966)
9.4	3.2	$2.69^{+0.26}_{-0.21}$	Stokes, Partridge and Wilkinson (1967)
10	3.0	2.62 ± 0.06	Kogut *et al.* (1990)
10.7	2.8	2.730 ± 0.014	Staggs *et al.* (1996a,b)
19.0	1.58	$2.78^{+0.12}_{-0.17}$	Stokes *et al.* (1967)
20	1.5	2.0 ± 0.4	Welch *et al.* (1967)
24.8	1.2	2.783 ± 0.025	Johnson and Wilkinson (1987)
31.5	0.95	2.83 ± 0.07	Kogut *et al.* (1996b)
32.5	0.924	3.16 ± 0.26	Ewing, Burke and Staelin (1967)
33.0	0.909	2.81 ± 0.12	De Amici *et al.* (1985)
35.0	0.856	$2.56^{+0.17}_{-0.22}$	Wilkinson (1967)
53	0.57	2.71 ± 0.03	Kogut *et al.* (1996b)
90	0.33	$2.46^{+0.40}_{-0.44}$	Boynton, Stokes and Wilkinson (1968)
90	0.33	2.61 ± 0.25	Millea *et al.* (1971)
90	0.33	2.48 ± 0.54	Boynton and Stokes (1974)
90	0.33	2.60 ± 0.09	Bersanelli *et al.* (1989)
90	0.33	2.72 ± 0.04	Kogut *et al.* (1996b)
90.3	0.332	< 2.97	Bernstein *et al.* (1990)
113.6	0.264	2.70 ± 0.04	Meyer and Jura (1985)
113.6	0.264	2.74 ± 0.05	Crane *et al.* (1986)
113.6	0.264	2.75 ± 0.04	Kaiser and Wright (1990)
113.6	0.264	2.75 ± 0.04	Kaiser and Wright (1990)
113.6	0.264	2.834 ± 0.085	Palazzi *et al.* (1990)
113.6	0.264	2.807 ± 0.025	Palazzi, Mandolesi and Crane (1992)
113.6	0.264	$2.279^{+0.023}_{-0.031}$	Roth, Meyer and Hawkins (1993)
154.8	0.194	< 3.02	Bernstein *et al.* (1990)
195.0	0.154	< 2.91	Bernstein *et al.* (1990)
227.3	0.132	2.656 ± 0.057	Roth *et al.* (1993)
227.3	0.132	2.76 ± 0.20	Meyer and Jura (1985)
227.3	0.132	$2.75^{+0.24}_{-0.29}$	Crane *et al.* (1986)
227.3	0.132	2.83 ± 0.09	Meyer, Cheng and Page (1989)
227.3	0.132	2.832 ± 0.072	Palazzi *et al.* (1990)
266.4	0.113	< 2.88	Bernstein *et al.* (1990)
Broad range	Broad range	2.728 ± 0.002	Fixsen *et al.* (1990)
300	0.1	2.736 ± 0.017	Gush, Halpern and Wishnow (1990)

An important feature of these data is an extremely low absolute measurement error, which makes possible the calculation of the amplitude of today's temperature of the microwave background at the 95% confidence limit:

$$T_0 = 2.7356 \pm 0.038 \text{ K.} \tag{1.1}$$

It is well known that this temperature (T_0) determines all spectral characteristics of radiation (see, for example, Landau and Lifshits (1984)). For instance, the spectral intensity of radiation, defined as energy per unit area element in unit solid angle and unit frequency interval, is given by the expression

$$I_\nu = \frac{2h\nu^3}{c^2} n_\nu, \tag{1.2}$$

where h is Planck's constant, c is the speed of light in vacuum, ν is frequency and n_ν is the spectral density of the number of quanta. For the Planck radiation, n_ν is a function of only one parameter, namely temperature:

$$n_\nu = \left(e^{h\nu/kT} - 1\right)^{-1}, \tag{1.3}$$

where k is the Boltzmann constant, and the corresponding spectral brightness is given by

$$B_\nu(T) = \frac{2h\nu^3}{c^2} \left(e^{h\nu/kT_0} - 1\right)^{-1}. \tag{1.4}$$

Note that the dependence of I_ν on frequency will be different for non-equilibrium radiation, but in general it should not necessarily be characterized by a single universal parameter, i.e. temperature. Equation (1.4) readily leads to asymptotics for $B_\nu(T)$ in the limit $\left(\frac{h\nu}{kT} \ll 1\right)$

$$B_\nu^{\text{RJ}}(T) \simeq \frac{2\nu^2}{c^2} kT \tag{1.5}$$

for the Rayleigh–Jeans interval, and, for high energies of quanta $\left(\frac{h\nu}{kT} \gg 1\right)$,

$$B_\nu^{\text{W}}(T) \simeq \frac{2h\nu^3}{c^2} e^{-h\nu/kT} \tag{1.6}$$

for Wien's interval. We see that $B_\nu^{\text{RJ}}(T)$ describes the classical (non-quantum) part of the spectrum, which is independent of the value of Planck's constant. The Rayleigh–Jeans formula is well known in radioastronomy for determining the brightness temperature of a radiation source with spectral intensity I_ν:

$$T_{\text{A}} = \frac{c^2}{2k\nu^2} I_\nu(T). \tag{1.7}$$

As we see from Eq. (1.7), the relation between the thermodynamic and brightness temperatures for blackbody radiation has the form

$$T_{\text{A}}(\nu) = T_0 \frac{x^2 e^x}{(e^x - 1)^2}, \tag{1.8}$$

where $x = \frac{h\nu}{kT}$. Therefore, in the low-frequency limit, $x \ll 1$, Eq. (1.8) immediately implies the equality $T_{\text{A}} = T_0$, and if $x \gg 1$ then the brightness temperature is found to be systematically below the thermodynamic temperature. In what follows we require integral characteristics of the CMB in addition to spectral ones: energy density, ε_γ; concentration of quanta, n_γ;

entropy density, S_γ; and quantum energy averaged over the spectrum, $\bar{E}_\gamma$. These quantities are defined for the CMB in the standardized manner (Landau and Lifshits, 1984), regardless of its cosmological nature:

$$\varepsilon_\gamma = \sigma T_0^4 = 4.24 \times 10^{-13} \text{ erg cm}^{-3}; \quad n_\gamma = 0.244 \left(\frac{kT_0}{\hbar c}\right)^3 = 414 \text{ cm}^{-3};$$

$$S_\gamma = \frac{4}{3}\frac{\varepsilon_\gamma}{kT_0} = 1.496 \times 10^3 \text{ cm}^{-3}; \quad \bar{E}_\gamma = \frac{\varepsilon_\gamma}{n_\gamma} = 1.02 \times 10^{-15} \text{ erg},$$

(1.9)

where $\sigma = (\pi^2 k^y/15\hbar^3 c^3) = 7.5640 \times 10^{-15}$ erg cm^{-3} K^{-1} is the emission constant and $\hbar = h/2\pi$.

1.2.1   Electromagnetic emission from space

We mentioned at the beginning of this chapter that the pioneers of CMB research considered various types of electromagnetic emission coming from space as sources of very undesirable noise. However, in contrast to the CMB, electromagnetic radiation in the optical, ultraviolet, x-ray, γ and also long-wavelength ranges ($\lambda > 1$ m) are of non-cosmological origin. The most important characteristics of these electromagnetic backgrounds are, as in the case of the CMB, the intensity and degree of anisotropy of distribution over the sky. In this section we are mostly interested in the isotropic extragalactic component which is obtained by subtracting the component generated by the activities within the Milky Way Galaxy from the total signal. Figure 1.2 (Halpern and Scott, 1999) shows the combined distribution of various electromagnetic backgrounds published in Dwek and Arendt (1998), Hauser *et al.* (1998), Kappadath *et al.* (1999), Lagache *et al.* (1998), Miyaji *et al.* (1998), Pozzetti *et al.* (1998), and Sreekumar *et al.* (1998). In the long-wavelength limit ($\lambda > 10^3$ mm), we clearly see a contribution from extragalactic radio sources that is characterized by a power-law spectrum:

$$I_\nu \simeq 6 \times 10^3 \left(\frac{\nu}{1\text{GHz}}\right)^\alpha \text{ Jy ster}^{-1},$$

(1.10)

with the spectrum exponent $\alpha = -0.8 \pm 0.1$ (Longair, 1993) and 20% uncertainty in amplitude. The total contribution of this component to the total energy density of the radiation is extremely small, but the role of this background is found to be very significant in clarifying the origin of the so-called superhigh-energy cosmic rays ($E \geq 10^{20}$ eV) (Bhattacharjee and Sigl, 2000; Blasi, 1999; Doroshkevich and Naselsky, 2002).

Note, however, that as $\nu \to 0$, the intensity increases ($I_\nu \propto \nu^{-0.8}$) only up to frequencies $\nu \sim 1$–3 MHz. The data of Clark, Brown and Alexander (1970), Longair (1993) and Simon (1978) point to this behaviour. The slope I_ν changes at $\nu \leq 3$ MHz and the effective exponent becomes $\alpha \simeq 1$. The causes of this behaviour may be traced to synchronous self-absorption of radiation in the sources responsible for the formation of long-wavelength radio background (Longair, 1993).

Let us return, however, to discussing background radiation outside the range in which the CMB dominates. The most complete review of the available observational data in the infrared (IR) range of wavelengths from 1 mm to 10^{-3} mm is given in Hauser (1998) and Gispert, Lagache and Puget (2000). Note that the study of the defuse cosmic IR radiation is relatively recent, even though the data on the intensity of this background radiation make it possible to extract unique information on the evolution of pregalactic matter and on the dynamics of the formation of galaxies and stars. It appears that the first indications of the

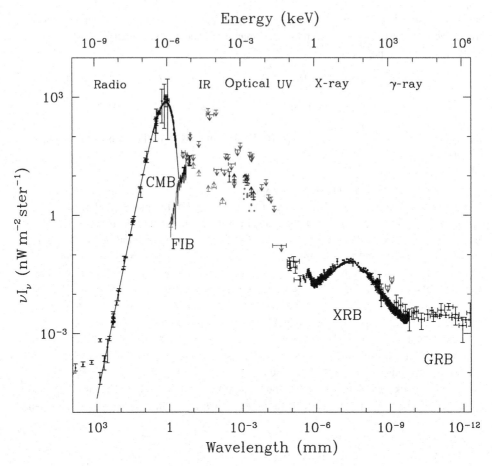

Figure 1.2 Spectral density of extragalactic electromagnetic radiation in the Universe. From Scott (1999a).

existence of this background were obtained in rocket experiments (see, for example, Hauser *et al.* (1991)). The IR background was later studied specifically using the DIRBE tool in the framework of the Cosmic Background Explorer (COBE) project that we have mentioned earlier. In combination with FIRAS – an instrument in the same project (Gispert *et al.*, 2000) – it was possible to obtain unique data on the spectral characteristics of IR radiation in the range from 100 μm to 1 cm, as shown in Fig. 1.3. The same figure shows the data for the optical and ultraviolet (uv) ranges that follow the IR range in the order of increasing energy of quanta. An important feature of these ranges, as in the case of the IR background, is their genetic relation to young galaxies being formed in the process of evolution of the Universe (the optical range 0.15–2.3 μm), to the diffuse thermal emission of intergalactic medium and to the integral ultraviolet luminosity of galaxies and quasars (UV range; $\lambda \simeq 1000$–2500 Å) (see Gispert *et al.* (2000) and the relevant references therein).

In the optical range in the interval $\lambda \simeq 3200$–24 000 Å, the intensity distribution is described sufficiently well by the following expression:

$$\lambda F_\lambda \simeq A^{(\lambda)} 10^{-6} \, \text{erg} \ \text{cm}^{-2} \, \text{s}^{-1} \, \text{ster}^{-1}, \tag{1.11}$$

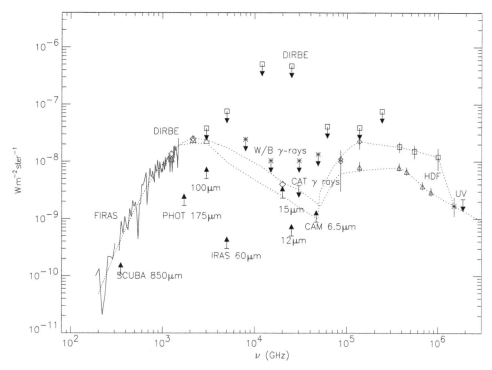

Figure 1.3 Spectrum of extragalactic radiation in the ultraviolet to millimeter wavelength ranges. From Gispert *et al.* (2000).

where the coefficient is given by $A = 2.5^{+0.07}_{-0.04}$ for $\lambda = 3600\,\text{Å}$, $A = 2.9^{+0.09}_{-0.05}$ for $\lambda = 4500\,\text{Å}$ and $\lambda = 6500\,\text{Å}$, and $A = 2.6^{+0.3}_{-0.2}$ for $\lambda = 9000\,\text{Å}$ (Gispert *et al.*, 2000). We see from these data that A can be considered to be practically independent of wavelength in a wide range of λ. From 2 μm (22 000 Å) we observe that the amplitude $A(\lambda)$ is almost doubled to $A = 7 \pm 1$ (Gispert *et al.*, 2000). In contrast to the optical range, this situation in the (UV) range is not as obvious. Here it is very difficult to separate the galactic and extragalactic components. It is assumed (see, for example, Henry and Murthy (1996) and Jakobsen *et al.* (1984)), that UV observations at high galactic elevations mostly single out the extragalactic component, even though it is not clear to what extent it is distorted by the influence of our Galaxy. The anticipated limits and observational data on the spectrum of extragalactic UV background may be found in Andersen *et al.* (1979), Fix, Craven and Frank (1989), Henry and Murthy (1996), Hurwitz, Bowyer and Martin (1990), Jakobsen *et al.* (1984), Joubert *et al.* (1983), Martin and Bowyer (1990), Onaka (1990), Parese *et al.* (1979), Tennyson *et al.* (1988), Weller (1983).

Moving further on along the scale of wavelengths of cosmic background, we reach, after the UV range, the region of diffuse x-ray radiation within wavelengths from 10^{-9} to 10^{-6} mm (see Fig. 1.4). Note that this range of wavelengths was an object of study even before the discovery of the cosmic microwave background. Even in 1962, in the course of rocket experiments, a diffuse x-ray component was detected, combined with simultaneously discovered powerful discrete sources of x-ray emission (Gehrels and Cheng, 1996; Zamorani, 1993). An x-ray survey of the sky followed, using the satellites UHURU, ARIEL V, EINSTEIN,

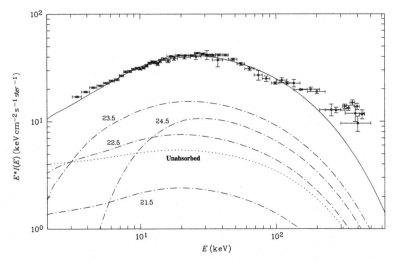

Figure 1.4 The spectrum of x-ray background in the range 1–10³ keV. Lines correspond to models of generation of background radiation (see Gehrels and Cheng (1996) and Zamorani (1993)). Adapted from Hasinger and Zamorani (1997).

ROSAT, GINGA, etc.; this made it possible to identify with certainty the spectrum of the diffuse component that reaches maximum at quantum energies $\overline{E} \simeq 25\,\text{keV}$ and manifests power-law asymptotics at $E < \overline{E}$ and $E > \overline{E}$ with exponents $\alpha_1 \simeq 0.4$ and $\alpha_2 \simeq 1.4$, respectively.

Figure 1.5 plots the data on the spectrum of diffuse x-ray and γ-ray backgrounds according to Strong, Moskalenko and Reimer (2004) and Zamorani (1993). The results of observation of the role of Seyfert galaxies of types I and II, taking into account the role of quasars with x-ray luminosities $L_x \geq 5 \times 10^{44}\,\text{erg s}^{-1}$, show that this component of diffuse cosmic background was formed at relatively low redshifts, $z < 3$. Note that excessive intensity of quanta at $E > 10^2\,\text{keV}$ up to 1 MeV shows a knee in the range $E \geq 1\,\text{MeV}$, which is clearly seen in Fig. 1.5. In the γ part of the spectrum, the intensity $I(E) \propto E^{-\alpha}$ for quanta with energy $E > 1\,\text{MeV}$ can be approximated, by following the work of Gehrels and Cheng (1996) by a set of power-law functions with exponents $\alpha \simeq 0.7$ for $E \simeq 1\,\text{MeV}$ and $\alpha \simeq 1.7$ for $2\,\text{MeV} < E < 10\,\text{MeV}$. The distribution of quanta over energy for the function $EI(E)$ is clearly manifest in the shape of the peak (Fig. 1.5). Note that the nature of this diffuse background remained unclear for a long time, despite numerous attempts to identify possible sources of its formation. Only a few powerful sources of gamma radiation were identified up to the end of the 1990s in the gamma range, such as 3C273, CenA, NGC4151 and NGC8-11-11 (Bassani *et al.*, 1985). The situation with the gamma background changed radically after the successful launch of the COMPTON satellite. The all-sky map obtained by the Energetic Gamma Ray Experiment Telescope (EGRET) on board the COMPTON Gamma Ray Observatory is shown in Fig. 1.6.

To conclude this section, it will be useful to summarize the observational data regarding the intensity distribution of cosmic rays (CRs) with energy above $10^2\,\text{keV}$ and up to the maximum energies that can be currently detected, $\sim 10^{21}\,\text{keV}$. Figure 1.7 shows the energy distribution of the flux of the CRs (no special effort was made to separate the electromagnetic

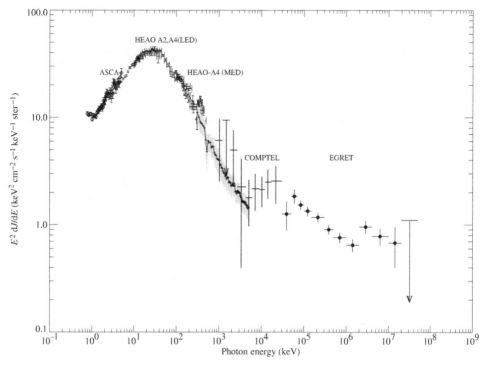

Figure 1.5 The spectrum of the diffuse x-ray and γ-ray backgrounds. Adapted from Strong, Moskalenko and Reimer (2004).

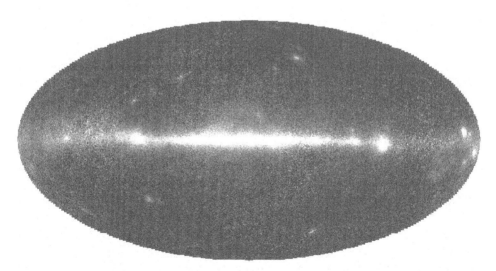

Figure 1.6 EGRET all-sky map of $E_\Gamma > 100\,\mathrm{MeV}\,\gamma$-ray intensity in galactic coordinate Aitoff projections. Adapted from Willis (2002).

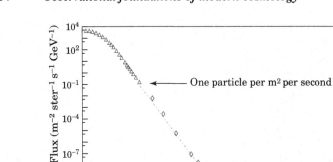

Figure 1.7 Flux of cosmic rays in the range 10^8 eV $< E < 10^{21}$ eV. From Sigl (2001a).

components). It is generally accepted that the main sources of formation of the CR spectrum in the energy range 10^{17}–10^{18} eV are pulsars, nuclei of active galaxies, galaxy clusters and a number of other non-cosmological sources of particle acceleration. However, in the range above 10^{18} eV, especially $E \geq 10^{20}$ eV, the situation is less trivial.

The spectrum of the so-called ultrahigh-energy cosmic rays (UHECR) composed from a number of sets of experimental data (Ave *et al.*, 2000; Hayashida *et al.*, 1994; Lawrence, Reid and Watson, 1991; Takeda *et al.*, 1998; Yoshida and Dai, 1998; Yoshida *et al.* 1995), is shown in Fig. 1.8. The fact of special importance is that several dozens of events were recorded in the energy range above the so-called Greisen–Zatsepin–Kuzmin limit (Greisen, 1966; Zatsepin and Kuzmin, 1966), $E_{\text{GZK}} \simeq 7 \times 10^{19} \left(\ddot{E}/10^{-3} \text{ eV} \right)^{-1}$ eV, where $\ddot{E}$ is the mean energy of the cosmic microwave background. The gist of the UHECR problem lies in that the characteristic free path length of nucleons in the cosmic background ($\gamma_{\text{CMB}} + \text{p} \rightarrow \text{p} + \text{e}^+ + \text{e}^- + \gamma$) is found to be ~ 20 Mpc (Greisen, 1966). In this case, the observed flux of CRs near the Earth must be characterized by a considerable correlation between the direction of arrival of the CRs and the expected sources that generate them. However, experimental data point to a high degree of isotropy of the background; this is the reason why the hypothesis of its cosmological nature deserves attention.

Table 1.2. *Energy distribution over various components of cosmic background*

Frequency range	Intensity ($m^{-2}\ sr^{-1}$)	Fraction of energy density
Radio	1.2×10^{-12}	1.1×10^{-6}
CMB	9.96×10^{-7}	0.93
IR	$4\text{--}5.2 \times 10^{-8}$	0.04–0.05
Optical	$2\text{--}4 \times 10^{-8}$	0.02–0.04
X-rays	2.7×10^{-10}	2.5×10^{-4}
Gamma radiation	3×10^{-11}	2.5×10^{-5}

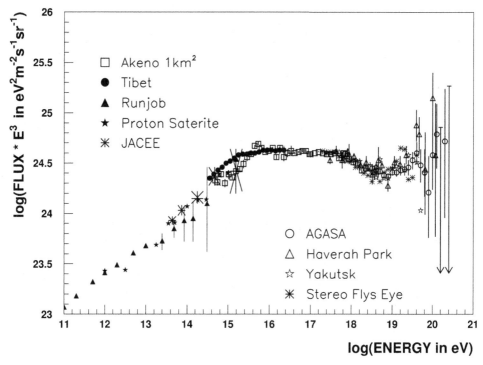

Figure 1.8 UHECR spectrum according to the observations by the facilities shown in the figure. From Sigl (2001b).

To conclude the survey of the current data on the distribution of cosmic radiation from the radio range to UHECR particle energies, we give in Table 1.2 a summary of the intensities of various components and their contribution to the total density of electromagnetic energy in the Universe. As we see from this table, 93% of the total energy density of electromagnetic radiation comes from CMB radiation, while the optical and infrared ranges constitute most of the remaining 7%. Taking into account the fact that diffuse components are formed at redshifts $z \leq 3$, we arrive at a result quite familiar in cosmology: the electromagnetic component of matter in the early Universe at $z \gg 3$ consisted of CMB only.

As the Universe continued expanding, the maximum of the spectrum shifted towards lower energies, in accordance with the law of temperature decrease $T_R(z) = T_0(1 + z)$ (Zeldovich and Novikov, 1983) and the quanta of CMB were undergoing the Doppler frequency shift. In this process, the energy density of radiation, ε_γ, the quantum concentration, n_γ, and the density of entropy, S_γ, changed with z in the following manner:

$$\varepsilon_\gamma = \overline{\varepsilon}_\gamma(1 + z)^4, \qquad n_\gamma = \overline{n}_\gamma(1 + z)^3, \qquad S_\gamma = \overline{S}_\gamma(1 + z)^3, \qquad (1.12)$$

where $\overline{\varepsilon}_\gamma$, $\overline{n}_\gamma$ and $\overline{S}_\gamma$ correspond to the current values for $z = 0$ (see Eq. (1.9)).

1.3   The baryonic component of matter in the Universe

In Section 1.2 we summarized the main parameters of the electromagnetic component of the current density of matter, in the Universe. However, in addition to this electromagnetic radiation, today's Universe is filled with conventional baryonic matter, which provides the original material for star formation and later serves as nuclear fuel that sustains their luminosity. An important feature of this component of matter is typically a very low temperature of matter, much lower than the relativistic limit, $T_p \simeq m_p c^2/k \sim (10^{13})$ K, where, m_p is the proton mass. Therefore, as the Universe expands, the baryonic component of matter changes following a law that differs from that for the primordial electromagnetic radiation,

$$\rho_b = \overline{\rho}_b(1 + z)^3, \qquad (1.13)$$

where $\overline{\rho}_b$ is the current value of the baryonic density at $z = 0$. We know that this fraction exists in the form of various structural forms, beginning with the condensed state and ending with plasma. It is mostly concentrated in clouds of gas and dust, in planets, stars and stellar remnants. In their turn, these younger components are building material for galaxies, groups of galaxies and galaxy clusters. Therefore, in contrast to the electromagnetic component, the baryonic matter is now very highly structured. In fact, by analysing the observational manifestations of these structural units, we can make a judgement about the content of baryons in them and, therefore, about their cosmological abundance. Following Fukugita, Hogan, and Peebles (1998), we evaluate the baryonic density of various structural forms of condensation of matter, using the standard normalization of the mean baryon density, $\Omega_b = \overline{\rho}_b/\rho_{cr}$, to the critical density, $\rho_{cr} = 3H_0^2/8\pi G \simeq 1.8 \times 10^{-29}h^2$, where h is the Hubble constant in units of $100\,\mathrm{km\,s^{-1}\,Mpc^{-1}}$.

1.3.1   Stars and stellar remnants in galaxies

Two subsystems of stars and their remnants must be distinguished in order to characterize the role of stars and stellar remnants in galaxies; these are connected to the structure of spiral and elliptical galaxies, namely the spherical population of old stars and the disk population that contains younger stars. The contributions of these two subsystems to the total mass of stars may differ for each type of galaxy. For instance, the spherical stellar population is most pronounced in elliptic galaxies, while the spherical component in irregular galaxies is either much less pronounced or is completely absent. Evaluations of baryon density in these two basic types of galactic population yield the following values for the parameter Ω_b (Fukugita *et al.*, 1998):

$$\Omega_{sp}h = 0.0018^{+0.0012}_{-0.0009}$$
$$\Omega_\alpha h = 0.0006^{+0.0003}_{-0.0002}. \qquad (1.14)$$

The estimate for irregular galaxies is given by

$$\Omega_{\text{Ir}} h = 0.0005^{+0.0003}_{-0.0002}.$$

1.3.2 Atomic and molecular gaseous components

The data for this fraction were obtained from HI 21 cm surveys (Rao and Briggs, 1993; Roberts and Haynes, 1994). For atomic hydrogen we have

$$\Omega_{\text{H}} h = 0.00025 \pm 0.00006; \tag{1.15}$$

$$\Omega_{\text{H}_2} h \simeq 0.00020 \pm 0.00006.$$

1.3.3 Baryons in galaxy clusters

Evaluations for Ω_b from the data of matter density concentrated in galaxy clusters are based on the distribution of the number of clusters as a function of their mass, suggested in Bahcall and Chen (1993):

$$n_{\text{cl}}(>M) = 4 \times 10^{-5} h^3 \left(\frac{M}{M_*} \right)^{-1} \exp \left(-\frac{M}{M_*} \right) \text{Mpc}^{-3}, \tag{1.16}$$

where $M_* = (1.8 \pm 0.3) \times 10^{14} h^{-1} M_\odot$ and M is the total mass of matter inside a sphere of radius $1.5 h^{-1}$ Mpc, enclosing the cluster. The distribution of matter within this radius is close to dynamic equilibrium. Following Fukugita *et al.* (1998), we define a galaxy cluster as an object with mass $M > 10^{14} h M_\odot$. Then the integral $\int dM\, M dn_{\text{cl}}/dM = \rho_{\text{cl}}$ corresponds to the average density of the baryonic component in the cluster:

$$\rho_{\text{cl}} = \left(7.7^{+2.5}_{-2.2} \right) \times 10^9 h^2 M_\odot \text{Mpc}^{-3}. \tag{1.17}$$

Normalizing ρ_{cl} to the critical matter density, we obtain

$$\Omega_{\text{cl}} = 0.028^{+0.009}_{-0.008}. \tag{1.18}$$

Note that the mass of the gas in the space between the galaxy clusters is reliably identified with the data of x-ray observations (Fabricant *et al.*, 1986; Hughes, 1989; White, Efstathiou and Frenk, 1993). A recalculation of the contribution of this component to the parameter Ω_b points to an extremely small contribution of the intercluster gas to the aggregate density of baryons (Mayers *et al.*, 1997; White and Fabian, 1995) as compared to Eq. (1.18):

$$\Omega_{\text{HII,cl}} h^{3/2} = 0.0016^{+0.001}_{-0.0007}. \tag{1.19}$$

1.3.4 Plasma in groups of galaxies

Evaluations of the density of the baryonic fraction in groups of galaxies are based on the observations of hard x-ray radiation made with the ROSAT satellite (Mulchaey *et al.*, 1996). According to Fukugita *et al.* (1998) it was possible to evaluate the density of the baryonic component for 17 groups of galaxies using the measurements of fluxes of soft x-ray radiation:

$$\Omega_{\text{HII,group}} h^{3/2} \simeq 0.003^{+0.004}_{-0.002}. \tag{1.20}$$

1.3.5 Massive compact halo objects (MACHOs)

Immediately after the discovery of the effect of gravitational lensing of starlight in the larger Magellanic Cloud (Alcock *et al.*, 1997), the nature of this galactic component

Table 1.3.

	Component	Mean value	Maximum value	Minimum value
1	Stars in spherical subsystems	$0.0026h_{70}^{-1}$	$0.0043h_{70}^{-1}$	$0.0014h_{70}^{-1}$
2	Stars in the disk	$0.00086h_{70}^{-1}$	$0.00129h_{70}^{-1}$	$0.00051h_{70}^{-1}$
3	Stars in irregular galaxies	$0.000069h_{70}^{-1}$	$0.000116h_{70}^{-1}$	$0.0000331h_{70}^{-1}$
4	Neutral atomic gas	$0.00033h_{70}^{-1}$	$0.00041h_{70}^{-1}$	$0.00025h_{70}^{-1}$
5	Molecular gas	$0.00030h_{70}^{-1}$	$0.00037h_{70}^{-1}$	$0.00023h_{70}^{-1}$
6	Plasma in clusters	$0.0026h_{70}^{-1.5}$	$0.0044h_{70}^{-1.5}$	$0.0014h_{70}^{-1.5}$
7	Plasma in groups	$0.014h_{70}^{-1}$	$0.030h_{70}^{-1}$	$0.0072h_{70}^{-1}$
	Gas component	at $z \simeq 3$		
10	Lyman-alpha clouds	$0.04h_{70}^{-1.5}$	$0.05h_{70}^{-1.5}$	$0.01h_{70}^{-1.5}$

h_{70} – Hubble constant
in units of 70 km s^{-1} Mpc^{-1}

attracted widespread attention. Judging by the data of Alcock *et al.* (1997), we can state that these are manifestations of objects whose masses are comparable to the solar mass, i.e. $M_{\mathrm{MACHO}} \simeq 0.5^{+0.3}_{-0.2} M_{\odot}$. Nevertheless, their nature remains problematic. Fukugita *et al.* (1998) note that if MACHOs consist of baryons, then the maximum of the parameter Ω_b may reach $\Omega_{b,\mathrm{MACHO}} \simeq 0.25$. However, this evaluation only points to an upper bound, and its reliability is uncertain. As a counter-example, we may cite the hypothesis that these objects are massive black holes (Ivanov, Naselsky and Novikov, 1994) formed at the earliest stages of the expansion of the Universe. Then the fraction of baryons in these objects should be negligibly small, $\Omega_b \simeq 0$ (see the discussion in de Freitas-Pacheco and Peirani (2004)).

1.3.6 *Ly-α 'forest' for redshifts* $z \simeq 3$

In contrast to the current epoch, in which the main representatives of the baryonic fraction of matter are stars, an analysis of the Ly-α lines in absorption spectra of quasars at redshifts $z \simeq 3$ makes it possible to evaluate the density of baryonic matter in the gaseous phase. The abundance of such clouds and the density contrast in them depend on a specific model of structure formation in the expanding Universe. It was shown in Rauch *et al.* (1997) that for the theory to fit the observational data on Ly-α absorption lines, the baryon fraction in clouds must be above $\Omega_{\mathrm{Ly}\text{-}\alpha}h^2 \geq 0.017 - 0.021$. However, this estimate depends greatly on the choice of the cosmological model (Fukugita *et al.*, 1998). Hui *et al.* (2002) came to a similar conclusion, showing that the baryon density may reach $\Omega_b h^2 \simeq 0.045$. In this case, we speak about uncertainty characterized by a factor of 2, even though it could be possible that all subsequent improvements of the models would lead to a significantly reduced estimate.

The summary of the results of this subsection are given in Table 1.3 for the expected values of density of the baryonic fraction of matter based on the above-listed observational tests and on their theoretical interpretation.

Assuming $\Omega_b h^2 \simeq 0.02$ in order to estimate the total density of the baryonic fraction, it is not difficult to evaluate today's concentration of baryons: $n_b \simeq 2 \times 10^{-7} \left(\frac{\Omega_b h^2}{0.02} \right)$ cm^{-3}. For comparison, the concentration of CMB quanta is $412 \, \mathrm{cm}^{-3}$, and therefore

$$\xi_{10} = 10^{10} \frac{n_{\mathrm{b}}}{n_\gamma} = 274\Omega_{\mathrm{b}} h^2. \tag{1.21}$$

1.3.7 Cosmological nucleosynthesis and observed abundance of light chemical elements

We mentioned in Section 1.1 that the effort to try to explain why the current chemical composition of matter in the Universe is as we observe today was the starting point for creating today's cosmology and for expanding it. Beginning with the pioneering paper of George Gamow and his colleagues, the theory of the cosmological synthesis of light chemical elements was gradually improved, acquiring ever greater predictive power. We also mentioned that the blackbody (Planckian) character of the spectrum of primordial radiation is an indication that radiation and the $\mathrm{e}^+\mathrm{e}^-$ plasma were in electrodynamic equilibrium at some point in the past. Inevitably this equilibrium had to break down after the $\mathrm{e}^+\mathrm{e}^-$ annihilation, when the characteristic plasma temperature became comparable to $T \simeq m_{\mathrm{e}}c^2/k \sim 10^{10} \, \mathrm{K}$. Until that moment, the high concentration of electron–positron pairs, comparable to that of gamma quanta, sustained the equilibrium not only between them, but also between the electron neutrinos, ν_{e}, and antineutrinos, $\overline{\nu}_{\mathrm{e}}$. In its turn, the presence of electron neutrinos ($\nu_{\mathrm{e}}\overline{\nu}_{\mathrm{e}}$) in the cosmological plasma maintains equilibrium between neutrons and protons in weak interaction reactions (Hayashi, 1950; Olive, Steigman and Walker, 2000; Wagoner, 1973; Wang, Tegmark and Zaldarriaga, 2002):

$$\mathrm{n} + \mathrm{e}^+ \leftrightarrow \mathrm{p} + \overline{\nu}_{\mathrm{e}}; \qquad \mathrm{n} + \nu_{\mathrm{e}} \leftrightarrow \mathrm{p} + \mathrm{e}^-; \qquad \mathrm{n} \leftrightarrow \mathrm{p} + \mathrm{e}^-\overline{\nu}_{\mathrm{e}}. \tag{1.22}$$

Since the typical weak interaction reaction rates, $\Gamma = \langle \sigma_{\nu\mathrm{p},\mathrm{n}} n_\nu c \rangle$, where n_ν is the neutrino concentration, are proportional to T^5, and the plasma temperature decreases with progressive expansion of the Universe, it is clear that beginning with a certain moment, t_*, the equilibrium between protons and neutrons in weak interaction reactions should break down.[2] Formally, the moment of 'quenching' of Eq. (1.22) can be found from the condition $\Gamma(t_*) \cdot t_* = 1$. Detailed calculations show (Olive *et al.*, 2000) that the plasma temperature corresponding to time t_* is close to $T(t_*) = 10^{10} \, \mathrm{K} \sim 1 \, \mathrm{MeV}$. The residual ratio of neutron to proton concentrations is given by the Boltzmann factor, $(n/p) \simeq \exp(-\Delta mc^2/kT_*)$, where Δm is the difference between the proton and neutron masses. Immediately after quenching of the weak interaction reactions, the merger of a neutron and a proton into a deuteron nucleus, $\mathrm{n} + \mathrm{p} \leftrightarrow \mathrm{D} + \gamma$, becomes energetically favoured.

However, owing to a large number of quanta with energy $E \sim 2.7kT$ ($n_\gamma/n_\nu \sim 10^{10}$), the deuterium photodissociation reactions become extremely efficient, and the equilibrium deuterium concentration at the moment of quenching is extremely low. We have mentioned earlier that, as the Universe expands, the maximum of the primordial radiation spectrum shifts to lower temperatures. Note that in Wien's segment of the spectrum the quantum concentration decreases as $\exp\left(-\frac{E}{kT}\right)$. As the bounding energy of the deuterium nucleus is $E_{\mathrm{D}} = 2.2 \, \mathrm{MeV}$, it is not difficult to find the critical temperature $T = T_{\mathrm{D}}$, beginning with which

---

[2] The dependence $\Gamma \propto T^5$ is readily obtained using the following argument. With particle energy $\sim 1 \, \mathrm{MeV}$, the cross-sections of the processes (1.22) are $\sigma \propto E^2$, where $E \sim kT$ is the mean neutrino energy. The neutrino concentration, n_ν, in equilibrium with the plasma is close to the concentration of γ quanta and, therefore, n_ν is proportional to T^3.

the photodissociation process becomes inefficient. This value of temperature corresponds to the condition $\xi^{-1} \exp\left(-\frac{E_D}{kT_D}\right) \simeq 1$, where $\xi = \frac{n_N}{n_\gamma} \sim 10^{-10}$ and yields the value $T_D \simeq 0.1$ MeV (Wagoner, 1973; Wang *et al.*, 2002). The characteristic time counted off the start of the Universe's expansion is then close to $t_D \simeq 10^2$ s, which is of critical importance for subsequent estimates of the upper bound of the abundance of cosmic He^4. The point is that between the 'quench' moment ($t_* \simeq 1$ s) of the weak interaction reactions (1.22) and the 'quench' moment of deuterium photodissociation reactions, neutrons decay freely with characteristic time $\tau_N \simeq 887 \pm 2$ s (Olive *et al.*, 2000). Therefore, the quenched concentration of neutrons by the moment t_D decreases to

$$\left(\frac{n}{p}\right)_{t_D} \simeq \left(\frac{n}{p}\right)_{t_*} \exp\left(-\frac{t_D}{\tau_N}\right). \tag{1.23}$$

Even if all the free neutrons bind later into He^4 nuclei, their mass concentration will be given by

$$X_{He^4}^{max} \simeq \frac{2\left(\frac{n}{p}\right)_{t_D}}{1 + \left(\frac{n}{p}\right)_{t_D}} \simeq 0.26. \tag{1.24}$$

In reality, the He^4 content predicted in the course of cosmological nuclear synthesis is even lower. The point is that immediately after 'quenching' of photodissociation processes, the deuterium synthesis reaction $n + p \rightarrow D + \gamma$ leads rapidly to formation of a deuterium peak in the concentration of light chemical elements. In addition, reactions that transform deuterium into tritium and He^3 are immediately triggered:

$$\begin{aligned}
&D + D \leftrightarrow p + T; \quad D + D \leftrightarrow He^3 + n; \\
&D + n \leftrightarrow T + \gamma; \quad D + n \leftrightarrow He^3 + \gamma; \\
&He^3 + n \leftrightarrow T + p;
\end{aligned} \tag{1.25}$$

followed by transformation to He^4:

$$\begin{aligned}
&D + D \leftrightarrow He^4 + \gamma; \quad D + He^3 \leftrightarrow p + He^4; \quad D + T \leftrightarrow He^4 + n; \\
&He^3 + He^3 \leftrightarrow He^4 + 2p; \quad T + p \leftrightarrow He^4 + \gamma; \quad He^3 + n \leftrightarrow He^4 + \gamma.
\end{aligned} \tag{1.26}$$

Owing to the absence of nuclei with atomic numbers $A = 5$ and $A = 8$, the synthesis of Li, Be and heavier elements proceeds through the following channels:

$$He^3 + He^4 \leftrightarrow Be^7 + \gamma; \quad T + He^4 \leftrightarrow Li^7 + \gamma; \ldots \tag{1.27}$$

Figure 1.9 plots the dynamics of synthesis of light chemical elements as a function of temperature as it decreases in the course of expansion of the Universe. Figures 1.10(a) and (b) plot mass concentration of Be^9 and B^{10-11}, respectively, as a function of the parameter η. Today's values of mass concentration are given, without taking into account their possible transformation in the course of formation and evolution of stars (Olive *et al.*, 2000).

Roughly, these are the predictions of the current theory of cosmological nuclear synthesis based on the 'hot' model of uniform and isotropic Universe. As we see from Fig. 1.10 (a) and (b), the predictions of the theory with regard to current concentrations of He^4, and especially deuterium, are very sensitive to the current density of baryons, provided the temperature and concentration of CMB quanta are known. Therefore, observational cosmology offers us a new

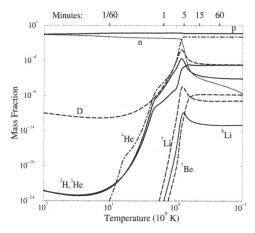

Figure 1.9 Dynamics of synthesis of light chemical elements in the hot Universe. Adapted from Taytler *et al.* (2000).

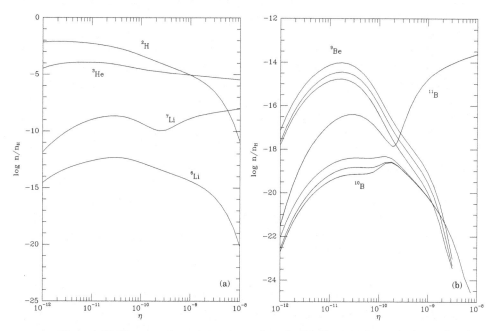

Figure 1.10 The mass element abundances from the Big Bang nucleosynthesis as a function of η. (a) For He^4, He^2, Li^7; (b) for Be^9, B^{10} and B^{11}. Solid lines for Be^9 and B^{10} correspond to uncertainties in the theoretical predictions. Taken from Esposito (1999) and Thomas *et al.* (1993).

possibility of higher accuracy of calculation of the current density of baryonic matter based on analysing the cosmic abundance of He^4, D and Li^7. Note, however, that this method is not direct, mostly because light chemical elements are either synthesized (for example He^4) or will burn out in the course of evolution of the stellar populations of galaxies. Subsequently, a detailed analysis of the possible channels of transformation of the primordial chemical composition of baryonic matter to its current state is one of the most important problems of today's astrophysics, attracting more and more attention from researchers. An additional important factor that strengthens the hope of successful implementation of the program of establishing the primordial chemical composition of matter is the following familiar observational fact: different chemical elements are contained in different types of objects in different ratios. For instance, an analysis of UV absorption from the ground level (see references in Olive *et al.* (2000)) detects deuterium mostly in cold clouds of neutral gas (HI regions). At the same time, the abundance of He^3 can be determined by radioastronomical techniques (similar to observing the 21 cm line) that detect He^{3+} that concentrates in clouds of hot ionized hydrogen, HII. However, tritium is observed in absorption spectra of hot low-mass stars. Naturally, the two most abundant isotopes of hydrogen, He^4 and D, play the main role in determining the present value of the baryonic density in the Universe. Let us consider this aspect of the problem in more detail.

Cosmic He^4

We know that in addition to the cosmological nucleosynthesis channel, the He^4 isotope can be synthesized in the process of stellar evolution, though in considerably smaller amounts. Therefore, for greater certainty in identifying the upper bound of its cosmological abundance, it was suggested in Izotov and Thuan (1998), Izotov, Thuan and Lipovetsky (1994), Olive and Steigman (1995), Pagel *et al.* (1992), Skillman and Kennicutt (1993), Skillman *et al.* (1994) that attention should be focused on analysing the He^4 content of extragalactic HII regions characterized by undoubtedly low metal content. Since a sample of such regions is composed of about 40 areas, the accuracy of determining X_{He^4} is sufficiently high ($\sim 1\%$) (Olive *et al.*, 2000). In fact, we are talking here, and later when discussing lithium abundance, about He generated in nuclear reactions in primordial matter within the first five minutes of the life of the Universe. Furthermore, as the lowest-metallicity areas of the sample contain 2–3% of the solar abundance of He^4, the recalculation of the mass concentration of He^4 to its mean abundance has an uncertainty at the same error level ($\sim 2\%$). For instance, the estimate of the mass concentration of He^4 made in Olive and Steigman (1995) using low-metallicity HII regions (Pagel *et al.*, 1992) yielded $Y_{He^4} \simeq 0.234 \pm 0.003$. At the same time, Izotov and Thuan (1998) used a somewhat bigger sample and arrived at $Y_{He^4} \simeq 0.244 \pm 0.002$. It is pointed out in Skillman, Terlevich and Terlevich (1998) that the cause of the discrepancy between these two estimates may be the insufficient attention paid to the collisional excitation of recombination lines of He^4 as it could result in reduced $Y_{He^4} \simeq 0.241 \pm 0.002$. Added to this must be the effects related to the systematic effects, for instance the uncertainty in the absorption estimates that result in increased He^4 content in analysed areas. In view of these factors, we can arrive at a sufficiently reliable evaluation of the cosmic abundance of He^4: $0.228 \leq Y_{He^4} \leq 0.248$ (Olive and Steigman, 1995).

Cosmic deuterium

It is generally accepted (see, for example, Olive and Steigman (1995) and references therein) that the cosmic abundance of deuterium is one of the more reliable tests for identifying the current density of the baryonic fraction of matter. First of all, in contrast to He^4, the cosmological abundance of deuterium essentially depends on the parameter ξ, which allows us to narrow down the range of possible values of $\Omega_b h^2$ and achieve an agreement between the predictions of the theory of cosmological nucleosynthesis and the observational data.

Furthermore, in contrast to He^4, the cosmological deuterium can only decay in the course of star and galaxy formation (Olive and Steigman, 1995). Therefore, those observations that identify the maximum possible deuterium content (Epstein, Lattimer and Schramm, 1976) are of special interest. It is curious that without taking into account the evolution of the chemical composition of our Galaxy, deuterium content relative to hydrogen is evaluated using the UV observation data (Linsky, 1998) as $D/H \simeq (1.5 \pm 0.1) \times 10^{-5}$. We also need to mention recent measurements of deuterium abundance in Jupiter's atmosphere (Mahaffy *et al.*, 1998) that yielded an estimate $D/H \simeq (2.6 \pm 0.7) \times 10^{-5}$. An important place among the newer methods of evaluating the cosmic abundance of deuterium is occupied by observations of low-metallicity clouds at high ($z \sim$ 2–4) redshifts. Spectra of the absorption system Q1937-1009 ($z \simeq 3.572$) were studied in Tytler, Fan and Burles (1996). It was shown that, in this system, $D/H = (2.3 \pm 0.3) \times 10^{-5}$. For a similar combined object Q1009 + 2956, the value $D/H = (4.0 \pm 0.7) \times 10^{-5}$ was obtained. Summarizing the results of this study, we calculate the 95% significance level interval for deuterium abundance to be as follows:

$$2.9 \times 10^{-5} \leq D/H \leq 4 \times 10^{-5}. \tag{1.28}$$

The paper by O'Meara *et al.* (2001) that reports the measurement of deuterium in QSO HS0105 + 1619 for $z = 2.536$ should also be mentioned. The abundance was calculated for $D/H = (2.54 \pm 0.23) \times 10^{-5}$. We can say, with a degree of caution, that these estimates result in placing the parameter $\eta_{10} = 10^{10}\eta$ in the range $4.2 \leq \xi_{10} \leq 6.3$. Taking into account the dependence of η_{10} on the density of the baryonic fraction, we obtain

$$0.015 \leq \Omega_b h^2 \leq 0.023. \tag{1.29}$$

Note that the uncertainty in choosing the range of the parameters ξ_{10} and Ω_b could, in principle, be minimized relative to cosmic lithium (Olive *et al.*, 2000).

Lithium
The abundance of the Li^7 isotope is evaluated from the data of observations of about 100 hot stars of population II. For such stars with high surface temperature, $T > 5500$ K, and relatively low metal content ($\sim 0.05 z_\odot$), it is possible to determine the Li^7 content Li$^7/H = (1.6 \pm 0.1) \times 10^{-10}$ (Molaro, Primas and Bonifacio, 1995). However, it was mentioned earlier that the observed content of Li^7 cannot in any way be interpreted as the primordial level. We refer those who are interested in specific details to the original publications (Cayre *et al.*, 1999; Molaro *et al.*, 1995; Ryan, Norris and Beers, 1999; Ryan *et al.*, 2000). Recapitulating on the results of theoretical predictions of the mass content of light chemical elements and observational data concerning their abundance, we provide a summarizing diagram (Fig. 1.11a) in which the evaluation errors are marked as grey rectangles. It is clear from this figure that the optimum range for the parameter η_{10} is $1.5 \leq \eta_{10} \leq 6.3$, which means that the current density of the baryonic fraction of matter does not exceed $\Omega_b h^2 = 0.023$. Recalculated to the Hubble constant, $H_0 = 70$ km s^{-1} Mpc^{-1} ($h = 0.7$), this result is in excellent agreement with the results of evaluating the baryonic density in the Universe from the data on Ly-α clouds.

1.3.8 Global parameters of the present-day Universe
We have mentioned several times that the considerable progress made in cosmology has been, driven not only by the detailed study of the spectrum of primordial electromagnetic

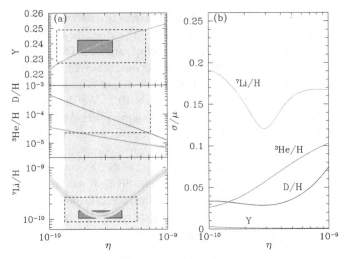

Figure 1.11 (a) Theoretical predictions and (b) observational constraints on the abundance of light chemical elements in the Universe as a function of the parameter η_{10}. Adapted from Olive (2000).

background, but also by the discovery of a number of novel phenomena and the improvement in the already classic tests of observational cosmology. In this context, questions about what the 'dark matter' (which provides most of the density of matter in space),[3] really is what its physical nature is, what its properties are and to what extent it determines the age of the Universe, are still topical and important, and, furthermore, lead to a number of questions of principal importance for astrophysicists. It is natural that the two most important parameters, namely the current density of the 'dark matter' and the value of the Hubble constant, cannot be treated separately from one another. On the other hand, the fraction of density carried by the hidden mass, Ω_{dm}, and the Hubble constant, H_0, are the main parameters that characterize the current expansion of the Universe and furthermore dictate the possible evolution scenarios for its future. Moreover, we have already seen in the preceding sections that not only does the value of the Hubble constant determine the age of the Universe, but that it is also a normalizing parameter through which the current density of the baryonic fraction of matter expresses itself.

In this subsection we mostly discuss the observational aspect of the problem of measuring two global parameters – the density of the dark matter, Ω_{dm}, and the Hubble constant, H_0. Each of these parameters has a long and fairly dramatic history. We can recall that between the mid 1940s and the end of the 1950s the calculated value of the Hubble constant was higher than its current value by a factor of 5 to 10, which implied an anomalously low age of the Universe. The 'dark matter' problem in astrophysics was formulated slightly earlier, in the mid 1930s. On one hand, this matter manifested itself as a stabilizer of the visible ('luminous') component of galaxies (Zwicky, 1957), but, on the other hand, in the form of uniformly distributed component (the cosmological constant), this

---

[3] We include in the density of the dark matter not only the invisible matter concentrated in galaxies and their clusters, but also the uniformly distributed component – the cosmological constant, or quintessence (the so-called dark energy; see below).

matter increased the theoretically predicted age of the Universe (Kardashev, 1967; Shklovsky, 1965).

These problems are still relevant and important today, they and stimulate progress in novel techniques that improve the accuracy of the evaluations of H_0 and Ω_{dm}. In this subsection we briefly discuss the current observational status of these two most important global parameters of the present Universe.

The Hubble constant

It is amusing that the entire 'dramatic clash of ideas' of the last 20 years in connection with the determination of the value of the Hubble constant takes root at the very beginning of the 1980s (but was discernible even before that), when Sandage and Tammann, and practically at the same time de Vaucouleurs, arrived at two mutually exclusive estimates of the value of H_0. According to Sandage and Tammann (1982), $H_0 = 50$ km s^{-1} Mpc^{-1}. According to de Vaucouleurs (1982), the constant was twice as large, $H_0 = 90$–100 km s^{-1} Mpc^{-1}. Curiously enough, astrophysicists of the day were joking that if one wanted to know the value of the Hubble constant, they should take the result of Sandage and Tammann, add it to the result of de Vaucouleurs and divide by 2.

It was later found that this joke, as often happens, contained an element of, if not the entire, truth. Namely, the current data relevant to the value of the Hubble constant and based on the models of expanding photospheres of type II supernovas yield the value $H_0 = 73 \pm 9$ km s^{-1} Mpc^{-1} (Schmidt, Eastman and Kirshner, 1994). The method of determining H_0 from the retardation of the signal in gravitational lensing of quasars yielded very similar values. Franx and Tonry (1999) found for the system $0957 + 561$ the value $H_0 = 71 \pm 7$ km s^{-1} Mpc^{-1}. The effect of retardation was measured quite recently for three lenses, namely B0218 + 357, B1608 + 656 and PKS 1830 + 211. Retardation time for the system B0218 + 357 (Biggs *et al.*, 1999) yielded $H_0 = 69^{+13}_{-19}$ km s^{-1} Mpc^{-1}. In B1608 + 656 the corresponding value was $H_0 = 64 \pm 7$ km s^{-1} Mpc^{-1} for $\Omega_{dm} = 0.3$, and 59 ± 7 km s^{-1} Mpc^{-1} if the total density of matter in the Universe corresponds to $\Omega_{dm} = 1$ (Koopmans and Fassnacht, 1999). We must emphasize that the dependence of retardation time on Ω_{dm} in this technique is fairly weak, but must nevertheless be taken into account when processing experimental data (Fukugita, 2000). In addition to the methods cited above for evaluating H_0, we also need to mention such astronomical methods as observation of Cepheids, planetary nebulae and type I supernovas using both land-based telescopes and, above all, the Hubble Space Telescope.

Table 1.4 summarizes the data on the evaluation of the Hubble constant based on using the above-described methods (see references in the table). The data are combined in the table with the results obtained from gravitational lensing of quasars, yielding as a result the current value of the Hubble constant (Fukugita, 2000):[4]

$$H_0 = (71 \pm 7) \times_{0.95}^{1.15} \text{ km s}^{-1} \text{ Mpc}^{-1}, \tag{1.30}$$

where the 10% error source in parentheses originates with the error of measuring distances to the Large Magellanic Cloud (LMC), $L^{LMC} \simeq 50$ kpc, which is the standard 'unit' for measuring distances to cosmic objects. As we see from Eq. (1.30), the upper bound on H_0 reaches the already mentioned value $H_0 \simeq 90$ km s^{-1} Mpc^{-1}, and the lower bound is

---

[4] The upper index (1.15) corresponds to choosing the + sign and the lower index (0.45) to choosing the minus sign.

Table 1.4.

Secondary indicators	References	Hubble constant (km s^{-1} Mpc^{-1})
Tally–Fisher test	HST-KP (Sakai *et al.*, 2000)	$71 \pm 4 \pm 7$
Fundamental plane	HST-KP (Kelson *et al.*, 1999)	$78 \pm 8 \pm 10$
SBF	HST-KP (Ferrarese *et al.*, 1999)	$69 \pm 4 \pm 6$
SBF	Tonry *et al.* (2000)	$77 \pm 4 \pm 7$
SNeIa	Riess, Press and Kirshner (1995)	67 ± 7
SNeIa	Hamuy *et al.* (1996)	$63 \pm 3 \pm 3$
SNeIa	Jha *et al.* (1999)	$63^{+5.6}_{-5.1}$
SNeIa	Suntzeff *et al.* (1999)	65.6 ± 1.8
SNeIa	HST-KP (Gibson *et al.*, 2000)	$68 \pm 2 \pm 5$
SNeIa	Saha *et al.* (1999)	60 ± 2
Combined value (see text)		$(64-78) \pm 7$

$H_0 \simeq 60$ km s^{-1} Mpc^{-1}. Note that the values of H_0 close to $H_0 = 58 \pm 6$ km s^{-1} Mpc^{-1} were obtained in Tammann (1999) from the data of the velocity distribution in Type Ia supernovae (SNeIa) using the Hubble Space Telescope. (For a review of the problem, see in Jansen, Tonry and Blakeslee (2004). For new data on H_0 from the observation of the CMB anisotropy, see Chapter 8.)

The Ω_{dm} parameter

The evaluation of the total density of matter in the Universe includes the estimation of both the clusterized component, Ω_{dm}, and a possible diffusely distributed component that is not included in the galaxies and their clusters and is spread uniformly through the Universe. Following the tradition, we assign this component to the cosmological constant Λ (and the corresponding parameter Ω_Λ), the background of low-mass particles (for example, neutrinos) or other physical fields whose presence in the current Universe is being actively discussed in the literature (the so-called 'quintessence'). Differences in the characteristics of the special distribution of 'dark' matter dictate the differences in methods of detecting it. Note that in this subsection we mostly concentrate on considering the observational manifestations of the 'hidden mass', leaving the discussion of various models of its physical nature to later sections. In this context the 'hidden mass' interests us only as another source of gravitational field in addition to baryonic matter and electromagnetic radiation, acting as a stabilizing factor for the structural forms in the mass distribution that we observe.

Hidden mass in galaxies and clusters

Historically, the first objects that clearly demonstrated that there was a 'virial paradox' between the luminous and gravitating masses were the spiral galaxies, for which the rotation curves $V(r)$ are non-decreasing functions of distance from the centre (Fig. 1.12). This behaviour of $V(r)$ is an important observational confirmation of the hypothesis that a massive component exists in these galaxies, which extends to scales that exceed the visual size of the galaxies. The natural cause of this behaviour of rotation curves $V(r)$ is the presence of a massive galactic halo containing a low-luminosity component. The existence of a

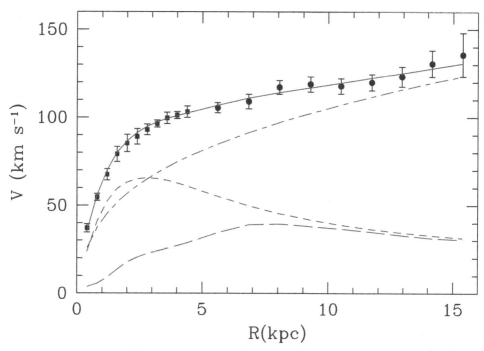

Figure 1.12 Rotation curves for the galaxy M33. Dots with vertical bars are observational data; solid curves trace the best-fit approximation of observations in models with different contributions of stars of the disk (short dashed line), halo (dot–dash) and the gas (long dashed line). Adapted from Corbelli and Salucci (1999).

background of weakly luminous matter (or matter emitting no quanta at all) also manifests itself in the so-called 'mass–luminosity' test. The idea of this method is based on the relation $\rho_G \simeq I_G \cdot \langle \frac{M}{L} \rangle$, where ρ_G is the density of matter in the galaxy, I_G is the specific luminosity per unit volume and $\langle \frac{M}{L} \rangle$ is the mean ratio of mass to luminosity averaged over a given class of galaxies. The specific luminosity, I_G, in this formula is approximated fairly well by an empirical relation $I_G = (2.0 \pm 0.4) \times 10^8 h L_\odot \, \text{Mpc}^{-3}$ (Fukugita, 2000). For galaxies with characteristic scales $\delta \leq 100$ kpc, the value $\langle \frac{M}{L} \rangle$ approximately equals $(1–2) \times 10^2 h^{-1} \langle \frac{M_\odot}{L_\odot} \rangle$, where the symbol $\odot$ stands for solar units (Bahcall, Lubin and Dorman, 1995; Zaritsky *et al.*, 1997). Note that at the end of the 1970s and the beginning of the 1980s, it seemed that the value of $\langle \frac{M}{L} \rangle$ grows as the mass of the object increases. Modern observations show that this growth does indeed take place but, beginning with characteristics scales 200–300 kpc corresponding to typical sizes of coronas of massive galaxies (Bahcall *et al.*, 1995; Calberg *et al.*, 1996), the curve reaches a 'plateau'.

For galaxies with typical sizes below 100 kpc, the virial radius evaluated using the spherical collapse model is given by $r_v = 0.13 \Omega_{\text{dm}}^{-0.15} (M/10^{12} M_\odot)^{1/2}$ (Fukugita, 2000). It is typically assumed that the distribution of 'dark matter' inside r_v is isothermal. Then, $\langle \frac{M}{L} \rangle \simeq 150$–400. Note that this value is practically the same as the value one finds for groups and clusters of galaxies (Fukugita, 2000). Using the relationships presented above, we can evaluate the parameter Ω_{dm} in the framework of the method outlined above. According to Fukugita (2000) and Fukugita *et al.* (1998), it is close to $\Omega_{\text{dm}} \simeq 0.2^{+0.2}_{-0.1}$. A more reliable estimate of $\Omega_{\text{dm}} \simeq$

0.19 ± 0.06 was obtained in Calberg, Yee and Ellingson (1997), where the galactic field method was used. However, the error depends essentially on assumptions about the nature of the dark matter distribution outside galaxy clusters. We see that uncertainty of evaluations is fairly high; nevertheless, the fact that the value of Ω_{dm} does not exceed 40% of the critical density is fairly significant.

Another method of finding the density of dark matter is the so-called 'peculiar velocity versus density' test based on the relationship of a perturbation of the Hubble velocity of matter ($\vec{v}$) and a perturbation of matter density, $\delta = \delta\rho/\rho$, in the expanding Universe (Peebles, 1983):

$$\Delta\vec{v} + H_0\Omega_{dm}^{0.6}\delta = 0. \tag{1.31}$$

Small-scale peculiar motion of matter ($r < 1$ Mpc) and the density, δ, evolve in a non-linear mode. For these scales, an analysis of the relation '$\vec{v} - \delta$' is based on the so-called cosmic virial theorem, in which it is assumed that the peculiar accelerations occurring on small scales between pairs of galaxies are balanced out by their relative motion velocities. Using this approximation, it is possible to evaluate rather crudely the parameter $\Omega_{dm}(10\,\text{kpc} < r \le 1\,\text{Mpc}) \simeq 0.15 \pm 0.10$ (Peebles, 1999a). On scales exceeding 1 Mpc, density perturbations and matter velocities evolve linearly ($|v| \ll c$ and $\delta \ll 1$). In this case, the value $\Omega_{dm} \simeq 0.2$ was obtained using, for instance, the data on the velocity field of galaxies in the neighbourhood of the Virgo cluster (Davis and Peebles, 1983). However, it is necessary to emphasize that attempts to evaluate Ω_{dm} from the data on large-scale matter–velocity field typically result in large systematic errors and often yield mutually exclusive results (see, for example, Dekkel *et al.* (1999) and Hamilton (1998)).

One of the factors leading to this situation is the fact that the field of velocities, $\vec{v}$, and density perturbations, δ_G, are detected using the luminous matter in galaxies and clusters, δ_g, while Eq. (1.31) deals with the total density, Ω_{dm}, and its perturbations, δ. By introducing a normalizing parameter $b = \delta_G/\delta$, we find from Eq. (1.31) that the velocity field is directly related not only to Ω_{dm}, but also to the parameter b that indicates to what extent luminous matter represents the distribution of the hidden mass. Another important factor is found in the errors of determining the field of peculiar velocities and, as we have already mentioned, uncertainties in determining the distance scale.

In addition to the methods outlined above, techniques for indirect determination of Ω_{dm} are rapidly progressing; they are based on models of specific effects. Among them we find first of all the direct analysis of the spectrum of density and velocity of matter perturbations on scales up to 10–100 Mpc (Peacock and Dodds, 1996), an analysis of formation of the current abundance of galaxy clusters (Viana and Liddle, 1999; White *et al.*, 1993), an analysis of the power spectrum of density fluctuations on small scales $r \le 3$ Mpc (Peacock, 1997), and some others.

At the same time, a very important factor that gives a measure of today's density of not only the hidden mass, but also of the diffusely distributed component is the evaluated age of the Universe, t_U. In this case, it is not enough to know the density of matter confined to galaxies and clusters; we need to take into account all possible types of diffusely distributed matter. Traditionally the nature of this 'hidden mass' of the Universe was identified with the cosmological term Λ that characterizes the energy density and pressure of the vacuum.

The idea that is being actively discussed these days is that the age of the Universe can be affected not only, and not so much by, the vacuum; in principle, any highly uniform and isotropic physical fields may act as sources of 'negative' pressure. We leave these novel

Table 1.5.

Ω_{dm}	0.2	0.3	0.2	0.3	1.0
Ω_Λ	0	0	0.8	0.7	0
$t_U(10^9 \text{ years})$	12	11	15	13.5	9

cosmological hypotheses to subsequent chapters and concentrate now on discussing the cosmological model with the Λ term, nowadays a standard model, which reflects the main features of the effect of diffusely distributed dark matter on the age of the Universe.

Within the model of the on-average uniform and isotropic Universe, its current age is related to the parameter Ω_{dm} as follows (Kolb and Turner, 1989; Zeldovich and Novikov, 1983):

$$t_U = \frac{\Omega_{dm} H_0^{-1}}{2(\Omega_{dm}-1)^{3/2}} \left[\cos^{-1}\left(2\Omega_{dm}^{-1}-1\right) - \frac{2}{\Omega_{dm}}(\Omega_{dm}-1)^{1/2} \right] \qquad \text{for } \Omega_{dm} > 1,$$

$$t_U = \frac{\Omega_{dm} H_0^{-1}}{2(1-\Omega_{dm})^{3/2}} \left[\frac{2}{\Omega_{dm}}(1-\Omega_{dm})^{1/2} - \cos h^{-1}\left(2\Omega_{dm}^{-1}-1\right) \right] \qquad \text{for } \Omega_{dm} < 1, \qquad (1.32)$$

$$t_U = 2/(3H_0) \qquad \text{for } \Omega_{dm} = 1.$$

These expressions hold if the dynamics of expansion of the current Universe is dictated by the dark matter, $\Lambda \equiv 0$. For a non-zero cosmological term and assuming $\Omega_\Lambda + \Omega_{dm} = 1$, the expression for t_U changes to (Kolb and Turner, 1989):

$$t_U = \frac{2}{3H_0\Omega_\Lambda^{1/2}} \ln \frac{1+\Omega_\Lambda^{1/2}}{1-\Omega_\Lambda^{1/2}}. \qquad (1.33)$$

Assuming $H_0 = 70\,\text{km s}^{-1}\,\text{Mpc}^{-1}$ to be the value of the Hubble constant in the evaluations to follow, we immediately obtain from Eq. (1.32)–(1.33) several characteristic values of the theoretically predicted age of the Universe.

Let us compare these theoretical predictions with estimates of the age of the Universe. One of the classical, but unfortunately still imprecise, tests is the time scale of decay of heavy isotopes: $Th^{232}(\tau \simeq 20.3 \times 10^9$ years); $U^{235}(\tau \simeq 20.3 \times 10^9$ years); $U^{238}(\tau \simeq 6.8 \times 10^9)$ years; and $Rb^{87}(\tau \simeq 69.2 \times 10^9$ years). The crucial point of radiocosmochronology is the prediction of the possible initial contents of these elements or their ratios using models of stellar evolution and the comparison of the observed abundances with the predictions of the theory of nuclear decay. Unfortunately, the age of the Universe is evaluated by this technique very imprecisely: $t_U \simeq (10–20) \times 10^9$ years (Kolb and Turner, 1989), which, in fact, supports any of the models given in Table 1.5. Thus this test proves unsuitable for comparing the value of Ω_{dm} with the hidden mass density found from brightness curves of SNeIa Supernovas (Riess *et al.*, 1998; Schmidt *et al.*, 1998).

If Euclidean geometry is applicable and the Universe is stationary, the distance to the supernova is evaluated from its own luminosity, I, and the observed flux, F:

$$D_L = \left(\frac{I}{4\pi F} \right)^{1/2}. \qquad (1.34)$$

In terms of the observed (m) and absolute (M) stellar magnitudes, this distance corresponds to

$$\mu = m - M = 5\log D_L + 25. \qquad (1.35)$$

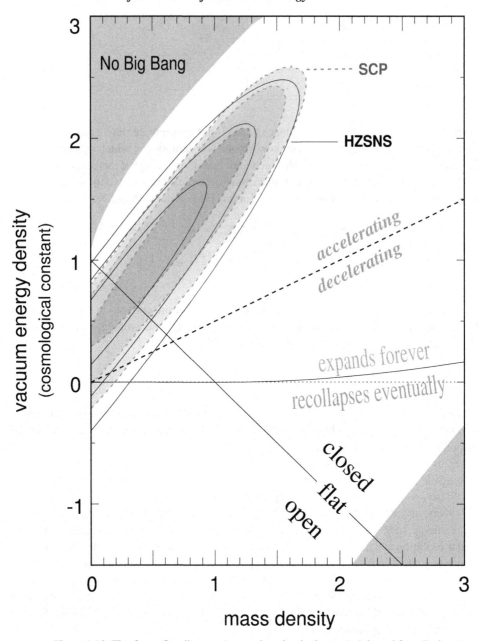

Figure 1.13 The $\Omega_\Lambda - \Omega_M$ diagram (see explanation in the text). Adapted from Perlmutter and Schmidt (2003).

On the other hand, in a cosmological model with non-zero vacuum constant $\Omega_\Lambda \neq 0$, the same distance is given by

$$D_{\rm L} = cH_0^{-1}(1+z)|\Omega_{\rm t}|^{-\frac{1}{2}} Sinn \left\{ |\Omega_{\rm t}|^{\frac{1}{2}} \int {\rm d}z \left[(1+z)^2(1+\Omega_{\rm U}z) - z(2+z)\Omega_\Lambda\right]^{-\frac{1}{2}} \right\},$$

$$(1.36)$$

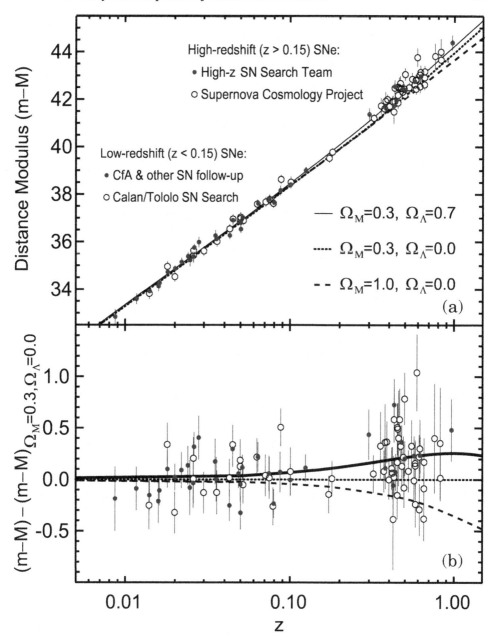

Figure 1.14 (a) $m - M$(mag) as a function of z for a number of models of Ω_M and Ω_Λ. (b) Difference $\Delta(m - M))/M$(mag). Adapted from Perlmutter and Schmidt (2003).

where $\Omega_t = 1 - \Omega_M - \Omega_\Lambda$ and $Sinn = \sinh$ for $\Omega_t \geq 0$ and $Sinn = \sin$ for $\Omega_t \leq 0$. A comparison of Eq. (1.34) with Eqs (1.32)–(1.33) makes it possible to extract from the SNeIa data in an independent manner both Ω_U vs Ω_Λ and the value of H_0. The corresponding Ω_m vs Ω_Λ diagram is plotted in Fig. 1.13. In Fig. 1.14 we plot $m–M$ as a function of z.

Table 1.6.

	Ω
Primary sources	
'Dark energy' (Λ-term, quintessence)	$10^{-0.1\pm0.1}$
Non-baryonic hidden mass in galaxies and clusters	$10^{-0.75\pm0.25}$
Baryonic fraction of matter	$10^{-1.3\pm0.1}$
Primordial neutrinos	$10^{-2.4\pm0.8}$
Thermal radiation	$10^{-4.15}$
Energy of gravitational bonding	$\sim-10^{-6}$
Energy release during structure formation	
Energy of gravitational bonding:	
Relativistic	$\sim-10^{-5.4}$
Stellar	$\sim-10^{-7.8}$
Galactic	$\sim-10^{-8.3}$
Nuclear bonding energy:	
Helium	$10^{-5.6\pm0.5}$
Heavy elements	$10^{-5.9\pm0.3}$
X-ray radiation	$\sim-10^{-8.5}$
Optical/near IR	$\sim-10^{-6}$
Far IR/ submillimeter radiation	$\sim-10^{-6}$

To conclude this chapter, we briefly recapitulate the main results of determining the values of the most important parameters using classical astronomical methods. These methods do not include those techniques that are based on using the primordial background radiation as a kind of probe of the early stages of cosmological expansion. For this purpose, we use the data of Table 1.6 taken from Peebles (1999b) (see also de Freitas-Pacheco and Peirani, 2004). In fact, these are the very values of parameters with which modern cosmology enters the era of 'precision cosmology', increasing the accuracy of classical methods and adding a new tool: measuring anisotropy and polarization of background radiation in order to construct a realistic theory of structure and evolution of the Universe. Chapter 2 is devoted to describing this approach to cosmology research.

2

Kinetics of electromagnetic radiation in a uniform Universe

2.1   Introduction

When discussing the observational status of modern cosmology in Chapter 1, we emphasized that the spectrum of primordial background radiation corresponds in a high degree to the equilibrium Planck distribution of quanta with temperature $T_0 = 2.735$ K (Fixsen *et al.*, 1996). Although possible deviations from the equilibrium distribution function of quanta must be extremely small ($\leq 10^{-4}$), they may contain information on processes of energy release both in the early Universe and at stages that are closer to the current phase of cosmological expansion. Let us recall that the blackbody spectrum of primordial background radiation provides evidence that in the past there existed a local thermodynamic equilibrium between plasma and photons, at least at temperatures 10^{10} K (Kolb and Turner, 1989; Zeldovich and Novikov, 1983). An indirect confirmation of this is found in the predictions of the theory of cosmological nucleosynthesis. However, this test is only sensitive to substantial deviations of the spectrum from Planck's curve $\left(\frac{\delta f(\nu)}{f(\nu)} \sim 0.1–1 \right)$ while the current experiments predict the level of deviations by almost three orders of magnitude smaller than that indicated above.

The factor that exerts the decisive influence on the spectral characteristics of primordial background radiation is the interaction of quanta with the electron–positron plasma (at plasma temperatures above the threshold of electron and positron creation, $T_{cr} \gg 2m_e c^2/k$) and with a background of random electrons (at relatively low temperatures $T < \frac{m_e c^2}{K}$).

Possible deviations from the equilibrium energy distribution of quanta are damped in the range $T_{cr} \leq \frac{m_e c^2}{K} \simeq 5 \times 10^9$ K owing to the Compton scattering ($\gamma + e \leftrightarrow \gamma + e$) and the double Compton scattering ($\gamma + e \rightarrow \gamma + e + \gamma$). Moreover, processes of electron scattering on nuclei ($e + A \leftrightarrow e + A + \gamma$), play an important role and help maintain the equilibrium nature of the background radiation spectrum (Rephaeli, 1995; Sunyaev and Zeldovich, 1970a; Zeldovich and Novikov, 1983; Zeldovich and Sunyaev, 1969). It is natural that the method of 'freezing' Planck's frequency distribution function for quanta should be valid only if the thermal balance between electrons, positrons and radiation is in fact a thermodynamical equilibrium so that plasma at large redshifts lacks efficient sources for 'pump-feeding' energy to the plasma. We can point to decays of long-lived massive particles (decay halftime $> 10^2$ s), the dissipation of adiabatic perturbations of density and velocity of plasma, the evaporation of primary black holes with masses $10^9 < M < 10^{13}$, and some other processes, such as hypothetical sources at high $z \sim 10^6–10^7$.

As sources of spectral distortions of primordial radiation background at relatively small redshifts, $z \ll 10^3$, we can point to young galaxies and quasars, pregalactic massive black holes and some others. Finally, at $z \leq 5–10$ and up to our epoch, spectral distortions of primordial radiation background in the direction of a galaxy cluster are formed as a result of interaction with the hot gas at relativistic temperatures, $\sim 10^{7-8}$ K. These distortions are experimentally observed and lie at the foundation of one of the novel methods of determining the Hubble constant (see references in Fukugita, Hogan and Peebles (1998)). Consequently, this chapter is devoted to analysing Compton distortions of the spectrum of primordial radiation background at various stages of cosmological evolution and to comparing theoretically predicted variations in frequency distribution function of quanta with observational data.

2.2   Radiation transfer equation in the Universe

When using a kinetic description of the spectral properties of the cosmic microwave background radiation in the expanding Universe, we make use of the quantum transfer equation in its most general symbolic form,

$$\frac{df}{dt} = S_t[f] + I_n, \tag{2.1}$$

where $f(t, \vec{x}, p^i)$ is the distribution function, $S_t[f]$ is the collision integral describing the transformation of the distribution function as a result of interaction with electrons, I_n is the quantum source function, $\vec{x}$ are the space variables and p^i is the energy–momentum four-vector $(i = 0, 1, 2, 3)$.

For a uniform and, on average, isotropic Universe, the geometric properties of spacetime are completely characterized by fixing the interval

$$ds^2 = g_{ik} \, dx^i \, dx^k = -dt^2 + (a^2(t))\xi_{\mu\nu} \, dx^\mu \, dx^\nu, \tag{2.2}$$

where g_{ik} is the metric tensor of four-dimensional space, $\xi_{\mu\nu}$ is the metric tensor of the three-dimensional space and $a(t)$ is the scale factor; the Latin indexes run through the values from 0 to 3, and the Greek indexes run from 1 to 3.

The left-hand side of Eq. (2.1) describes the free distribution of photons in the absence of collisions and external sources:

$$\frac{df}{dt} = \frac{\partial f}{\partial t} + \frac{\partial f}{\partial x^\mu}\frac{dx^\mu}{dt} + \frac{\partial f}{\partial \gamma^\mu}\frac{d\gamma^\mu}{dt} + \frac{\partial f}{\partial p_0}\frac{dp^0}{dt}. \tag{2.3}$$

The following notation was used in this equation: $\gamma^\mu = a(p^\mu/p)$, $p^2 = p_i p^i$ and $dx^\mu/dt = p^\mu/p^0$; $dp^0/dt = -(\dot{a}/a)p$. For a uniform and, on average, isotropic Universe, only the first and the last terms on the right-hand side of Eq. (2.3) are non-zero. Let us turn to analysing the collision integral for Compton processes. We assume, first of all, that the temperature of the electron gas is definitely below the relativistic limit $T_{cr} \simeq 5 \times 10^9$ K. Furthermore, if $z \ll 10^9$, the radiation temperature is also below T_{cr} and, therefore, the Compton limit for the interaction cross-section can be used when describing the e–γ scattering. Also, the energy transfer from electrons to radiation at $T_e \ll T_{cr}$ changes the frequency of quanta by the quantity

$$\frac{\Delta \nu}{\nu} \simeq \frac{KT_e}{m_e} \sim \frac{T_e}{T_{cr}} \ll 1 \left(\frac{\Delta P}{P} \ll 1 \right).$$

Under these assumptions, the collision integral in Eq. (2.1) is, as shown in Hu (1995), Hu and Silk (1993) and Hu, Scott and Silk (1994), given by

$$S_t[f] = \frac{1}{16(2\pi)^5 E(p)} \int \frac{d^3q \, d^3q' \, d^3p'}{E(q)E(q')E(p')} |M(p, q, q', p')|^2$$

$$\times \delta^{(4)}(p + q - p' - q') \tag{2.4}$$

$$\times \left\{ f_e(t, \vec{x}, \vec{q}')f(t, \vec{x}, \vec{p}')[1 + f(t, \vec{x}, \vec{p})] - f_e(t, \vec{x}, \vec{q})f(t, \vec{x}, \vec{p})[1 + f(t, \vec{x}, \vec{p}')] \right\},$$

where $|M|^2$ is the matrix element for the Compton scattering of quanta by electrons, $\delta^{(4)}(p)$ is the Dirac delta function, $f(t, \vec{x}, \vec{p})$ is the photon distribution function, and $f_e(t, \vec{x}, \vec{q})$ is the electron distribution function. Again following the work of Hu and colleagues (Hu, 1995; Hu and Silk, 1993; Hu, Scott and Silk, 1994) we consider the equilibrium distribution of electrons over momentum q in the neighbourhood of a certain mean value of $m_e \vec{v}_e$ that describes possible large-scale matter fluxes. Obviously, no such direction of plasma flow exists in the cosmological formulation of the problem ($\vec{v}_e \equiv 0$ owing to the isotropy and uniformity of the Hubble motion of the medium). However, when analysing various types of non-equilibrium sources of plasma heating, and also in analysing the motion of the hot gas in galaxy clusters, situations when $\vec{v}_e \neq 0$ become possible.

In the general case, therefore, the Maxwellian momentum distribution of electrons has the following form:

$$f_e(t, \vec{x}, \vec{q}) = (2\pi)^3 n_e (2\pi m_e T_e)^{-3/2} \exp\left[-\frac{(\vec{q} - m_e \vec{v}_e)^2}{2m_e T_e} \right], \tag{2.5}$$

where m_e and T_e are the electron mass and temperature, respectively.[1]

In the electron's rest reference frame, the scattering matrix element has the following form after averaging over the photon's polarization:

$$|M|^2 = 8(2\pi)^2 \alpha^2 \left[\frac{\widetilde{p}'}{\widetilde{p}} + \frac{\widetilde{p}}{\widetilde{p}'} - \sin^2 \widetilde{\beta} \right]. \tag{2.6}$$

The tilde denotes the choice of the reference frame, $\alpha = \frac{1}{137}$ is the fine structure constant, $\widetilde{\beta}$ is the scattering angle in the chosen reference frame,

$$\widetilde{p} = \frac{1 - \vec{p} \cdot \vec{q}/pm_e}{\sqrt{1 - q^2/m^2}} \cdot p \tag{2.7}$$

and the relation $\widetilde{p}_\mu \cdot \widetilde{p}'^\mu = p_\mu \cdot p'^\mu$ fixes the dependence of the scattering angle on the particles' momenta.

As we see, the matrix element of scattering of quanta by electrons in the form of Eq. (2.6) describes only the first term of the expansion of $|M|^2$ in the parameter α^2 that takes into account only the Coulomb and Compton processes. An analysis of inelastic processes that emerge in the order α^3 was given in Bernstein and Dodelson (1990), Hu and Silk (1993), Lightman (1981). Following Hu and Silk (1993), we use the expansion of the matrix element

---

[1] Here and later in this section we use the $\bar{h} = c = k = 1$ system of units.

of $|M|^2$ in the parameter $q/m_e \ll 1$:

$$|M|^2 = 8(2\pi)^2\alpha^2 \sum_{i=0}^{4} I_i + O\left(\frac{q}{m_e}\right)^3, \qquad (2.8)$$

where

$$I_0 = 1 + \cos^2\beta; \quad I_1 = -2\cos\beta(1-\cos\beta)\left[\frac{\vec{q}\vec{p}}{m_e p} + \frac{\vec{q}\vec{p}'}{m_e p'}\right];$$

$$I_2 = \cos\beta(1-\cos\beta)\frac{q^2}{m_e^2}; \qquad (2.9)$$

$$I_3 = (1-\cos\beta)(1-3\cos\beta)\left[\frac{\vec{q}\vec{p}}{m_e p} + \frac{\vec{q}\vec{p}'}{m_e p'}\right]^2 + 2\cos\beta(1-\cos\beta)\frac{(\vec{q}\vec{p})(\vec{q}\vec{p}')}{m_e^2 pp'};$$

$$I_4 = (1-\cos\beta)^2\frac{p^2}{m_e^2}.$$

Likewise, the following expression for the electron energy in Eq. (2.4) is obtained by taking into account Eq. (2.7):

$$\frac{1}{E(q')} = \frac{E(q)}{m_e^2}\left[1 - \frac{q^2}{m_e^2} - \frac{(\overline{p}-\overline{p}')\cdot\overline{q}}{m_e^2} - \frac{(\overline{p}-\overline{p}')^2}{2m_e^2}\right] + O\left(\frac{q}{m_e}\right)^3. \qquad (2.10)$$

Finally, the Dirac δ-function that we find in Eq. (2.4) can also be written as a Taylor series expansion in the parameter $q/m_e \ll 1$, as follows:

$$\delta^{(4)}(p+q-p'-q') = \delta(p-p') + G(\overline{p},\overline{p}',\overline{q})p\left[\frac{\partial}{\partial p'}\delta(p-p')\right]$$

$$+ \frac{1}{2}G^2(\overline{p},\overline{p}',\overline{q})p^2\left[\frac{\partial^2}{\partial p'^2}\delta(p-p')\right] + O\left(\frac{q}{m_e}\right)^3, \qquad (2.11)$$

where $G(\overline{p},\overline{p}',\overline{q}) = \frac{1}{m_e p}[(\overline{p}-\overline{p}')\cdot\overline{q} + (\overline{p}-\overline{p}')^2]$. Integrating Eq. (2.4) over the momenta of the electronic component and taking into account the normalizations,

$$\int \frac{d^3\vec{q}}{(2\pi)^3}f_e(\vec{q}) = n_e; \qquad \int \frac{d^3\vec{q}}{(2\pi)^3}q^i f_e(\vec{q}) = m_e v_e^i n_e;$$

$$\int \frac{d^3\vec{q}}{(2\pi)^3}q^i q^j f_e(\vec{q}) = m_e v_e^i v_e^j n_e + m_e T\delta^{ij}n_e, \qquad (2.12)$$

where δ^{ij} is the Kronecker delta, we arrive at the following expression for the collision integral (Hu and Silk, 1993):

$$S_t[f] = \frac{d\tau}{dt}\int dp' \frac{p'}{p}\int \frac{3d\Omega}{16\pi}\sum_{i=0}^{4}H_i(f), \qquad (2.13)$$

where the function $H[f]$ describes the following processes.

Thomson scattering

$$H_0[f] = \delta(p-p')(1+\cos^2\beta)[f(t,\overline{x},\overline{p}') - f(t,\overline{x},\overline{p})]. \qquad (2.14)$$

Linear and quadratic Doppler effects

$$H_1[f] = \left\{ \left[\frac{\partial}{\partial p'} \delta(p - p') \right] (1 + \cos^2 \beta) \, \vec{v}_e(\vec{p}\,\vec{p}') \right.$$

$$\left. - \delta(p - p') \cdot 2\cos\beta(1 - \cos\beta) \left[\frac{\vec{v}_e \vec{p}}{p} + \frac{\vec{v}_e \vec{p}'}{p'} \right] \right\} \cdot F_1(t, \vec{x}, \vec{p}, \vec{p}'), \quad (2.15)$$

$$H_2[f] = \left\{ \frac{1}{2} \left[\frac{\partial^2}{\partial p'^2} \delta(p - p') \right] (1 + \cos^2\beta)[\vec{v}_e(\vec{p} - \vec{p}')^2] \right.$$

$$- \left[\frac{\partial}{\partial p'} \delta(p - p') \right] \cdot 2\cos\beta(1 - \cos\beta) \left[\frac{\vec{v}_e \vec{p}}{p} + \frac{\vec{v}_e \vec{p}'}{p'} \right] \cdot (\vec{v}_e(\vec{p} - \vec{p}')) \bigg\} F_1(t, \vec{x}, \vec{p}, \vec{p}')$$

$$+ \delta(p - p') \left\{ -(1 - 2\cos\beta + 3\cos^2\beta)v_e^2 + 2\cos\beta(1 - \cos\beta)\frac{(\vec{v}_e \cdot \overline{p})(\vec{v}_e \cdot \overline{p}')}{pp'} \right.$$

$$\left. + (1 - \cos\beta)(1 - 3\cos\beta) \left[\frac{\vec{v}_e \cdot \overline{p}}{p} + \frac{\vec{v}_e \cdot \overline{p}'}{p'} \right]^2 \right\} F_1(t, \overline{x}, \overline{p}, \overline{p}'), \quad (2.16)$$

where $F_1(t, \overline{x}, \overline{p}, \overline{p}') \equiv f(t, \overline{x}, \overline{p}) - f(t, \overline{x}, \overline{p}')$.

Thermal Doppler effect and 'recoil' effect
In the absence of a directional electron flux $\vec{v}_e$, the thermal velocities of the electrons of order $(q/m_e)^2$ result in the same dependence of $H_2(f)$ on thermal energy, $\langle v_T^2 \rangle = \frac{3T_e}{m_e}$, as does the quadratic Doppler effect, Eq. (2.16). For the isotropic photon distribution, this effect is known as the Zeldovich–Sunyaev effect (Zeldovich and Sunyaev, 1969). The corresponding expression for the function $H_3^T[f]$ has the following form:

$$H_3[f] = \left\{ \left[\frac{\partial^2}{\partial p'^2} \delta(p - p') \right] (1 + \cos^2\beta)\frac{(\vec{p} - \vec{p}'')^2}{2} - 2\cos\beta(1 - \cos^2\beta) \right.$$

$$\left. \times (4\cos^3\beta - 9\cos^2\beta - 1)(p - p') \times \left[\frac{\partial}{\partial p'} \delta(p - p') \right] \right\} \frac{T_e}{m_e} F_1(t, \vec{x}, \vec{p}, \vec{p}').$$

$$(2.17)$$

Subsequent sections within this chapter treat the Zeldovich–Sunyaev effect in astrophysics in more detail. The effect of 'recoil' of electrons plays an important role not only in analysing the quadratic effect, but also in e–γ scattering at thermal energies close to the energies of photons. In this case, according to Hu and Silk (1993), the corresponding term in Eq. (2.13) is given by

$$H_4[f] = - \left[\frac{\partial}{\partial p'} \delta(p - p') \right] (1 + \cos^2\beta)\frac{(\vec{p} - \vec{p}'')^2}{2m_e} - F_2(t, \vec{x}, \vec{p}, \vec{p}'), \quad (2.18)$$

where $F_2(t, \vec{x}, \vec{p}, \vec{p}') = f(t, \vec{x}, \vec{p}) + f(t, \vec{x}, \vec{p}') + 2f(t, \vec{x}, \vec{p})f(t, \vec{x}, \vec{p}')$. Therefore Eqs (2.13–2.18) provide an exhaustive mathematical formulation of the problem of finding the form of the collision integral in the first order in the parameter α and up to the second order in the parameter $q/m_e \ll 1$.

In this chapter we are mostly interested in applications of the theory of radiation transfer, dealing with spectral distortions of the initial Planck function of quantum distribution. The mathematical formulation of the problem is discussed in more detail in the following section.

2.3   The generalized Kompaneets equation

We shall consider the interaction of uniformly distributed radiation with electron plasma, assuming deviations of the distribution function from the equilibrium value to be small. In this approximation, Eqs (2.4)–(2.18) yield the transfer equation for quanta in the absence of sources which, after integration over momenta p', takes the form (Hu and Silk, 1993; Zeldovich and Sunyaev, 1969, 1970):

$$
\frac{\partial f}{\partial t} - \frac{\dot{a}}{a} p \frac{\partial f}{\partial p^0} = \tau_T' \left\{ -\vec{\gamma} \vec{v}_e p \frac{\partial f}{\partial p} + [(\vec{\gamma} \vec{v}_e)^2 + v_e^2] \right.
$$
$$
\times p \frac{\partial f}{\partial p} + \left[\frac{3}{20} v_e^2 + \frac{11}{20} (\vec{\gamma} \vec{v}_e)^2 \right] p^2 \frac{\partial^2 f}{\partial p^2}
$$
$$
\left. + \frac{1}{m_e p^2} \frac{\partial}{\partial p} \left[p^4 \left\{ T_e \frac{\partial f}{\partial p} + f(1+f) \right\} \right] \right\}. \tag{2.19}
$$

Here, $\tau_T' = \sigma_T n_e$, $\sigma_T = 8\pi \alpha^2 / 3 m_e^2$, is the Thomson cross-section. This equation is obviously the generalized Kompaneets equation that includes possible microscopic fluxes in the medium in addition to the thermal motion of electrons.

Note that microscopic motions of matter in a uniform and, on average, isotropic Universe are either completely absent ($\vec{v}_e \equiv 0$) or are of random nature such that $\langle \vec{v}_e \rangle \neq 0$ but $\langle |\vec{v}_e|^2 \rangle = 0$. In this case, Eq. (2.19) takes the following form (after averaging over the scales of possible peculiar motions):

$$
\frac{\partial f}{\partial t} - \frac{\dot{a}}{a} p^0 \frac{\partial f}{\partial p^0} = \tau_T' \left\{ \frac{\langle \vec{v}_e^2 \rangle}{3} \frac{1}{p^2} \frac{\partial}{\partial p} \left(p^4 \frac{\partial f}{\partial p} \right) + \frac{1}{m_e p^2} \frac{\partial}{\partial p} \left[p^4 \left\{ T_e \frac{\partial f}{\partial p} + f(1+f) \right\} \right] \right\}. \tag{2.20}
$$

Let us first consider the effect of heating of electrons on the CMB spectrum assuming $\langle \vec{v}_e^2 \rangle = 0$. After a substitution $x = p^0 / T_e$ on the left-hand side of Eq. (2.20), we make use of the condition $\dot{T}_\gamma / T_\gamma = -\dot{a}/a$, where T_γ is the radiation temperature. Then we ultimately obtain

$$
\frac{\partial f}{\partial t} = y' \frac{1}{x^2} \frac{\partial}{\partial x} \left[x^4 \left(\frac{\partial f}{\partial x} + f(f+1) \right) \right] + x \frac{\partial f}{\partial x} \frac{\partial}{\partial t} \ln \frac{T_e}{T_0(1+z)}, \tag{2.21}
$$

where $y' = n_e \sigma_T (T_e m_e)$. When analysing Eq. (2.21), we shall neglect the last term. This approximation can be used because the characteristic time of the Doppler shift in the frequency of quanta, $\tau \sim a/\dot{a}$, is practically identical to the cosmological times, whereas the processes of heating of electrons and energy exchange between electrons and quanta have considerably shorter characteristic times. This aspect is discussed in more detail in Section 2.4. To conclude this section, we note that in Eq. (2.21) we can convert from differentiating with respect to the variable t to differentiating with respect to y:

$$
y = \int n_e \sigma_T \frac{T_e}{m_e} \, dt. \tag{2.22}
$$

It then follows from Eqs (2.20) and (2.21) for the quantum distribution function $f(x, y)$ that

$$\frac{\partial f}{\partial y} = x^{-2} \frac{\partial}{\partial x} \left[x^4 \left(\frac{\partial f}{\partial x} + f(1 + f) \right) \right]. \tag{2.23}$$

This equation was first derived by Kompaneets (1957), and its astrophysical applications were studied in detail in Sunyaev and Zeldovich (1970a,b, 1972) and later in Illarionov and Sunyaev (1975a,b). Let us emphasize two important features of Compton interactions between quanta and electrons. First, as follows from Eqs (2.20)–(2.22), this process conserves the total number of quanta. Multiplying the left- and right-hand sides of Eq. (2.20) by x^2 and integrating over x from 0 to ∞, we easily obtain

$$\frac{d}{dt}(n_\gamma a^3) \propto \int dx \cdot x^2 S_t[f] = 0, \tag{2.24}$$

where n_γ is the photon concentration.

This result allows very clear interpretation (see Zeldovich and Sunyaev (1969), (1970)). Since the dynamics of the process is accompanied by a redistribution of quanta, it is clear that the reduction in the number of quanta in one frequency range results in the emergence of quanta in another range, so that the total concentration does not change. The second important consequence follows from Eq. (2.20) if it is multiplied by x^3 and integrated over x:

$$\frac{1}{\varepsilon_\gamma a^4} \frac{\partial}{\partial t}(\varepsilon_\gamma a^4) = 4\tau_T' \frac{T_e}{m_e} \left[1 - \frac{T_e^4}{4\pi^2 \varepsilon_\gamma} \int_0^\infty dx \, x^4 f(1 + f) \right]. \tag{2.25}$$

Here, ε_γ is the radiation energy density. The first term in the square brackets corresponds to the thermal Compton effect, and the second describes the 'recoil' effect (Hu and Silk, 1993).

2.4 Compton distortion of radiation spectrum on interaction with hot electrons

In this section we discuss one of the most important applications of the theory of the Zeldovich–Sunyaev effect to the interaction model of cosmological 'hot' electrons with quanta of primordial background radiation. In the $T_e \gg T_\gamma$ approximation, it is convenient to transform Eq. (2.23) from variable $x = p/T_e$ to $\xi = p/T_\gamma$ and to ignore on the right-hand side those terms that are proportional to $f(f + 1)(T_\gamma/T_e)$. This operation transforms Eq. (2.23) as follows:

$$\frac{\partial f}{\partial y} = \xi^{-2} \frac{\partial}{\partial \xi} \left[\xi^4 \frac{\partial f}{\partial \xi} \right]. \tag{2.26}$$

Following Sunyaev and Zeldovich (1970a), we can apply perturbation theory in order to find a solution of this equation in the limit $y \ll 1$, after substituting the non-perturbed Planckian expression for $f_0(x) = (e^x - 1)^{-1}$ into the right-hand side of Eq. (2.26). After this, we obtain from Eq. (2.26) the following expression for the perturbation, Δf, of the distribution function:

$$\Delta f \simeq \frac{e^x \cdot xy}{(e^x - 1)^2} \left\{ \frac{x}{\tanh\left(\frac{x}{2}\right)} - 4 \right\} \tag{2.27}$$

and

$$\frac{\Delta f}{f_0} = \frac{xe^x}{(e^x - 1)^2} \left\{ \frac{x}{\tanh\left(\frac{x}{2}\right)} - 4 \right\}. \tag{2.28}$$

In the Rayleigh–Jeans asymptotics $x \ll 1$, Eq. (2.27) immediately yields

$$\frac{\Delta f}{f_0} \simeq \frac{\delta T_{\mathrm{RJ}}}{T_{\mathrm{RJ}}} = -2y. \qquad (2.29)$$

In the general case, the deformation of the quantum spectrum can be presented in the integral form for any value of the parameter y and for any x (Sunyaev and Zeldovich, 1970a):

$$f(x, y) = \frac{1}{\sqrt{4\pi y}} \int_0^\infty \frac{d\xi}{\xi} f_0(0, \xi) \exp\left[-\frac{(\ln x - \ln \xi + 3y)^2}{4y} \right], \qquad (2.30)$$

where, as before, $f_0(0, \xi) = (e^\xi - 1)^{-1}$. Multiplying $f(x, y)$ by x^3 and integrating in the entire domain of variation of x, we arrive at the well known expression for the radiation energy density,

$$\varepsilon_\gamma(y) = \sigma T_{0,\gamma}^4 e^{4y}, \qquad (2.31)$$

where $T_{0,\gamma}$ is the non-perturbed value of temperature. Since the Compton effect results in reduced temperature in the Rayleigh–Jeans range, and therefore lowers energy, it is clear that the increase in quantum energy density, Eq. (2.30), corresponds to accumulation of quanta in the Wien range of the spectrum. This factor was first pointed out by Sunyaev and Zeldovich (1970a). Figure 2.1 shows the spectrum of the CMB in the Zeldovich–Sunyaev approximation for various values of the parameter y. Leaving aside a discussion of the possible mechanisms of electron heating, we can present the dependence of the effective temperature on frequency and the parameter y in the most general case, as shown in Fig. 2.1. An analysis of this plot will be conducted in the following section, in which we discuss relativistic corrections to the Zeldovich–Sunyaev effect.

2.5   Relativistic correction of the Zeldovich–Sunyaev effect

The effect of scattering of quanta by hot electrons in the approximation of diffusion of quanta in frequency from the Rayleigh–Jeans range to the Wien range, discussed in Section 2.2, is of utmost importance, both for the understanding of the mechanisms of spectrum transformation in the early Universe, and for describing the interaction of quanta of the CMB with the hot gas in galaxy clusters. The key element in the description of quantum diffusion is the use of the parameter y as the time variable (see Eq. (2.22)); y is a function both of the optical depth of the plasma and the electron temperature T_e.

In this section, we are interested in two aspects.

(1)   How correct is it to use the diffuse approximation in the case when the optical depth of the plasma relative to the Thomson scattering is small ($\tau \ll 1$)?

(2)   What are the quantitative changes in the predictions of the diffuse approximation as the temperature of the electrons increases to several tens of keV? In other words, what will the behaviour of the radiation spectrum be in the limit $y \ll 1$ if the parameter T_e/m_e is not too small?

Note that this formulation of the problem attracted careful scrutiny immediately after the publication of Zeldovich–Sunyaev papers (Zeldovich and Sunyaev (1969) and then Sunyaev and Zeldovich (1972)). Interest in the relativistic correction of the Zeldovich–Sunyaev effect was largely stimulated by the discovery of galaxy clusters with gas temperature up to 15 keV;

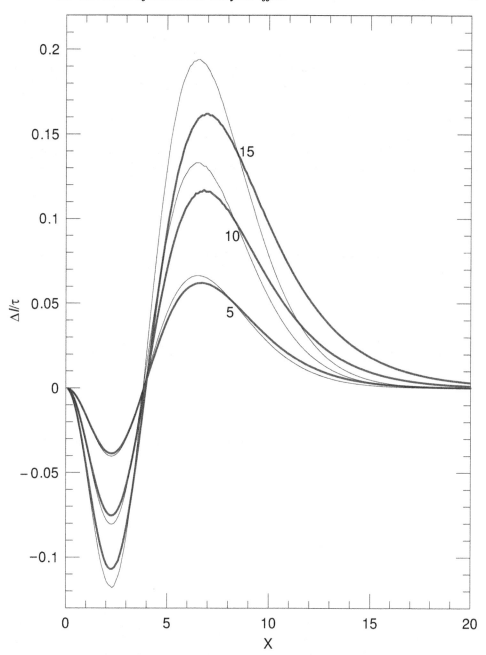

Figure 2.1 Comparison of the intensities $\Delta I/\tau$ (in units of $h^2c^2/2k^3T^3$) in the Zeldovich–Sunyaev approximation (thin solid lines), taking into account relativistic corrections (thicker curves). The numbers above the curves denote electronic temperatures in keV. The intensities $\Delta I/\tau$ are given in units of $(hc)^2/2(kT)^3$. Adapted from Rephaeli (2001).

on the other hand, the high precision of measurements of primordial radiation background using the DERBI and FIRAS instruments in the COBE project (Fixsen *et al.*,1996) makes it necessary to return again to the basics of the Compton theory of quantum interaction with electrons. As a result, we follow Rephaeli (1995) and Wright (1979) and assume that the smallness of the parameter y is especially due to the smallness of the optical depth of the plasma relative to the Thomson scattering, τ_T.

Following Wright (1979), we analyse this scattering of a quantum by an electron in the Chandrasekhar approximation (Chandrasekhar, 1950). The probability of scattering of a photon moving in direction θ ($\mu = \cos\theta$) before being scattered by an electron moving in direction θ' ($\mu' = \cos\theta'$) after the act of scattering, has the following form (in the reference frame of the electron at rest):[2]

$$\vartheta(\mu, \mu', \beta) = \frac{3}{8}\left[1 + \mu^2\mu'^2 + \frac{1}{2}(1 - \mu^2)(1 - \mu'^2)\right], \tag{2.32}$$

where $\beta = v/c$. The scattering by an electron results in a frequency change of the quantum, characterized by a parameter S,

$$S = \ln\frac{v'}{v} = \ln\left(\frac{1 + \beta\mu'}{1 - \beta\mu}\right), \tag{2.33}$$

where v is the frequency before scattering and v' is the frequency after scattering. After integration over the initial angles of arrival of quanta, we obtain from Eqs (2.31) and (2.32) the probability for this process to occur (Rephaeli, 1995):

$$P(S, \beta) = \frac{1}{2\gamma^4\beta}\int d\mu \frac{(1 + \beta\mu')\vartheta(\mu, \mu', \beta)}{(1 - \beta\mu)^3}, \tag{2.34}$$

where γ is the electron's gamma-factor. Integrating Eq. (2.33) over the electron distribution function, we arrive at the distribution function for the shift of the quantum frequency per scattering act (Rephaeli, 1995):

$$P_1(S) = \frac{\int d\beta \cdot \beta^2\gamma^5 e^{-\xi(\gamma-1)}P(S, \beta)}{\int \beta^2\gamma^5 e^{-\xi(\gamma-1)}d\beta}, \tag{2.35}$$

where $\xi \equiv m_e c^2/kT_e$. Taking into account the fact that $\tau \ll 1$, we can write the distribution of quanta over frequency in the following form:

$$P(S) = (1 - \tau)\delta(S) + \tau P_1(S) + O(\tau^2), \tag{2.36}$$

where the first term in Eq. (2.36) describes the effect of scattering with frequency unchanged and the second term describes the change in frequency in a single-scattering event. As a result, taking into account single Compton scattering, we have, for the spectrum of radiation,

$$\Delta I = I_0(x)\tau[\Phi(x, \xi) - 1], \tag{2.37}$$

---

[2] In this section, we use the conventional system of units.

where $I_0(x) = [2(kT_0)^3/(hc)^2]x^3(e^x - 1)^{-1}$ is the spectral energy density of blackbody radiation,

$$\Phi(x, \xi) = A(\xi)[\varphi_1(x, \xi) + \varphi_2(x, \xi)],$$

$$\varphi_1(x, \xi) = \int_0^1 \frac{dt\, t(e^x - 1)}{e^{xt} - 1} \int_{\beta_m}^1 d\beta\, \gamma e^{-\xi(\gamma-1)} \int_{\mu_m}^1 d\mu\, q(t, \mu, \beta), \qquad (2.38)$$

$$\varphi_2(x, \xi) = \int_0^1 \frac{dt(e^x - 1)}{t^3(e^{x/t} - 1)} \int_{\beta_m}^1 \gamma e^{-\xi(\gamma-1)}d\beta \int_{-1}^{\mu_M} d\mu\, q(t, \mu, \beta),$$

$$q(t, \mu, \beta) = \frac{\beta^{-2}(3\mu^2 - 1)\left[\left(1 - \frac{\beta\mu}{t}\right) - 1\right]^2 + (3 - \mu^2)}{(1 - \beta\mu)^2},$$

and

$$A(\xi) = \frac{3}{32}\left(\int_0^1 d\beta \cdot \beta^2\gamma^5 e^{-\xi(\gamma-1)}\right)^{-1};$$

$$\beta_m = \frac{1-t}{1+t}; \qquad \mu_m = \frac{t(t^{-1} - 1 - \beta)}{\beta}; \qquad \mu_M = \frac{t - 1 + \beta}{t\beta}.$$

Detailed information about the calculation of the function $\Phi(x, \xi)$ is given in Rephaeli (1995) along with the numerical calculation of the function $\Delta I(x, \xi)$. Note an important conclusion that follows from Eq. (2.37): in the general case, there is no self-similar solution for arbitrary values of the parameter x. Intensity perturbations, ΔI, of Eq. (2.37), as well as temperature perturbations,

$$\frac{\Delta T}{T_0} = \frac{e^x - 1}{xe^x}\tau[\Phi(x, \xi) - 1], \qquad (2.39)$$

are functions of τ and ξ; to be precise, functions of the combination $\tau f(\xi)$. This effect manifests itself best for $x > 5$–6 (see Fig. 2.1). Figure 2.1 plots the frequency distribution for the function $\Delta I/\tau$ calculated in the Zeldovich–Sunyaev approximation (Zeldovich and Sunyaev, 1969) and also from Eqs (2.37) and (2.38) (Rephaeli, 1995). As we see from this figure, deviations are considerable in the frequency range $x \geq 6$. At the same time, it follows from Eqs (2.38) and (2.39) that, as $x \to 0$, the function $\Phi(x, \xi) = -\xi^{-1}$ and, for $kT_e/m_ec^2 =$ constant, we arrive at Eq. (2.29).

A similar derivation of the relativistic correction of the Zeldovich–Sunyaev effect is given in Challinor and Lasenby (1997), where it is also shown that temperature deviations for the low-frequency part of the spectrum have the following form:

$$\frac{\Delta T_{RJ}}{T_0} = -2y\left[1 - \frac{17}{10}\xi^{-1} + \frac{123}{40}\xi^{-2} + O(\xi^{-3})\right]. \qquad (2.40)$$

As we see from this expression, the self-similarity effect breaks down in the first order of the parameter $\xi^{-1} \ll 1$. Detailed numerical calculations of correction terms were carried out in Dolgov *et al.* (2001) and Itoh *et al.* (2001).

The relativistic correction of the defuse approximation is thus found to be the most important aspect of the analysis of possible spectral distortions in the Wien range of the spectrum, where stronger distortions are formed as the temperature of electrons increases.

2.6 The kinematic Zeldovich–Sunyaev effect

In addition to cosmological applications to the effect of comptonization of the CMB spectrum when it interacts with hot electrons, its astrophysical aspect has an important practical significance. We speak here of the interaction of the quanta of the CMB with the hot gas in clusters of galaxies whose temperature may be as high as 10^8 K. The thermal Zeldovich–Sunyaev effect in the Rayleigh–Jeans range of the spectrum is then described by the following formula:

$$\frac{\Delta T_{\mathrm{RJ}}}{T_0} = -\frac{2k T_{\mathrm{e}}}{m_{\mathrm{e}} c^2} \cdot \tau, \tag{2.41}$$

where $\tau = \int dl\, \sigma_{\mathrm{T}} n_{\mathrm{e}}(r)$ is the optical depth relative to the Thomson scattering, measured along the line of sight. However, as shown by Sunyaev and Zeldovich (1980) when analysing the interaction of quanta of the CMB with the gas in galaxy clusters, it is necessary to take into account the Doppler shift of the frequency of quanta caused by the motion of a cluster as a whole relative to the background radiation. In this case, the change in the intensity of radiation is given by

$$\left.\frac{\Delta I}{I}\right|_{\mathrm{D}} = -\frac{x e^x}{e^x - 1} \frac{v_{\mathrm{r}}}{c} \cdot \tau, \tag{2.42}$$

where $x = h\nu / k T_\gamma$ and v_{r} is the radial component of the velocity of the cluster.

Recalculated to temperature perturbations, Eq. (2.42) immediately implies

$$\frac{\Delta T}{T} \simeq -\frac{v_{\mathrm{r}}}{c} \cdot \tau. \tag{2.43}$$

We see from Eq. (2.43) that the effect of increasing or reducing $\Delta T/T$ is frequency-independent and is only dictated by the direction of motion of the cluster. Temperature decreases if the cluster moves away from us, and increases if it moves towards us. Therefore, the fact of principal importance is that the data of the spectrum of CMB in the direction of galaxy clusters may yield radial components of their velocity of motion. In principle, however, it is also possible (but far from easy) to evaluate the tangential component from the data of polarization of the CMB in the direction of the galaxy cluster (Sunyaev and Zeldovich, 1980).

Since the optical depth of the best known clusters for which the kinematic Zeldovich–Sunyaev effect has already been measured does not exceed $\tau \simeq 0.02$–0.05, and is generally found to be even smaller, it is possible to use the single-scattering-event approximation to simulate the effects of polarization generation in scattering of quanta by hot electrons in the CMB (see Section 2.3).

According to the general theory of the Doppler effect (Landau and Lifshits, 1962), the temperature of radiation in scattering by an electron (in its rest frame) is given by the expression

$$T_0 = T_\gamma \frac{\sqrt{1 - v^2/c^2}}{1 + \frac{v}{c}\cos\theta}, \tag{2.44}$$

where θ is the angle between the directions of momentum of the quantum and the electron, and T_γ is the true temperature of the CMB. Assuming the motion of electrons to be non-relativistic, we can apply expansion in the small parameter $v/c \ll 1$ in Eq. (2.43) up to the second order of magnitude, $\sim (v/c)^2$,

$$T_0 = T_\gamma \left[1 - \beta \cos\theta + \beta^2 \left(\cos^2\theta - \frac{1}{3} \right) \right], \tag{2.45}$$

where $\beta = v/c$.

The intensity of radiation in the Rayleigh–Jeans range of the spectrum is connected to its temperature and possesses quadruple anisotropy in the order $(v/c)^2$. Since the differential cross-section of scattering of quanta by electrons is a familiar function of polarization (Chandrasekhar, 1950),

$$\frac{\mathrm{d}\sigma_T}{\mathrm{d}\Omega} \propto |\hat{\varepsilon} \cdot \hat{\varepsilon}'|^2, \tag{2.46}$$

where $\hat{\varepsilon}$ and $\hat{\varepsilon}'$ are the initial and final polarizations of photons, respectively; then, averaging over the initial values of polarization produces the resulting polarization (Sunyaev and Zeldovich, 1980):

$$\frac{I_\| - I_\perp}{I_\| + I_\perp} = P \simeq 0.1 \left(\frac{v_t}{c} \right)^2. \tag{2.47}$$

Note that this evaluation involves precisely the tangential component of velocity, v_t, while the contribution from the radio component is zero owing to Eq. (2.46). The polarization for galaxy clusters will be less by a factor of $\tau \ll 1$ than in Eq. (2.47) since the smallness of optical depth is related to the scattering probability via Eq. (2.36). Finally, we have

$$P_{cl} \simeq 0.1\tau \left(\frac{v_t}{c} \right)^2. \tag{2.48}$$

Note also that, in addition to Eq. (2.48), the polarization of radiation resulting from scattering by a moving cluster appears in the first order in v_t/c but only in the second order in τ^2. The relevant evaluations were made by Sunyaev and Zeldovich (1980):

$$\widetilde{P}_{cl} \simeq \pm \frac{xe^x}{4\theta(e^x - 1)} \frac{v_t}{c} \tau^2. \tag{2.49}$$

Before concluding this section, we should remark that, in addition to weakly linear effects emerging in the order $\tau(v_t^2/c^2)$ or $(v_t/c)\tau^2$, there is also a purely gravitational correction to the spectrum of the background radiation, independent of the optical depth of the plasma. This fact was first noticed by Gurvitz and Mitrofanov (1986), who discussed the effect of gravitational lensing of background radiation by a moving cluster. Perturbations of the background radiation intensity are evaluated within an order of magnitude as follows:

$$\frac{\Delta I}{I} \simeq \frac{xe^*}{e^* - 1} \frac{v_t}{c} \cdot \theta. \tag{2.50}$$

Here the beam deflection angle, $\theta \sim GM/Rc^2$, depends on the cluster mass M and its radius R. Assuming, for the sake of estimation, that $M \sim 2 \times 10^{15} M_\odot$, $R \sim 2\,\mathrm{Mpc}$ and $v_t \sim 2 \times 10^3\,\mathrm{km \cdot s^{-1}}$, we obtain from Eq. (2.49) that $\Delta T/T \simeq 10^{-6}$ in the Rayleigh–Jeans part of the spectrum.

On the whole, recapitulating this section, we need to acknowledge that even granting the importance of analysing non-linear corrections to the thermal and kinetic Zeldovich–Sunyaev effects, today's experimental capabilities are unfortunately quite far from achieving detection of high-order corrections. At the same time, the linear effects, especially when combined with x-ray and γ observations, make it possible not only to extract unique information on a type of motion of gas and its temperature in galaxy clusters, but also to map out approaches to independent determination of the Hubble constant, H_0, using the observed spectrum of the primordial background radiation. The following section is devoted to illustrating the general ideology of this technique.

2.7   Determination of H_0 from the distortion of the CMB spectrum and the data on x-ray luminosity of galaxy clusters

Novel methods of determining the value of the Hubble constant were discussed in Chapter 1; these methods used various objects – Cepheids, supernovas, etc. – as new 'standard references'. In this section we add to this list another 'standard candle' – the calculation of H_0 from the distortion of the background radiation spectrum via the thermal Zeldovich–Sunyaev effect, combined with the data on the x-ray luminosity of the gas in galaxy clusters. The gist of this idea of the combined use of two effects is extremely simple. X-ray luminosity yields a relation between temperature and luminosity of the cluster and density distribution within the cluster. Moreover, the Zeldovich–Sunyaev effect operates with practically the same parameters. Therefore, by combining the two effects, it is possible to express the angular size of the cluster in terms of the combination of x-ray and radio luminosities. The same angular size can be found using standard cosmological techniques (see Chapter 1) that relate the value of the Hubble constant H_0 with the radio and x-ray luminosities of clusters. A specific implementation of this algorithm was given in Birkinshaw (1999), Birkinshaw and Hughes (1994), Carlstrom, Joy and Greco (1996), Carlstrom *et al.* (2000, 2001), Reese (2004) and Udomprasert, Mason and Readhead (2001), where the latest results on H_0 obtained by this method are discussed. Let us look at the fundamental aspects of this approach.

The observed x-ray luminosity of the gas inside a cluster is described by a simple formula (Birkinshaw, 1999):

$$B_x \simeq \frac{\Lambda_0 n_0^2 d_A}{4\pi(1+z)^3} \int_0^{\theta_{cl}} \omega_n^2 \omega_\Lambda \, d\xi, \qquad (2.51)$$

where $n_0 \cdot \omega_n = n_e$ is the radial electron concentration distribution in the cluster, $\Lambda_0 \cdot \omega_\Lambda$ is the radial distribution of the x-ray luminosity of the hot gas, $\xi = r/d_A$, d_A is the distance to the cluster, r is the cluster radius, z is the redshift and θ_{cl} is the angular size of the cluster. The distortions of radiation intensity due to the Zeldovich–Sunyaev thermal effect are given by the formula

$$\Delta I = i_0 g(x) \frac{kT_{e0}}{m_e c^2} \sigma_T n_0 d_A \int_0^1 \omega_n \omega_T \, d\xi, \qquad (2.52)$$

where

$$i_0 = \frac{2(kT_0)^3}{(hc)^2}, \qquad g(x) = \frac{x^4 e^4}{(e^x - 1)^2} \left[\frac{x(e^x + 1)}{e^x - 1} - 4 \right],$$

and $T_e = T_{e0} \cdot \omega_T$ is the angular distribution of the electron temperature.

By combining Eqs (2.51) and (2.52), we finally obtain

$$d_A = \frac{1}{4\pi (1rz)^3} \left(\frac{\Lambda_0}{\sigma_T^2 B_x} \right) \left(\frac{\Delta I}{i_0 g(x)} \right)^2 \left(\frac{m_e c^2}{k T_{e0}} \right)^2 \left(\frac{Q_x}{Q_m^2} \right), \quad (2.53)$$

where $Q_x = \int d\xi \, \omega_n^2 \omega_\Lambda$ and $Q_m = \int \omega_n \omega_T \, d\xi$. A comparison of Eq. (2.53) and Eq. (2.51) allows us to express H_0 in terms of the parameters of the problem.

As we see from Eq. (2.53), the accuracy of determining the distance d_A to a cluster depends on the 'geometrical' parameters Q_x and Q_m, which are found by integrating the radial distribution functions of temperature and concentration of the gas and of its x-ray luminosity. In the 1990s, the decisive success in finding ω_n, ω_T and ω_Λ was achieved by using interferometric measurements in combination with high-angular-resolution x-ray data. Among such radio measurements we need to mention first of all the data of the BIMA and OVRO collaborations (Carlstrom *et al.*, 1996, 2000), which conducted observations of 35 galaxy clusters in the redshift range from 0.17 to 0.89.

Figure 2.2(a) shows a radio image of the cluster CL0016+16 obtained with the BIMA interferometer at 28 GHz. Shown in Fig. 2.2(b) is an x-ray photograph of the same cluster recorded by the ROSAT satellite. Note the high degree of correlation in the orientation of hot gas areas detected both in the radio and in the x-ray spectrum of emission from the cluster.

Radio luminosity distributions of the nine nearest galaxy clusters were measured recently by the CBI interferometer with angular resolution reaching $3'$ (Udomprasert *et al.*, 2000). A similar program was implemented in the framework of the observational programs SuZIE, PRONAOS and MITO. Compton distortions $y \simeq 1.2 \times 10^{-3}$ were measured in the observations of the cluster RXj1347 (Pointecouteau, 1999) using the 30 cm IRAM radio telescope. Especially important among the experiments on the observation of the Zeldovich–Sunyaev effect are the BIMA and ORVO data on measuring the Hubble constant. The results were summarized for 33 galaxy clusters (Carlstrom *et al.*, 2000); see Fig. 2.3. Since the expression for the distance, d_A, includes, in addition to the Hubble constant, the total density of matter and the cosmological constant, the results of calculation of H_0 are model-dependent. For instance, assuming $\Omega = 0.3$, the expected value of the Hubble constant is close to $60 \text{ km s}^{-1} \text{ Mpc}^{-1}$, and for $\Omega = 1$ ($\Lambda = 0$) it reduces to $H_0 = 58 \text{ km s}^{-1} \text{ Mpc}^{-1}$ with $\pm 5\%$ observational error. The level of systematic error is evaluated here to be almost six times higher ($\sim 30\%$) (Carlstrom *et al.*, 2000).

2.8   Comptonization at large redshift

This section is devoted to applications of the theory of Compton distortions of the spectrum of the CMB radiation caused by energy releases in the early Universe. Note that this aspect was first studied in Sunyaev and Zeldovich (1972), where the main features of the transformation of the background radiation spectrum upon heating of cosmological electrons were listed for times long before the epoch of the formation of the galaxy and galaxy clusters. In this section we return to this problem predominantly because the latest data from the DERBI and FIRAS instruments, obtained in the framework of the COBE project (Fixsen *et al.*, 1996), impose strict observational constraints on the chemical potential of photons and the degree of spectrum non-equilibrium.

An analysis of the temperature regime of electrons affected by possible energy releases in the early Universe will be conducted under the assumption of Maxwellian velocity distribution

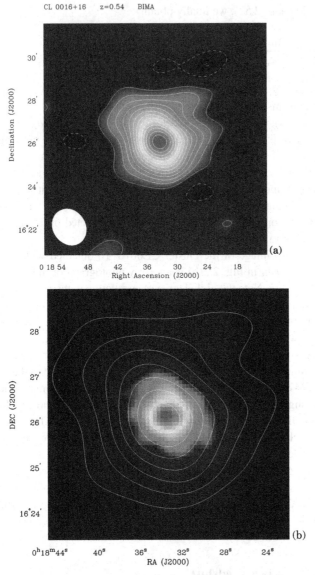

Figure 2.2 (a) Radio (b) and x-ray (ROSAT) images of CL0016+16, obtained by the BIMA collaboration (for details, see Carlstrom *et al.* (2000)).

and equality of the temperatures of electron and proton components. This hypothesis assumes that the characteristic electron–electron collision time, τ_{ee}, and electron–proton collision time, τ_{ep}, must be considerably shorter than the Compton time of energy 'pumping' from electrons to $\tau_{e\gamma}$ (see Zeldovich and Novikov (1983)).

When the condition $\tau_{ee}, \tau_{ep} \ll \tau_{e\gamma}$ is satisfied, the thermal balance in the electron gas in the presence of sources of electron heating is found from the first law of thermodynamics

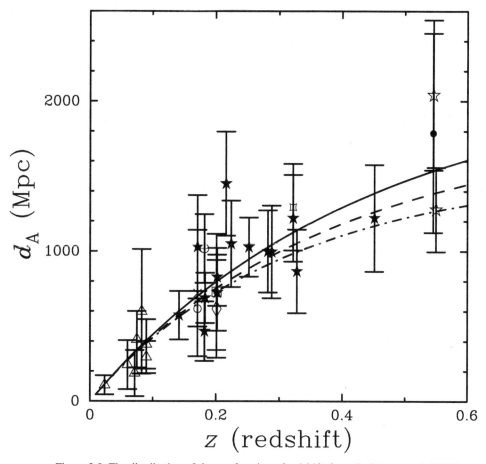

Figure 2.3 The distribution of d_A as a function of redshift, from Carlstrom *et al.* (2000).

to be

$$d(\rho_{tot}a^3) + P_{tot}\,da^3 = dQ, \tag{2.54}$$

where a is the scale factor, dQ is the energy release from heat sources, and ρ_{tot} and P_{tot} are the total density and total pressure of the electron–proton gas and radiation, respectively. Taking into account the relations between energy density, pressure and temperature for each component,

$$P_\gamma = \frac{1}{3}\rho_\gamma; \qquad \rho_e c^2 = m_e n_e c^2 + \frac{3}{2}n_e k T_e; \qquad P = k n_e T_e, \tag{2.55}$$

where n_e is the electron concentration and T_e is the electron temperature, we can obtain from Eqs (2.54) and (2.55) the following equations of thermal equilibrium for hydrogen and helium:

$$\frac{dT_e}{dt} + 2\frac{\dot{a}}{a}T_e = \frac{2}{3(n_e + n_H + n_{He})}\left[\frac{\dot{Q}}{ka^3} - \frac{d}{a^4 dt}(\rho_\gamma a^4)\right]. \tag{2.56}$$

From Eq. (2.25), in the approximation of Compton heating of radiation by hot electrons, we find

$$\frac{d}{a^4 dt}(\rho_\gamma a^4) = \rho_\gamma \cdot 4\tau_C \frac{kT_e}{m_e c^2}\left[1 - \frac{\sigma T_e^4}{4\pi^2 \varepsilon_\gamma}\int dx \cdot x^4 f(1+f)\right], \qquad (2.57)$$

where $\varepsilon_\gamma = \rho_\gamma c^2$, $x = h\nu/kT_e$ and f is the frequency distribution function of quanta.

Integrating now in Eq. (2.57) over a new variable $x' = h\nu/kT_\gamma$, we finally obtain

$$\frac{d(T_e a^2)}{a^2 dt} + \frac{1}{\tau_{e\gamma}}\left(T_e - \frac{15}{4\pi^4}\frac{\sigma T_\gamma^5}{\varepsilon_\gamma}\int_0^\infty dx \cdot x^4 f(x)(1+f)\right) = \frac{2q}{3n_{tot}a^3}, \qquad (2.58)$$

where

$$n_{tot} = n_e + n_H + n_{He}; \qquad q \equiv \dot{Q}; \qquad \tau_{e\gamma} = \frac{3}{8}\frac{m_e c}{\sigma_T \varepsilon_\gamma}\cdot\frac{n_{tot}}{n_e} = \frac{3}{8}\frac{m_e c}{\sigma_T \varepsilon_\gamma}x_e^{-1}, \qquad (2.59)$$

where the degree of plasma ionization is $x_e = n_e/n_{tot}$. Recalling here that the cosmic plasma is totally ionized at redshifts $z \gg 10^3$, and also that the characteristic time of Compton interaction is much shorter than the cosmological time scale, we can derive from Eq. (2.58) the following:

$$T_e \simeq \frac{15}{4\pi^4}\frac{\sigma T_\gamma^5}{\varepsilon_\gamma}\int_0^\infty dx \cdot x^4 f(x)(1+f) + \frac{2}{3}\tau_{e\gamma}\frac{q}{n_{tot}}a^3. \qquad (2.60)$$

If we assume that deviations from the equilibrium form of the distribution function for quanta are small, we can reduce the second term on the right-hand side of Eq. (2.60) to the radiation temperature, T_γ, so that the corresponding value of the parameter y is found to be

$$y \simeq \int \frac{k(T_e - \tau)}{m_e c^2} \simeq \frac{2}{3}\int d\tau \cdot \frac{\tau_{e\gamma}}{m_e c^2}\frac{qk}{n_{tot}a^3}. \qquad (2.61)$$

Further evaluation of the parameter y depends on the relation between the characteristic variation time of source parameters, $t_S \simeq (Q/\dot{Q})$, and the characteristic variation time of the plasma optical depth, $t_{opt} \simeq \tau/\dot{\tau}$. If we ignore for the moment the period of hydrogen recombination during which t_{opt} is small, we can assume that $t_{opt} \sim t_{exp}$ for $z \gg 10^3$, where $t_{exp} \simeq a/\dot{a}$ is the characteristic time of expansion. Then the parameter y for $t_S \ll t_{exp}$ is on the order of

$$y \simeq \frac{2}{3}\tau_T' \cdot \frac{\tau_{e\gamma}}{m_e c^2}\frac{Q}{n_{tot}a^3}. \qquad (2.62)$$

In the opposite case, when the evolution of the heating source is slow ($t_S \geq t_{exp}$), we can use the following approximate expression for y:

$$y \simeq \frac{2}{3}\tau \cdot \frac{\tau_{e\gamma}}{m_e c^2}\frac{Qk}{n_{tot}a^3} = \frac{2}{3}\tau \frac{\tau_{e\gamma}}{m_e c^2}\frac{q}{n_{tot}a^3}. \qquad (2.63)$$

For the sake of certainty, we refer to the regime described by Eq. (2.62) as the regime of explosive energy release, and to the regime of Eq. (2.63) as the quasi-stationary one. Forgetting for the moment concrete details of possible mechanisms of implementing the explosive and quasi-stationary modes, we can say that, to within a coefficient ~ 1, they correspond to approximately the same value of y. To be specific, let us consider the mode given by Eq. (2.62). In view of the equality $\tau'_T = \sigma_T n_e c$ and Eq. (2.59), we arrive at the following expression for y:

$$y \simeq \frac{1}{4} \frac{Q}{\varepsilon_\gamma a^3}. \tag{2.64}$$

The quantity $\varepsilon_S = \frac{Q}{a^3}$ in Eq. (2.64) identifies with the density of energy released by external sources to heat electrons. Then Eq. (2.64) implies a simple estimate of the ratio $\varepsilon_S/\varepsilon_\gamma$:

$$\frac{\varepsilon_S}{\varepsilon_\gamma} \simeq 8 \times 10^{-5} \left(\frac{y}{2 \times 10^{-5}} \right) \sim 10^{-4}, \tag{2.65}$$

where the parameter y was normalized using the data of Fixsen *et al.* (1996).

Note that this evaluation is of sufficiently general nature and can be applied successfully to evaluate energy release regardless of the value of the redshift. In addition to the y-distortions of the primordial background radiation spectrum discussed above, it is also necessary to look at another mechanism of principal importance that arises if the optical plasma depth is large. By this we mean the formation of the Bose–Einstein spectrum of the CMB:

$$f(x, \mu) = \left[\exp\left(\frac{h\nu + \mu}{kT_e} \right) - 1 \right]^{-1}. \tag{2.66}$$

Formally, this spectrum satisfies the Kompaneets equation, Eq. (2.22), for $\partial f/\partial y = 0$. This means in turn that the parameter y must be sufficiently large, i.e.

$$y \sim \tau \cdot \frac{k(T_e - T_\gamma)}{m_e c^2} \gg 1. \tag{2.67}$$

We can expect in this case that the asymptote of the solution of Eq. (2.23) will tend to the mode described by Eq. (2.66) and that the chemical potential μ will be (Sunyaev and Zeldovich, 1972)

$$\mu = 1.4 \, kT_\gamma \delta_S, \tag{2.68}$$

where δ_S is the energy release.

To recapitulate the results of this section, we can point to two basic types of spectrum distortions for the primordial background radiation: y-distortions and the chemical potential, μ. Both these parameters are dependent on specific features of electron heating and are calculated with certain assumptions concerning the properties of the sources of a non-equilibrium energy release. In Chapter 3 we analyse the ionization history of the Universe and specifically the mechanisms of reionization of the cosmic plasma, and treat in detail a number of such mechanisms, paying attention to possible observational manifestations of non-equilibrium energy release processes. At the same time, the most important factor that defines constraints on the limits of applicability of the theory is the experiment. For spectral distortions of background radiation, the role of such 'critical' experiments is played by the data of the DERBI and FIRAS instruments of the COBE project, which limit the observational values of the parameters y and μ as follows:

$$y_0 \leq 1.5 \times 10^{-5} \quad (95\% \text{ CL}),$$
$$\mu < 9 \times 10^{-5} \quad (95\% \text{ CL}).$$

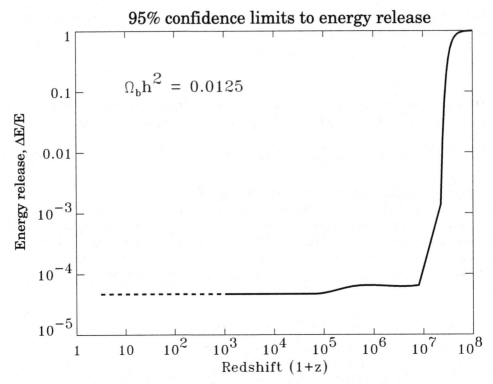

Figure 2.4 Upper confidence limits (95%) on fractional energy release, $\Delta E / E_{\mathrm{CMB}}$ from different epochs. Adapted from Smoot and Scott (1997).

The constraints on the possible energy release in the early Universe are plotted in Fig. 2.4 for these parameter values as functions of redshift (Fixsen *et al.*, 1996).

As we see, the level of energy release does not exceed 10^{-3}–10^{-4} in practically the entire range of values of redshift, $10^3 < z \le 10^6$. If $z \ge 10^7$, the 'degree of non-equilibrium' in the Universe, $\Delta\varepsilon/\varepsilon_\gamma$, does not exclude the possibility of considerable energy release, $\Delta\varepsilon/\varepsilon_\gamma \sim 1$. However, this 'softening' of the limitation fails to highlight the fact that the cosmic plasma needs to be in non-equilibrium. What we mean is that the Compton mechanism of distortion formation is found not to be sensitive to considerable distortions in the spectrum of the CMB. It should be remembered that the epoch of cosmological nucleosynthesis is not immediately followed by the $z \sim 10^7$ epoch, and that the survivors of nucleosynthesis (He^4 and D) are indicators of the equilibrium nature of the spectrum of quanta up to $z \sim 10^9$–5×10^9.

3

The ionization history of the Universe

3.1 The inevitability of hydrogen recombination

The physics of the cosmic microwave background is tightly related to the kinetics of interactions between quanta and electrons because electrons are the lightest charged particles that annihilated with positrons in the process of cooling of the Universe at the time when its temperature was falling to 10^9 K. The remaining electrons, by now non-relativistic, constitute the most important factor of possible distortions of the primordial microwave background during the $z < 10^7$ epoch when Compton scattering became the dominant mechanism of the e–γ interaction. Within the 'y-theory' of Comptonization of the CMB, we find that this parameter, which characterizes the degree of 'non-equilibrium' of electrons with respect to radiation, depends not only on temperature, T_e, but also on the plasma's optical depth in terms of Thomson scattering, τ. In turn, the rate at which the optical depth changes in time, $\dot{\tau} = \sigma_T n_e c$, is determined by two very important factors: the expansion of the Universe and the dynamics of evolution of electron concentration. As a rule, this process is described in terms of the degree of ionization of the plasma, as follows:

$$x_e = \frac{n_e}{n_{tot}}, \tag{3.1}$$

where n_{tot} is the total concentration of baryons in the plasma. If the plasma temperature definitely exceeds 10^5 K, electrons must be free (i.e. not bound to protons) since otherwise a gigantic number of ionizing quanta would immediately destroy hydrogen atoms. In other words, the efficiency of the reaction $H + \gamma \rightarrow p + e$ is so high that it is beyond any doubt that the amount of neutral hydrogen in cosmological matter is inconsequential. In this case, therefore, the degree of ionization, x_e, equals unity with high accuracy, and changes in the optical depth of the plasma relative to Thomson scattering are caused only by the expansion of the Universe. Note that even when the degree of plasma ionization does not change with time ($x_e = 1 = $const.), the anticipated plasma depth continues to diminish anyway as a result of the cosmological expansion (Hu, 1995):

$$\tau = \int_t^{t_{now}} \sigma_T n_b c \, dt \simeq 4.1 \times 10^{-2} \frac{\Omega_b}{\Omega_m} h \left\{ \left[\Omega_\Lambda + \Omega_m (1 + z)^3 \right]^{1/2} - 1 \right\}, \tag{3.2}$$

where we have used the same notation as in Chapter 2 and $\Omega_\Lambda + \Omega_m = 1$. Equation (3.2) clearly shows that if $z > z_{cr}$, where z_{cr} is found from the condition $\Omega_\Lambda \simeq \Omega_m (1 + z_{cr})^3$, the behaviour of the optical depth follows the relation $\tau \propto (1 + z)^{3/2}$, while for $z \rightarrow 0$ we have $\tau(z) \propto \frac{3}{2}\Omega_m \cdot z \rightarrow 0$.

In fact, the first important conclusion that follows from this analysis of the extremal asymptote $x_e = 1$, regardless of the value of the redshift, is that, relative to the Thomson scattering, today's Universe must be optically thin – to within $\leq 1\%$. As we see from Eq. (3.2), the behaviour of the optical depth for $z \gg 1$ is independent of Ω_Λ:

$$\tau(z) \simeq 4.1 \times 10^{-2} \Omega_b \Omega_m^{-1/2} h (1 + z)^{3/2}, \qquad (3.3)$$

and formally the zone of the 'last scattering' of quanta by electrons ($\tau(z) = 1$) corresponds to the redshift

$$z_* \simeq 8.4 \Omega_b^{-2/3} \Omega_m^{1/3} h^{-2/3}. \qquad (3.4)$$

To be specific, we assume $\Omega_b h^2 \simeq 0.02$, $\Omega_m \simeq 0.3$ and $h \simeq 0.7$ (see Chapter 1) and we finally obtain $z \simeq 60$. Therefore, with the cosmic plasma completely ionized, the maximum redshift after which the CMB radiation propagates freely is a relatively low at $z \simeq 60$. There remains the question of whether the total hydrogen ionization can be self-maintained down to this redshift. This question can be answered using the following qualitative reasoning.

To maintain the degree of ionization at the level $x_e = 1$, it is necessary for the fraction of quanta having energy above the hydrogen ionization potential, $I \simeq 13.6\,\text{eV}$, to reach approximately one quantum per baryon. As in cosmological nucleosynthesis (see Chapter 1), this formally leads to an estimate of

$$\xi^{-1} \exp\left(-\frac{I}{kT(z)}\right) \sim 1, \qquad (3.5)$$

where $\xi = \xi_{10} \times 10^{10}$. We can evaluate the optimum range for ξ from the data on the abundance of the cosmic He^4 and D: $\xi_{10} \sim 5$. The substitution of this estimate into Eq. (3.5) yields

$$T(z) \sim T_i \ln^{-1}(\xi^{-1}) \simeq 3.8 \times 10^3 \text{ K}, \qquad (3.6)$$

where $T_i = \frac{I}{k} \simeq 1.5 \times 10^5$ K is the temperature corresponding to the ionization energy. Taking into account $T(z) = T_0(1 + z)$, where $T_0 = 2.736$ K is the current temperature of the CMB radiation, we see that Eq. (3.6) implies that the ionizing (Wien) part of the spectrum cannot sustain the degree of ionization at the level $x_e \simeq 1$ at redshifts $z < 1400$. Therefore, to sustain the $x_e = 1$ mode at $z < 1400$, it is necessary that a powerful ionizing component of matter is present because there is simply not enough quanta of the primordial background radiation!

The estimates given above yield the obvious conclusion that the ionization history of the cosmic plasma is one of the most important probes for studying the properties of cosmic matter in the epoch of redshift $z \leq 1400$. Any information on the degree of plasma ionization in this period is inevitably tied to testing the processes of energy release and, therefore, to the identification of the possible sources of this energy release. In fact, the situation becomes even more dramatic if we take into account the observation of the hydrogen line $\lambda = 21$ cm and the Ly-α absorption in the spectra of remote quasars: these show that the cosmological hydrogen must already be ionized up to $x_e \simeq 1$ at redshifts $z \sim 5\text{--}6$ (see Section 3.8). Therefore, the idea of non-equilibrium sources of energy release is directly confirmed, but unfortunately for small redshifts only. What is the situation in the range $60 < z < 1400$? What can we say about the presence or absence of sources of non-equilibrium ionization (non-equilibrium relative to the primordial background radiation)? It appears that Zeldovich and Sunyaev (1969) were the first to attempt to justify the inevitability of a neutral hydrogen period in the Universe, at least over a limited range of redshifts z. The key element in their work was the idea that hydrogen ionization must be accompanied by heating of electrons to temperatures $T_e \geq 10^4$ K. At such temperatures the plasma would emit quanta (the free–free emission) with the emission coefficient given by

$$E_{ff}(\nu) = 5.4 \times 10^{-39} g T_e^{1/2} e^{-\frac{h\nu}{kT_e}} n_e^2 \text{ erg cm}^{-3} \text{ s}^{-1} \text{ ster}^{-1} \text{ Hz}^{-1} \qquad (3.7)$$

where g is the Gaunt factor ($g = 1$ for $h\nu \gg kT_e$ and $g = \frac{\sqrt{3}}{\pi} \ln[(4kT_e/h\mu) - 0.577]$ in the limit $h\nu \ll kT_e$ (Karzas and Latter, 1961)). As we see from Eq. (3.7), the emission spectrum in the long-wavelength range is practically independent of frequency. Therefore, in this wavelength range we should expect peculiarities in today's spectrum of the background radiation due to the heating of cosmic plasma at high redshifts.

When describing the spectrum of cosmic electromagnetic radiation in the Universe in Chapter 1, we pointed out that the radiation flux in the radio wavelength range ($\nu \simeq 1$–10 GHz) is definitely below $J_R \simeq 10^{-23}$ erg s^{-1} ster^{-2} Hz^{-1} for $\nu \sim 10$ GHz (see Fig. 3.1). Note that the value for J_R chosen in the original paper (Zeldovich and Sunyaev, 1969) was almost four orders of magnitude higher than the limit given above, even though this was at a different frequency ($\nu \simeq 0.6$ GHz; $\lambda \simeq 50$ cm). In view of the 'flat' behaviour of the spectrum of free–free emission, it is clear that the values $E_{ff}(\nu \simeq 0.6 \text{ GHz})$ and $E_{ff}(\nu \simeq 10 \text{ GHz})$ should remain constant to within an order of magnitude. At the same time, the observed flux at the frequency 600 MHz decreases in comparison with that at $\nu \simeq 10$ GHz by approximately another order of magnitude (see Fig. 3.1), reaching a local minimum. Consequently, the bounds on the temperature of electron heating can be derived from the condition (Zeldovich and Sunyaev, 1969)

$$\int \frac{E_{ff} \, dl}{(1+z)^3} < J_R(\nu), \qquad (3.8)$$

where $dl = c \, dt \simeq (c/H_0)(\Omega_m)^{-1/2} z^{-5/2}$ and $z \gg 1$. Combining Eqs (3.7) and (3.8), we finally obtain

$$\frac{(\Omega_b h^2/0.02)^2}{(\Omega_m h^2)^{1/2}} \int_0^{z_{max}} dz \sqrt{1+z} T_e^{-1/2}(z) \leq 0.5 j_R, \qquad (3.9)$$

where $j_R \equiv J_R f(\nu)/10^{-24}$ erg s^{-1} cm^{-2} Hz^{-1} ster^{-1}. As is clear from Eq. (3.9), the constraints on $T_e(z)$ depend on the dynamics of variation of the electron temperature as the redshift diminishes.

Let us consider a model in which the electron temperature depends on z in a power-law fashion, $T_e(z) \simeq 10^4 (1 + z)^\xi / (1 + z_*)^\xi$, where the parameter $\xi \leq 0$ and z_* defines the moment when heating starts. Clearly, as z changes at $\xi > 0$, the electron temperature should decrease, and therefore the ionizing power of the source would become insufficient for sustaining the $x_e = 1$ mode.

Let us consider the limiting case of $T_e = \text{const.} \simeq 10^4$ K, when T_e is independent of z ($\xi = 0$). We immediately obtain from Eq. (3.9) that

$$z_{max}^{3/2} \leq 75 j_R (\Omega_m h^2)^{1/2} \left(\frac{\Omega_b h^2}{0.02} \right)^{-2} \left(\frac{T_e}{10^4} \right)^{1/2}. \qquad (3.10)$$

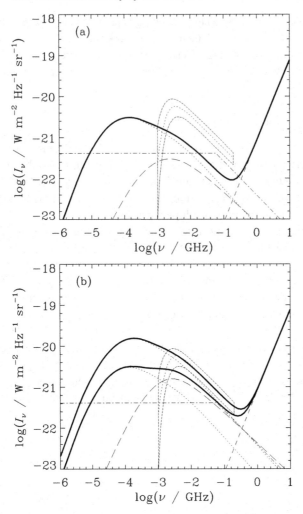

Figure 3.1 Contributions of normal galaxies (dotted curves), radio galaxies (long-dash curves) and the CMB (short-dash curve) to the extragalactic radio background intensity (thick solid curves) for the models (a) no evolution and (b) with evolution of galactic luminosity. Dotted lines depict an observational estimate of the total extragalactic radio background intensity and the dot–dash curve gives an earlier theoretical estimate. For details, see Protheroe and Biermann (1996).

So, electrons heated to 10^4 K must transfer energy to CMB quanta through the Compton mechanism. The corresponding value of the parameter y is then evaluated to be

$$y \sim \tau \frac{kT_e}{m_e c^2} \simeq 1.6 \times 10^{-9} \left(\frac{\Omega_b h^2}{0.02} \right) (\Omega_m h^2)^{-1/2} \left(\frac{T_e}{10^4} \right) z_{max}^{3/2}. \tag{3.11}$$

Taking into account Eqs (3.10) and (3.11), we find

$$y \leq 1.2 \times 10^{-7} \left(\frac{\Omega_b h^2}{0.02} \right)^{-1} \left(\frac{T_e}{10^4} \right)^{3/2} j_R, \tag{3.12}$$

which is definitely below the observational limit of COBE for $T_e < 6.5 \times 10^6$ K. Returning to Eq. (3.10) and using the value $T_e < 6.5 \times 10^6$ K as the temperature maximum, we arrive at $z_{max} \leq 150$. Note that a more detailed derivation of the quantity z_{max}, based on searching for today's minimum of the functional that takes into account constraints on the radio background and the y parameter that do not require the assumption $T_e(z) = $ const., was given in Zeldovich and Novikov (1983). In the framework of this generalized formulation, it is readily shown that quantitative conclusions on a requirement for the neutral hydrogen epoch in the Universe to have existed are only weakly affected. We can state, quite safely, that cosmic plasma must be neutral ($x_e \ll 1$) down to redshifts as low as $z \simeq 300$, and that its temperature must be low ($T_e < 10^4$ K).

It is therefore inevitable that the cosmological hydrogen must undergo recombination and, taking into account the Gunn–Peterson effect, be re-ionized later. The hidden plot behind the history of ionization of the Universe boils down to what the sources of this process could be; we discuss this aspect in the subsequent sections of this chapter.

3.2 Standard model of hydrogen recombination

In this section we turn our attention to the standard model of hydrogen recombination. Its fundamentals were formulated at the end of the 1960s and the 1970s in a number of pioneer publications (Peebles, 1968; Zeldovich, Kurt and Sunyaev, 1969). It is necessary to point out that during this period the role played by hidden mass (dark matter) in the kinematics and dynamics of the evolution of the Universe was underestimated. Therefore, all the results of the hydrogen recombination theory in the baryonic Universe needed certain corrections that would take into account the simple fact that the density of hidden matter exceeds that of baryonic matter. Consequently, the rate of expansion of the Universe should follow the law $a \propto t^{2/3}$ from the time of redshift $z_{eq} \simeq 1.2 \times 10^4 \omega^4 \Omega_m h^2$ when its density becomes equal to that of the radiation background. At the same time, the moment of equality for z_{eq} at low density of baryonic matter $\Omega_b h^2 \simeq 0.02$ would correspond to $z_{eq} \sim 240 (\Omega_b h^2 / 0.02)$ (with the hidden mass (dark matter) background neglected), and hydrogen recombination would be completed already at the radiation-dominated phase. The imbalance of hydrogen recombination reaction rates (affected by Ω_b) and the rate of cosmological expansion (dictated by Ω_m) is the principal distinctive feature of the 'standard' models when the hidden mass (dark matter) is taken into account. This factor was first pointed out by Zabotin and Naselsky (1982a) (see also Jones and Wyse, 1985; Krolik, 1990; Lubarsky and Sunyaev, 1983). At the same time as the hidden matter factor was being taken into account, Basko (1981), Krolik (1990) and Rybicki and Dell'Antonio (1994), improved the model of transfer of resonance quanta in the expanding Universe, and Lepp, Stancil and Dalgarno (1998) calculated the effect of the ionization regime on the molecular synthesis at later stages of hydrogen recombination ($z \ll 400$). A new element due to the unique accuracy of the future CMB experiments was that the effect of He^4 on the kinetics of the cosmological hydrogen recombination, and on its residual ionization, was to be taken into account. Seager, Sasselov and Scott (1999a,b) analysed a multilevel model of the hydrogen atom (~ 300 levels) and gave a systematic summary of the main achievements of the theory. This work reached its conclusion with the creation of a specialized programs package RECFAST, which at the moment is the most successful tool for calculating the dynamics of cosmological hydrogen recombination.

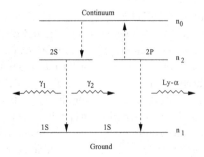

Figure 3.2 The three-level model of the hydrogen atom.

3.3   The three-level approximation for the hydrogen atom

In this section we consider the three-level model of the hydrogen atom comprising the ground state HI and the 2P and 2S states as the initial approximation for describing its recombination kinetics. We will neglect the contribution of He^4 atoms to this process, leaving it to the following section. In this formulation, the ideology of the model is in complete agreement with the approximations of Peebles (1968) and Zeldovich *et al.* (1969). The model is schematically presented in Fig. 3.2 and indicates the possible directions of electron transitions in the hydrogen atom.

Following the papers of Peebles (1968) and Zeldovich *et al.* (1969), we describe the recombination kinetics, taking into account the following specifics of the level structures of the hydrogen atom and also the properties of the plasma.

(1)   The plasma temperature is sufficiently low ($< 10^4$ K) for the collisional ionization to be negligible in comparison with radiative processes.
(2)   The time of transition for an electron to travel from the 2P level to the ground state is much shorter than the cosmological time.
(3)   The population of higher levels obeys equilibrium thermal distribution.
(4)   The population of the 2S level satisfies the condition $n_{2S} \ll n_1$, where n_1 is the hydrogen population at the ground level.
(5)   Each act of recombination to the 2P level creates a resonant Ly-α quantum, and each decay of the 2S level produces two low-energy photons.
(6)   Electrons, protons and radiation are in equilibrium at temperatures above 10^4 K.

This means that the reaction $e + p \leftrightarrow H + \gamma$ occurs in both directions, in such a way that the equilibrium concentrations of electrons, protons and neutral hydrogen atoms obey the Saha formula,

$$\frac{n_e n_p}{n_H} = \frac{g_e g_p}{g_H} \frac{(2\pi m_e kT)^{3/2}}{h^3} e^{-I/kT}, \tag{3.13}$$

where g_i are the statistical weights of each component. We now introduce the degree of ionization, $x_0 \equiv n_e/(n_p + n_H)$. Then the evolution of the equilibrium degree of ionization is described by the following equation (Zeldovich and Novikov, 1983):

$$\frac{x^2}{1-x} \simeq 4.4 \times 10^{22} \left(\frac{\Omega_b h^2}{0.02}\right)^{-2} (1+z)^{-3/2} \exp\left[-\frac{5.77 \times 10^4}{1+z}\right]. \tag{3.14}$$

Note, however, that the equilibrium degree of ionization determined from the Saha formula (3.14) does not provide the entire detailed picture of the formation of neutral hydrogen atoms. The point is that each act of recombination, $p + e \rightarrow H + \gamma$, is accompanied by the emission of one Ly-α quantum with energy $h\nu_\alpha = \frac{3}{4}I = 10.2\,\text{eV}$, which immediately excites a hydrogen atom, while quanta with $h\nu_\alpha = \frac{1}{4}I = 3.4\,\text{eV}$ that are abundant in the blackbody radiation immediately ionize this atom.

The absorption cross-section of neutral hydrogen for resonance quanta is found to be extremely large. The corresponding optical depth, $\tau_\alpha \simeq \sigma_1 n_b ct$, is found to be $\geq 4 \times 10^8$ (Peebles, 1968), and therefore Ly-α quanta generated in each recombination event should be immediately absorbed by the generated hydrogen atoms.

Relatively low-energy quanta are needed to ionize hydrogen atoms from the 2P level: $E = I - h\nu_\alpha = \frac{1}{4}I \simeq 3.4\,\text{eV}$. Owing to the Wien character of the spectrum of background photons, the number of such 'soft' quanta is found to exceed that of 'hard' quanta (with energy $E \simeq I$) by a factor of approximately $e^{-I/4kT}/e^{-I/kT} \simeq e^{\frac{3}{4}\frac{I}{kT}} \gg 1$; therefore, hydrogen atoms in the 2P state are immediately ionized by quanta from the 'soft' part of the CMB spectrum.

Therefore, the right-hand side branch in Fig. 3.2 that describes the dynamics of population of the hydrogen atom 2P state is, at the same time, a sort of 'engine' for producing – and, more importantly, accumulating – Ly-α quanta in the process of hydrogen recombination over their equilibrium concentration in the Wien range of the spectrum. The reaction channel on the left is directly responsible for the formation of neutral hydrogen – via the metastable 2S level, as was first shown in Peebles (1968) and Zeldovich *et al.* (1969). Let us consider the kinetics of this process in more detail.

3.3.1 Equations for the populations of hydrogen levels

Look at the diagram of electron transfers in the hydrogen atom as shown in Fig. 3.2. In the three-level atom approximation, this diagram corresponds to the processes $H_{n=2,l=2S} + \gamma_\alpha \leftrightarrow e + p^+$, $H_{n=1} + \gamma_\alpha \leftrightarrow H_{n=2,l=2P}$, $H_{n=2,l=2S} + \gamma_\alpha \leftrightarrow H_{n=1} + \gamma + \gamma$. For each state of an electron, we introduce the corresponding concentration of hydrogen atoms, n_i ($n_{1S}, n_{2S}, n_{3S}, \ldots$). This concentration is found from the following kinetic equation (Seager, Sasselov and Scott, 2000):

$$\frac{d(n_i(t)a^3)}{a^3 dt} = n_e n_p \alpha_{ic} - n_i \beta_{ic} + \sum_{j=1}^{N} n_j(+)\rho_{ji} - n_i \sum_{j=1}^{N} \rho_{ij}, \qquad (3.15)$$

where α is the coefficient of recombination to level ν from the continuum, β_{ic} is the corresponding coefficient of ionization from level i to continuous spectrum, ρ_{ij} are the coefficients of transition from level i to j, n_e is the electron concentration and n_P is the concentration of atoms in the ionized state.

The set of equations (3.15) must be supplemented with an equation for the concentration of free electrons; in fact, this equation describes the rate of change of the degree of ionization, x_e, as a function of time (Peebles, 1968) as follows:

$$\frac{d(n_e a^3)}{a^3 dt} = -\sum_{i>1} \left(\alpha_{ie} n_e^2 + \beta_{ie} n_e \right), \qquad (3.16)$$

where α_{il} and β_{il} are the coefficients of recombination and photoionization at the level i, and the index l describes the possible P and S states.

We have emphasized earlier that all high-energy levels of the hydrogen atom are in dynamic equilibrium with radiation. In this case (Peebles, 1968),

$$n_{ie} = n_{2S}(2l+1)e^{-\frac{(B_2-B_1)}{kT}}, \tag{3.17}$$

where B_i is the bonding energy of the ith level. Introducing the notation

$$\alpha_c = \sum_{i>1} \alpha_{ie}; \quad \beta_c = \sum_{i>1}(2l+1)\beta_{ie}e^{-\frac{(B_2-B_1)}{kT}} = \alpha_c \cdot e^{-\frac{B_2}{kT}} \frac{(2\pi m_e kT)^{3/2}}{h^3} \tag{3.18}$$

for free electron concentration, Eqs (3.16)–(3.18) yield the following equation:

$$\frac{\mathrm{d}(n_e a^3)}{a^3 \mathrm{d}t} = -\alpha_c n_e^2 + \beta_c \cdot n_{2S}. \tag{3.19}$$

We see from this equation that the dynamics of evolution of n_e is directly related to the population dynamics of the level 2S, found from the set of equations (3.15). Following Peebles (1968), we introduce the relative population of the 2S level:

$$R_{2S} = \frac{n_{2S}}{n_{1S}}. \tag{3.20}$$

Obviously this population is sustained by Ly-α quanta, whose dynamics is described by the following transfer equation (Peebles, 1968):

$$\frac{\mathrm{d}(n_\alpha^3)}{a^3 \mathrm{d}t} = -\nu_\alpha H[n_{(\nu+)} - n_{(\nu-)}] + R, \tag{3.21}$$

where $n_\alpha = \int_{\nu_-}^{\nu+} n_\nu \, \mathrm{d}\nu$ is the concentration of Ly-α quanta in the spectral line, ν_- and ν_+ are, respectively, the lower and upper bounds dictated by the profiles of the Ly-α line, R is the concentration of the Ly-α quanta fed into the plasma by each hydrogen recombination event per unit time, and H is the Hubble constant. Since the width of the Ly-α line is $\Delta\nu/\nu_\alpha = (\nu_t - \nu_-)/\nu_\alpha \sim 10^{-5}$ and $n_\alpha \sim n_{(-)}\Delta\nu$, Eq. (3.21) immediately implies that the right-hand side must be nearly zero. Therefore

$$n_{(\nu-)} \simeq n_{(\nu_t)} + \frac{R}{\nu_\alpha H}. \tag{3.22}$$

Introducing filling numbers for quanta, $\mathfrak{R} = n_\nu c/(8\pi\nu^2)$, we find from Eq. (3.22)

$$\mathfrak{R}_- = \mathfrak{R}_+ + \frac{R\lambda_\alpha^3}{8\pi H}, \tag{3.23}$$

where $\lambda_\alpha = c/\nu_\alpha$ and $\mathfrak{R}_+ = \exp(-(B_1 - B_2)/kT)$ correspond to the thermodynamic equilibrium value.

To find the intensity of Ly-α quanta that are newly fed into the process during recombination, it is necessary to remember that their number is determined using the following arguments. Each recombination event generates one Ly-α quantum which is absorbed by a hydrogen atom. At the same time the decay of the 2S level 'removes' Ly-α quanta from the plasma, converting them to 'soft' photons. Therefore,

$$R = \left(\alpha_c n_e^2 - \beta_c n_{2S}\right) - \Lambda_{2S-1S}\left(n_{2S} - n_{1S}e^{-\frac{(B_1-B_2)}{kT}}\right), \tag{3.24}$$

where $\Lambda_{2S-1S} = 8.224\,58 \text{ s}^{-1}$ (see Spitzer and Greenstein (1951)) is the rate of the two-atom decay of the 2S state. Equation (3.24) shows that in the equilibrium state $R \equiv 0$ because recombination and ionization events completely cancel out (the expression in the first brackets in Eq. (3.24)) and the equilibrium population of the 2S level ensures that the expression in the second pair of brackets is zero.

The last condition that closes the set of equations for the degree of ionization in the presence of Ly-α quanta reduces to the equality $R_{2S} = \mathfrak{R}_+$, which points to a simple fact: the population of the 2S level is dictated by the equilibrium value of Φ_+ plus the Ly-α quanta that the redshift pushed under the ionization threshold. Substituting $R_{2S} = \mathfrak{R}_-$ into Eq. (3.20) and taking into account Eq. (3.23), we arrive at an equation for R, and therefore also for R_{2S}, by expressing n_{2S} in terms of n_{1S} and other parameters of the problem. Introducing now the degree of ionization, $x_e = n_e/(n_p + n_{1S})$, we obtain from Eq. (3.19)

$$-\frac{dx_e}{dt} = D\left[\alpha_c n x_e^2 - \beta_c(1 - x_e)e^{-\frac{B_1 - B_2}{kT}}\right], \tag{3.25}$$

where

$$D = \frac{1 + K\Lambda_{2S,1S}n(1 - x_e)}{1 + K(\Lambda_{2S,1S} + \beta_c)n(1 - x_e)} \tag{3.26}$$

and $K = \lambda_\alpha^3/6\pi H(t)$, $B_1 - B_2 = \frac{1}{4}$. For the recombination coefficient we use the improved value (Hummer, 1994; Pequignot, Petitjean and Boisson, 1991; Verner and Ferland, 1996)

$$\alpha_c = 10^{-13}\frac{at^b}{1 + ct^d} \text{ cm}^3 \text{ s}^{-1}, \tag{3.27}$$

where $a = 4.309$, $b = -0.6166$, $c = 0.6703$, $d = 0.5300$ and $t = T_M/10^4$ K. Here, T_M is the plasma temperature, which is assumed to be equal to the CMB temperature in the three-level recombination model chosen here.

3.4 Qualitative analysis of recombination modes

In this section we give a qualitative description of the behaviour of the plasma ionization degree, x_e, in various ranges of redshift z, leaving detailed numerical calculations to the following section. The point is that the three-level hydrogen recombination model discussed above requires certain modifications. Namely, we need to take into account the role of He^4 and also a detailed description of the behaviour of Ly-α quanta and of the temperature of matter (it does not always follow the radiation temperature, especially for low redshifts, $z < 10^2$). Nevertheless, an analysis of the approximate model proves extremely useful for describing the dynamics, $\dot{x}_e$, in the range $700 \leq z \leq 1100$ if the factors listed above do not noticeably affect the ionization balance, and the result of the analytical solution of Eq. (3.23) is found to be applicable not only for a qualitative, but equally well for a quantitative description of the problem.

Therefore, for redshifts $z \geq 1400$ the degree of ionization, x_e, is given by the stationary solution of Eq. (3.25):

$$\frac{x_e^2}{1 - x_e} = \frac{\beta_c}{\alpha_c n}e^{-\frac{B_1 - B_2}{kT}}. \tag{3.28}$$

Taking into account the relationship between β_c and α_c (see Eq. (3.18)), we readily see that Eq. (3.28) corresponds to the equilibrium Saha formula. As the radiation temperature decreases, Eq. (3.25) predicts that this equilibrium is violated.

Taking into account the fact that in Eq. (3.25) $D \simeq (\Lambda_{2S,1S}/(\Lambda_{2S,1S} + \beta_c))$ and also that $\beta_c \gg \Lambda_{2S,1S}$, we obtain

$$\frac{dx_e}{dt} = \Lambda_{2S,1S} \left[\frac{\alpha_c n x_e^2}{\beta_c} - (1 - x_e) e^{-\frac{B_1 - B_2}{kT}} \right]$$

$$= \Lambda_{2S,1S} \frac{\alpha_c n}{\beta_c} \left[x_e^2 - (1 - x_e) \frac{x_e^{2(eq)}}{(1 - x_e^{(eq)})} \right]. \tag{3.29}$$

The equilibrium degree of ionization, $x_e^{(eq)}$, decreases exponentially with decreasing redshift. Therefore, beginning with a certain critical value $z = z_*$, for all $z < z_*$, in Eq. (3.29), we can neglect the quantity $\sim x_e^{2(eq)}$ and retain only the first term in square brackets. Since z_* and $z \gg 1$, we have

$$\frac{dx_e}{dt} = -\frac{dx_e}{dz} H_0 \sqrt{\Omega_m} z^{5/2} \tag{3.30}$$

and Eq. (3.29) takes the form (Longair and Sunyaev, 1969)

$$\frac{dx_e}{dt} = \frac{\Lambda_{2S,1S} n_{bar}^{(0)} e^{\frac{I}{4kT_0 z}} h_P^3}{(2\pi m_e k T_0)^{3/2}} H_0^{-1} \Omega_m^{-1/2} x_e^2 z^{-1}, \tag{3.31}$$

where T_0 is the current CMB temperature, $n_{bar}^{(0)} \simeq 2 \times 10^{-7} (\Omega_b h^2/0.02)$ is the current baryonic density, and h_P is the Planck constant. Introducing the notation $A = (\Lambda_{2S,1S} n_b^{(0)} h_P^3 H_0^{-1} \Omega_m^{-1/2})/(2\pi m_e k T_0)^{3/2}$ and $I/4kT_0 z \gg 1$, we derive from Eq. (3.31) (Zabotin and Naselsky, 1982a,b):

$$x_e \simeq \left[C + \frac{4kT_0 z}{I} A e^{\frac{I}{4kT_0 z}} \right]^{-1}. \tag{3.32}$$

The constant C is found from the condition $x_e(z = z_*) = x_e^{(eq)}(z_*)$. Let us look at the asymptotic behaviour in Eq. (3.32) for $z < z_*$. In this case, the second term in square brackets increases in comparison with the first and (see Zeldovich and Novikov (1983))

$$x_e \simeq \frac{I}{4kT_0 z} A e^{-I/4kt_0 z} \ll 1. \tag{3.33}$$

As this expression clearly shows, the degree of ionization of hydrogen considerably exceeds the equilibrium value given in Eq. (3.28); this occurs because of the exponential factor $I/4kT_0 z \gg 1$ obtained from Saha's formula, which is slower in comparison with $I/2kT_0 z \gg 1$. In fact, once recombination starts in the equilibrium mode ($z = z_* \sim 1400$), the generation of excessive Ly-α quanta and their redshifting in frequency result in an increased degree of ionization, the behaviour of which is described by Eq. (3.33).

We must emphasize that the solution for x_e in the form (3.33) describes the behaviour of the degree of ionization within a relatively small interval of redshifts z: $900 < z < 1400$. If $z < 900$, the function D in Eq. (3.25) approaches unity and the asymptotic behaviour of recombination begins to be dominated by the first term on the right-hand side of Eq. (3.25) with $D = 1$. At this stage, the plasma is sufficiently rarefied, and recombination quanta are incapable of sustaining the equilibrium between the excited levels of the hydrogen atom. In reality, each recombination event results in the emission of a Ly-α quantum, which is removed by the redshift beyond the half-width of absorption. In this mode, Eq. (3.25) implies

$$\frac{dx_e}{dz} = \Gamma(z) \cdot x_e^2, \tag{3.34}$$

where $\Gamma(z) = \alpha_c n_b^{(0)} z^{1/2} H_0^{-1} \Omega_m^{-1/2}$, and the solution for $x_e(z)$ has a simple form (see Eq. (3.25)) (Zabotin and Naselsky, 1982a,b; Zeldovich *et al.*, 1969)

$$x_e \simeq \left[a - \int \Gamma(z) \, dz \right]^{-1}, \tag{3.35}$$

where a is the integration constant found from the condition of matching to the preceding mode. Note that owing to the power-law dependence of $\Gamma(z)$, the decrease in ionization degree is fairly monotonic in contrast to the initial stages at $z \sim 10^3$–1.4×10^3. We must stress at the same time that the transition from one asymptotic behaviour $x_e(z)$ to another constantly requires matching the solutions in the transition zone, resulting in accumulation of errors. Indeed, it is impossible to delineate unambiguously the zones of influence of distinct ionization modes. Therefore in this section we have limited the discussion to the above-described characteristic modes of diminishing of plasma ionization degree, preferring to look at the quantitative side of the problem in the light of the numerical solutions given in the following section.

3.5 Detailed theory of recombination: multilevel approximation

We begin by formulating the main reasons for which it is necessary to analyse the kinetics of hydrogen recombination in detail, in contrast to the framework of the approximate three-level model of the hydrogen atom. One of the main reasons for this lies in the need to develop a modern theory of formation of the CMB anisotropy. We will see in Chapter 5 that fluctuations $\Delta T / T$ on a scale of several minutes of arc are formed at the very beginning of recombination, at $z \sim 1400$. In fact, it is this phase of hydrogen recombination that is fairly sensitive to the presence of He[4] (Seager *et al.*, 1999a,b). From this, a direct connection can be traced between cosmological nucleosynthesis (prediction of mass concentration of He[4]), hydrogen recombination and small-scale CMB anisotropy.

Another important reason follows from the high accuracy of measurements in the WMAP experiment and in the PLANCK mission planned for 2007. If we choose the relative error of determining ΔT to be 10% – a very conservative estimate of determining the characteristics of CMB anisotropy – we need to be absolutely sure that these errors are not 'accumulated' only as a result in inaccuracies in theoretical predictions of the dynamics of the thinning of cosmic plasma for primordial radiation. Therefore, a detailed theory of recombination must be capable of predicting the behaviour of $x_e(z)$ with an error $\leq 1\%$, and possibly even lower.

Finally, there is the third cause, which is mostly of predictive nature. Hydrogen recombination results in considerable distortions of the background radiation spectrum in the Ly-α frequency range at the expansion stage, $z \sim 10^3$. As a result of redshifting, these distortions must be represented today in the long-wavelength range $\lambda \simeq c z_{rec}/\nu_\alpha \sim 11.3 \times 10^{-2}$ cm (Peebles, 1968; Zeldovich, Kurt and Sunyaev, 1969) that is, in the near-infrared range of the cosmic radiation spectrum. Experimental detection of a specific electromagnetic 'echo' of the recombination epoch would be fantastically important for testing the properties of the cosmic plasma at redshifts $z \sim 10^3$. Unfortunately, the high level of infrared background in this range does not offer us any hope of rapid experimental solution of the problem. However, a detailed

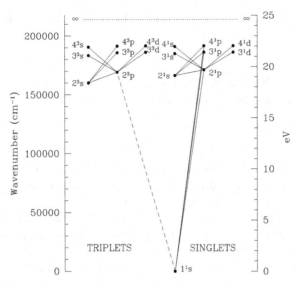

Figure 3.3 He atom levels included in the comparative analysis of models. Adapted from Seager *et al.* (1999a).

prediction of the shape of these distortions would be of extreme interest for understanding the mechanisms of formation of radiation in the near-infrared band.

It must be emphasized that the ionization history of the cosmic plasma at the stage $z \ll 10^3$ is extremely important for understanding the processes of formation of primary molecules of cosmological origin – those that could act as an efficient 'coolant' in dense clouds, facilitating the formation of first-generation stars. We must again emphasize that the development of a detailed theory of hydrogen recombination does not mean in any way a diversion from the simplified two-level recombination model, the fundamentals of which were created more than 30 years ago. Moreover, this simplified model reproduces qualitatively, and very often quantitatively, the main physical principles and processes that resulted in the transformation of ionized hydrogen to the neutral state. Assuming this model as a basis, we will focus mainly on discussing its more important additions that considerably increase the predictive accuracy of the theory. The paper by Seager *et al.* (1999a,b) that formed the basis for the development of a specialized package, RECFAST, for calculating the dynamics of evolution of $x_e(z)$, beginning from the earliest recombination phases at $z \sim 10^4$ up to $z = 0$, became an important step on this path, stimulating many years of research into the recombination kinetics of cosmological hydrogen. When describing important additions to the standard hydrogen recombination scheme below, we follow the ideology of this publication.

We will now list the main features of the detailed theory of cosmological recombination.

3.5.1 *Dynamics of excited states of hydrogen and helium*

Equations for the populations of hydrogen and helium atomic levels coincide, in the general case, with Eq. (3.15). The main difference lies in the increase in the number of hydrogen levels to $N = 300$ and in adding higher levels of He^4 to the analysis (see Fig. 3.3). A new term responsible for collisional ionization, and the excitation of hydrogen atoms is included in the reaction rates α_{ij} and β_{ij} in addition to radiative processes:

$$\alpha_{ij} = \alpha_{ij}^{\text{rad}} + n_{\text{e}} C_{ij}^{(\text{col})}. \tag{3.36}$$

3.5.2 Radiation kinetics

As we saw in the three-level model of the hydrogen atom, the kinetics of cosmo-logical hydrogen recombination is very sensitive to the behaviour of hydrogen resonance lines. Resonance quanta take part in absorption, scattering and emission by hydrogen and helium atoms, undergoing redshift because of the expansion of the Universe. The following kinetic equation is used to describe radiation transfer within lines for the directions-averaged intensity of quanta, instead of Eq. (3.21):

$$\frac{\partial y(\nu, t)}{\partial t} - \nu H(t) \frac{\partial y(\nu, t)}{\partial \nu} = -3H(t)y(\nu, t) + c\left[j(\nu, t) - k(\nu, t)y(\nu, t)\right]. \tag{3.37}$$

Here $j(\nu, t)$ is the spectral power of the sources, $k(\nu, t)$ is the radiation absorption coefficient and $H(t) = \dot{a}/a$ is the Hubble parameter at time t. The 'Sobolev approximation' is used for the coefficient $k(\nu, t)$ (see Dell'Antonio and Rybicki (1994)). For each cascade transition between the levels i and j ($i < j$) of the discrete spectrum, we can introduce photons concentration in the following equation:

$$\Delta R_{ij} = P_{ij} \left[n_j [A_{ij} + B_{ji} B(\nu_{ij}, t)] - n_i B_{ij} B(\nu_{ij}, t) \right], \tag{3.38}$$

where A_{ij}, B_{ji} and B_{ij} are the Einstein coefficients, $B(\nu_{ij}, t)$ is the spectral intensity of the CMB radiation at frequency ν_{ij} and P_{ij} is the probability that a quantum emitted in the $j \to i$ transition does not undergo a subsequent scattering or absorption. By the logic of the definition, $P_{ij} = 1$ corresponds to the probability of a quantum 'escaping' to infinity, and $P_{ij} = 0$ signifies that all photons of the $j \to i$ transition will be absorbed by atoms. As we have already seen for the Ly-α line of hydrogen, its dynamics in terms of the Sobolev probability of 'escaping' corresponds to the approximation $P_{ij} \ll 1$, while for all other lines we assumed $P_{ij} = 1$. Following Seager *et al.* (2000), we first consider the dynamics of transitions between the i and j levels, neglecting the cosmological solution. In this case the population dynamics for i and j levels is given by the equation

$$n_i R_{ij} = n_i B_{ij} \bar{y}; \quad n_j R_{ji} = n_j A_{ji} + n_j B_{ji} \bar{y}, \tag{3.39}$$

where

$$\bar{y} = \int_0^\infty Y(\nu, t) \Phi(\nu) \, d\nu$$

and $\Phi(\nu)$ is the line profile. To simulate $\Phi(\nu)$, we can use, depending on the accuracy required of the solution, the Voigt profile or the $\Phi(\nu) = \delta(\nu - \nu_\alpha)$ approximation.

In Sobolev's approximation for the probability P_{ij}, it is necessary to take into account the expansion of the Universe, which determines the matter velocity field, $\nu(r)$, at the scale $r \sim L$, comparable to thermal velocity ν_{T}: $L \simeq \delta_{\text{T}}/(d\nu/dr)$. On the other hand, the probability P_{ij} is a function of the optical depth of matter (Rybicki, 1984):

$$P_{ij} = \exp(-\tau(\nu_{ji})). \tag{3.40}$$

Here $d\tau(\nu_{ji}) = -\widetilde{K} \Phi(\nu_{ji}) \, dl$, τ is the optical thickness along the line of sight from the point of emission to the point of absorption, and $\widetilde{K}$ is the integral absorption coefficient:

$K = \widetilde{K}\Phi(\nu_{ji})$. Introducing the notation $x = (\nu - \nu_{ij})/\Delta\nu$, where $\Delta\nu$ is the line width (in units of the Doppler shift of the frequency of quanta) and ν_{ij} is the frequency corresponding to the line centre, we derive from Eq. (3.40)

$$d\tau_{ij} = -\frac{\widetilde{K}}{\Delta\nu}\Phi(x)\,dl. \tag{3.41}$$

According to Seager *et al.* (2000) the absorption coefficient, $\widetilde{K}$, in Eq. (3.41) has the form

$$\widetilde{K} = \frac{h_P\nu}{4\pi}(n_i B_{ij} - n_j B_{ji}). \tag{3.42}$$

Then, taking into account that $g_i B_{ij} - g_j B_{ji}$ and $A_{ji} = (2h\nu^3/c^2)B_{ji}$, we transform $\widetilde{K}$ to

$$\widetilde{K} = \frac{A_{ji}\lambda_{ij}^2}{8\pi}\left(n_i\frac{g_j}{g_i} - n_j\right). \tag{3.43}$$

Substituting Eq. (3.43) into Eq. (3.41) and taking into account Eq. (3.40) we obtain for the probability P_{ij}

$$P_{ij} = \exp\left[-\tau_s\int_{-\infty}^{\infty}\Phi(x')\,dx'\right], \tag{3.44}$$

where

$$\tau_s \equiv \frac{\widetilde{K}}{\Delta}\cdot L, \qquad L = \frac{\nu_T}{d\nu/dr} = \sqrt{\frac{3kT_M}{m_A}}\Big/\frac{d\nu}{dr}, \qquad \Delta = \frac{\nu_0}{c}\sqrt{\frac{3kT_M}{m_A}}\Big/\frac{d\nu}{dr},$$

ν_0 is the line centre frequency, m_A is the atomic mass and T_M is the temperature of the matter. Also taking into account that $d\nu/dr = H_0$ for the Hubble flow and choosing the normalization of the line profile integral $\int_0^\infty \Phi(x)\,dx = 1$, we derive from Eq. (3.44) expressions for τ_s and P_{ij}:

$$\tau_s = \frac{A_{ji}\lambda_{ij}^3[n_i(g_j/g_i) - n_j]}{8\pi H(t)}, \tag{3.45}$$

$$P_{ij} = \frac{1 - \exp(-\tau_s)}{\tau_s}. \tag{3.46}$$

It is easy to understand from Eq. (3.46) the dynamics of the sinking of the Ly-α line under the absorption threshold. At high redshifts, when $\tau_s \gg 1$ and $P_{\nu_\alpha} \simeq 1/\tau_s \ll 1$, practically all Ly-$\alpha$ quanta are absorbed by hydrogen atoms that are being formed. In the process of recombination, the Sobolev optical depth switches to the mode with $\tau_s \ll 1$ and $P_{\nu_\alpha} \simeq 1$. In the intermediate range of z, when $\tau \simeq 1$, the probability for this 'sinking' is $P_{\nu_\alpha} \simeq 0.6$ and rapidly diminishes to $\tau_s \ll 1$ for $z < 10^3$.

3.5.3 Thermal history of matter

Details of the thermal history of matter are extremely important when analysing the asymptotics of recombination ($z \ll 10^3$), when the Compton scattering of the CMB quanta fails to sustain the thermal contact, $T_M = T_R$. When constructing the thermal history of cooling of the matter, we need to take into account, along with the Compton process, the free–free transitions and the cooling via photorecombination radiation in the plasma. A detailed balance of these processes in a hydrogen–helium plasma was described in Seager *et al.* (2000). We give here the final equation for the temperature of the matter:

$$(1+z)\frac{dT_{\mathrm{M}}}{dz} = \frac{8\sigma_{\mathrm{T}}\varepsilon_{\mathrm{R}}}{3H(z)m_e c}\frac{n_e(T_{\mathrm{M}} - T_{\mathrm{R}})}{n_e + n_{\mathrm{H}} + n_{\mathrm{He}}} + 2T_{\mathrm{M}} + \frac{2}{3Kn_{\mathrm{tot}}H(z)}\sum_{i=1}^{6}\Lambda_i, \qquad (3.47)$$

where $n_{\mathrm{tot}} = n_e + n_{\mathrm{H}} + n_{\mathrm{He}}$, ε_{R} is the energy density of the primordial background radiation in the epoch with the redshift z and T_{R} is the background radiation temperature. The functions of heating and cooling, Λ_i, in Eq. (3.47) for various processes are as follows.

Free–free emission

$$\Lambda_1 = \Lambda_{\mathrm{ff}} = \frac{2^5 \pi e^6 z}{3^{3/2} h_{\mathrm{P}} m_e c^3}\left(\frac{2\pi k T_{\mathrm{M}}}{m_e}\right)^{1/2} g_{\mathrm{ff}} n_e (n_{\mathrm{p}} + n_{\mathrm{HeII}} + 4n_{\mathrm{HeIII}}), \qquad (3.48)$$

where g_{ff} is the Gaunt factor, n_{p} is the proton concentration, and n_{HeII} and n_{HeIII} are the concentrations of single- and double-ionized helium nuclei, respectively.

Photorecombination cooling

$$\Lambda_2 = +4\pi\sum_{j=1}^{N} n_e n_{\mathrm{p}}\left(\frac{n_i}{n_e n_{\mathrm{p}}}\right)^{\mathrm{LTE}} \times \int_{\nu_j}^{\infty} d\nu\, h_{\mathrm{P}}(\nu - \nu_0)\frac{\alpha_{ij}(\nu)}{h_{\mathrm{P}}\nu}$$

$$\times \left[\frac{2h_{\mathrm{P}}\nu^3}{c^3} + B(\nu, T_{\mathrm{R}})\right]\exp\left[-\frac{h_{\mathrm{P}}\nu}{kT_{\mathrm{M}}}\right], \qquad (3.49)$$

where α_j is the recombination rate to the jth level and the ratio $\left(n_i/(n_e n_{\mathrm{p}})\right)^{\mathrm{LTE}}$ (the superscript LTE stands for 'local thermodynamic equilibrium') is found from the equilibrium Saha formula as follows:

$$\left(\frac{n_j}{n_e n_{\mathrm{p}}}\right)^{\mathrm{LTE}} = \frac{g_i}{2g_c}\left(\frac{h^2}{2\pi m_e k T_{\mathrm{M}}}\right)^{3/2} e^{E_j/kT_{\mathrm{M}}}.$$

Photoionization heating

$$\Lambda_3 = -4\pi\sum_j n_j \int_{\nu_0}^{\infty}\frac{\alpha_{jc}(\nu)}{h_{\mathrm{P}}\nu}B(\nu, T_{\mathrm{R}})h_{\mathrm{P}}(\nu - \nu_0)\,d\nu, \qquad (3.50)$$

where $\alpha_{ij}(\nu)$ is the ionization rate from level j to continuum.

Cooling in lines

$$\Lambda_4 = -h_{\mathrm{P}}\nu_0[n_j R_{ij} - n_i R_{ij}]. \qquad (3.51)$$

Collisional cooling and heating

$$\Lambda_5 = h_{\mathrm{P}}\nu_0 C_{ic},$$
$$\Lambda_6 = -h_{\mathrm{P}}\nu_0 C_{ij}, \qquad (3.52)$$

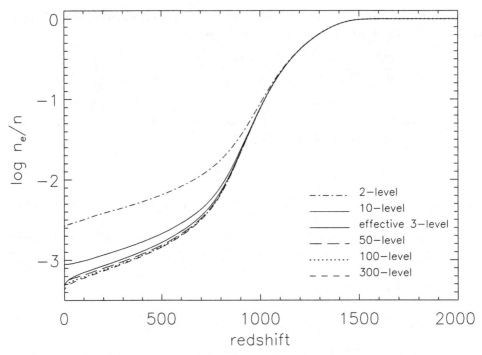

Figure 3.4 Multilevel model of hydrogen recombination. Upper set of curves: the standard CDM model with $\Omega_{tot} = 1.0$, $\Omega_b = 0.05$, $h = 0.5$. Lower set of curves: $\Omega_{tot} = 1.0$, $\Omega_b = 1.0$, $h = 1.0$. Both models assume $T_0 = 2.728$ K, $Y_p = 0.24$. From Seager *et al.* (1999a).

where C_{ic} and C_{ij} are the coefficients of collisional ionization and recombination, respectively (Mihalas, 1978).

3.6 Numerical analysis of recombination kinetics

We will now consider various modes of recombination of hydrogen and helium as functions of the parameters of the cosmological plasma in the framework of the mathematical formulation of the problem posed in the previous section. At the same time, following Seager *et al.* (2000), we will monitor the error level generated in analysing the dynamics of $x_e(z)$ in terms of the simplest model of the three-level recombination and with multilevel approximation. Figure 3.4 plots the dynamics of the decrease in the degree of ionization in two cosmological models with $\Omega_{tot} = 1$ but differing in the density of the baryonic fraction of matter. The upper curves correspond to the model $\Omega_b = 0.05$, $h = 0.5$, and the lower ones to the model $\Omega_b = 1$, $h = 1$. For each of these models, Seager *et al.* (2000) give the dependence of the degree of ionization, $x_e(z)$, for various sets of levels of the hydrogen atom. As we see from Fig. 3.4, the most 'sensitive' to the choice of model is the range $z \leq 10^3$, where the maximum difference is obtained (of approximately one order of magnitude) between the two-level and the 300-level approximations. We need to emphasize especially that in the model with the 10-level hydrogen atom we find a substantial difference in residual ionization at $z = 0$; it reaches a factor of 3.

An important element of numerical analysis of hydrogen recombination dynamics is the clarification of how the basic reaction rates of transitions $2P - 1S$, $2S - 1S$ and the rate of formation of neutral hydrogen atoms depend on the redshift, z. The dynamics of these processes is shown in Fig. 3.5 for the two cosmological models plotted in Fig. 3.4. Figures 3.5(b) and (c) show the corresponding reaction rates for the recombination of HeI and HeII. An important conclusion that follows from Seager *et al.* (2000) is that by the moment when hydrogen recombination started ($z \sim 1500$), helium recombination was practically completed. For the two cosmological models mentioned above, this conclusion follows from Fig. 3.6, which is a comparison of the results of detailed calculations of changes in the degree of ionization in the course of HeI recombination and of the dynamics of $x_e(z)$ in accordance with the Saha formulas for helium,

$$\frac{x_e(x_c - 1)}{1 + f_{He} - x_e} = 4\frac{(2\pi m_e kT + M)^{3/2}}{h_p^3 n_H}e^{-x_{HeI}/kT_M} \qquad \text{for} \quad \text{HeI} \rightarrow \text{HeII},$$

$$\frac{(x_e - 1 - f_{He})x_e}{1 + 2f_{He} - x_c} = \frac{(2\pi m_e kT + M)^{3/2}}{h_p^3 n_H}e^{-x_{YtHeII}/kT_M} \qquad \text{for} \quad \text{HeII} \rightarrow \text{HeIII},$$

(3.53)

where x_{HeI} and x_{HeII} are the corresponding helium ionization potentials, $f_{He} = n_{He}/n_H = Y_p/[4(1 - Y_p)]$, n_H is the hydrogen concentration and Y_p is the He^4 mass concentration.

On the basis of the available numerical data, Seager *et al.* (2000) proposed a generalized model of the three-level recombination that takes into account the contribution of helium recombination. This model is the basis of the program package RECFAST for computing the ionization history of cosmic plasma. The package is widely used in analysing the anisotropy of the CMB. In the framework of this generalized model, the dynamics of hydrogen and helium ionization degrees is described by the following set of equations:

$$H(z)(1 + z)\frac{dX_H}{dz} = D(x_H, x_{He}, x_0)\left[x \cdot x_H n_H \alpha_H - \beta_H(1 - X_H) \times e^{-\frac{h_p \nu_{HeII}}{kT_M}}\right],$$

(3.54)

$$H(z)(1 + z)\frac{dX_{HeII}}{dz} = G(x_H, x_{He}, x_0)\left[x_{He} \cdot x n_H \alpha_{HeI} - (f_{He} - x_{HeII})e^{-\frac{h_p \nu_{HeII}}{kT_M}}\right],$$

where $\alpha_H = 5 \times 10^{-13}\frac{at^b \cdot f}{1+ct^d}$ cm^3 s^{-1} (see Eq. (3.27) for values of b and d) and

$$\alpha_{HeI} = q \times 10^6 \left[\sqrt{\frac{T_M}{T_2}}\left(1 + \sqrt{\frac{T_M}{T_2}}\right)^{1-p}\left(1 + \sqrt{\frac{T_M}{T_1}}\right)^{1+p}\right]^{-1} \text{cm}^3 \text{ s}^{-1};$$

$x \equiv x_c = x_H + f_{He} \cdot x_{He}$, $q = 10^{-16.744}$, $p = 0.711$, $T_1 = 20^{5.114}$ K, $T_2 = 3$ K, $f = 1.14$.

The functions D and G that we find in these equations are given by the following expressions:

$$D = \frac{1 + \Lambda_{2S,1S}K_H n_H(1 - x_H)}{f^{-1}K_H(\Lambda_{2S,1S}n_H(1 - x_H)f^{-1} + K_H \beta_H n_H(1 - x)},$$

$$G = \frac{1 + K_{He}\Lambda_{He}n_{He}(1 - x_{He})e^{-h\nu_{PS}/kT_M}}{1 + K_{He}(\Lambda_{He} + \beta_{He})n_{He}(1 - x_{He})e^{-h\nu_{PS}/kT_M}},$$

(3.55)

where $K_{He} = \Lambda_{HeI}^3/8\pi H(z)$, $K_H = \Lambda_{H_{2p}}^3/8\pi H(z)$, $\Lambda_{HeI} \simeq 58.433$ nm is the wavelength of the quantum corresponding to the $2^1P \rightarrow 1^1S$ transition in the helium atom and, $\nu_{PS} = \nu_{HeI\ 2^1P} - \nu_{HeI\ 2^1S}$.

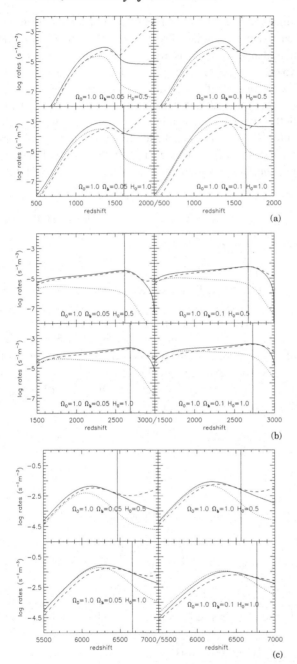

Figure 3.5 A comparison of reaction rates that dictate the hydrogen–helium recombination kinetics. (a) 2P to 1S transition rate (dashed curve) and 2S to 1S two-photon decay rate (dotted curve) against the recombination rate (solid curve) for CDM models with different Ω_b and h. The vertical line marks the level at which 5% of hydrogen atoms have already recombined. (b) Same as (a) but for the HeI recombination. Dashed curve: the rate of the $2'P$ to $1'S$ transition; dotted curve: the two-photon $2'S$ to $1'S$ decay; solid curve: recombination rate. (c) HeII recombination. Dashed curve: 2P to 1S transition rate; dotted curve: two-photon 2S to 1S decay; solid curve: recombination rate. From Seager *et al.* (1999a).

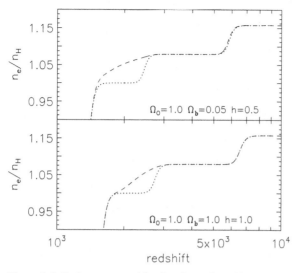

Figure 3.6 Hydrogen recombination dynamics taking into account He for two CDM models that differ in baryon density and Hubble constant. The dotted curve plots the data of Saha *et al.* (1999) and the dashed curve is an approximation from Seager *et al.* (2000). Adapted from Seager *et al.* (1999a).

The thermal contact between radiation and matter is described, as before, by Eq. (3.47), in which the main part is played by Compton processes and adiabatic cooling resulting from the expansion of the Universe. Using this modification of the three-level model of hydrogen recombination as a basis, we consider the dependence of ionization modes, $x(z)$, on the parameters of the cosmological model.

3.6.1   *The function* $x_e(\Omega_{dm})$

As our base model, we use the frequently cited ΛCDM cosmological model with the following set of parameters: $\Omega_m = 0.3, \Omega_b h^2 = 0.02, h = 0.65$ and $\Omega_{tot} 1 = \Omega_m + \Omega_b + \Omega_\Lambda$. To analyse the function $x_e(\Omega_m)$ we fix the parameters $\Omega_b h^2$ and h of this model, but, at the same time, we vary Ω_m and Ω_Λ, so that the condition Ω_{tot} continues to hold. Using RECFAST, we calculate the function $x_e(z)$, taking into account that the helium mass concentration is independent of the value Ω_m. The results of these calculations are plotted in Fig. 3.7. This plot shows that as the density of dark matter Ω_m decreases (while $\Omega_m + \Omega_\lambda \simeq 1$), the degree of ionization drops systematically in the whole redshift range $0 \leq z \leq 1500$. In Fig. 3.8 we illustrate the corresponding ratio for ionization degrees $x_e(\Omega_m)/x_l(\Omega_m = 0.3) = \Gamma(\Omega_m)$. Figure 3.8 demonstrates that differences in ionization degrees during the $z \sim 10^3$ epoch are at the 10–25% level while the residual degree of ionization (for $z = 0$) for $\Omega_m = 0.1$ is lower by a factor of approximately 1.5 than in the model with $\Omega_m = 0.3$. Qualitative arguments support this result. Namely, a decrease in Ω_m is accompanied by a rise in Ω_Λ and, as a consequence, by an increase in the age of the Universe. In its turn, the rate of expansion for $z > 1$ mostly depends on Ω_m and is practically independent of Ω_Λ: $t_{exp} \sim a/\dot{a} \sim H_0^{-1}\Omega_m^{-1/2}z^{-3/2}$. It is clear now that in models with lower values of Ω_m the age of the Universe for the same z is higher and, therefore, a larger fraction of hydrogen atoms have sufficient time to recombine.

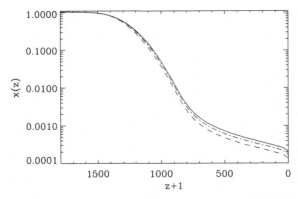

Figure 3.7 Degree of ionization, $x(z)$, as a function of redshift, z, for a number of values of the parameter Ω_m. The solid curve corresponds to $\Omega_m = 0.3$, the dash–dot curve to $\Omega_m = 0.2$ and the dashed curve to $\Omega_m = 0.1$.

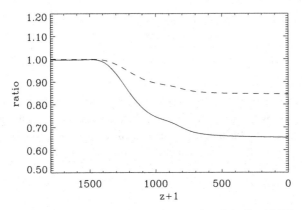

Figure 3.8 Parameter $\Gamma(\Omega_m)$ (ratio) as a function of redshift in the cosmological ΛCDM model. The solid curve corresponds to $\Omega_m = 0.1$, the dashed curve to $\Omega_m = 0.2$.

3.6.2   The function $x_e(\Omega_b h^2)$

The quantity $x_e(z)$ is plotted in Fig. 3.9 as a function of density of the baryonic fraction of matter in the base ΛCDM model. The main conclusion is that as $\Omega_b h^2$ increases, the recombination rate, α_H, grows, resulting in reduced ionization degree of the plasma. In Fig. 3.10 we plot the ratios of the corresponding ionization degrees, $\Gamma(\Omega_b) = x_e(\Omega_b h^2)/x_e(\Omega_b h^2 = 0.02)$ for the two values of the parameters $\Omega_b h^2 = 0.01$ and 0.03 in the entire range of variation of z. As we see from these curves, the residual degree of ionization varies roughly by a factor of 2 in comparison with the $\Omega_b h^2 = 0.02$ model.

3.6.3   The function $x_e(h)$

Figure 3.11 gives the results of numerical calculations of the ionization degree for three values of the Hubble constant: $H_0 = 50$, 65 and 100 km s^{-1} Mpc^{-1}. The effect of this parameter is not as trivial as that of Ω_m because, on one hand, it dictates the rate of expansion of the Universe, and on the other hand it determines the recombination rate via the

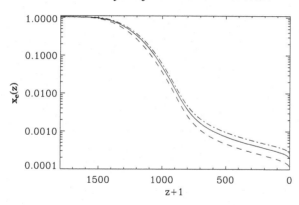

Figure 3.9 Degree of ionization as a function of redshift in the ΛCDM model for various densities of the baryonic fraction. Solid curve: $\Omega_b h^2 = 0.02$; dashed curve: $\Omega_b h^2 = 0.03$; dash–dot curve: $\Omega_b h^2 = 0.01$; $h = 0.65$.

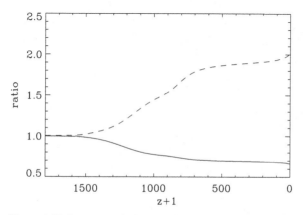

Figure 3.10 Parameter $\Gamma(\Omega_m)$ as a function of redshift. The solid curve corresponds to $\Omega_b h^2 = 0.03$, the dashed curve to $\Omega_b h^2 = 0.01$.

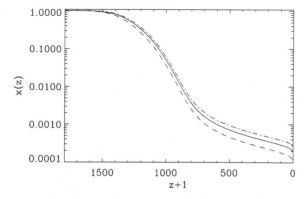

Figure 3.11 Degree of ionization as a function of z for various values of the Hubble constant. Solid curve: $h = 0.65$; dashed curve: $h = 1$; dash–dot curve: $h = 0.5$. The base model is ΛCDM with $\Omega_{tot} = 1$, and Ω_b is identical for all models.

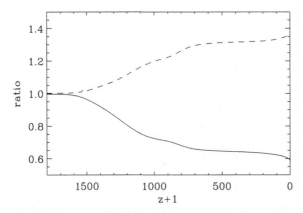

Figure 3.12 Ratio of ionization degree $\Gamma(h)$ as a function of z for $h = 0.5$ (solid curve) and $h = 2$ (dashed curve).

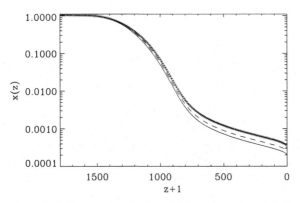

Figure 3.13 Degree of ionization $x(z)$ in an 'open' model: $\Omega_\Lambda = 0$, $\Omega_b h^2 = 0.02$, $h = 0.65$. Solid curve: $\Omega_m = 0.3$; dashed curve: $\Omega_m = 0.5$; marked curve: $\Omega_m = 1$.

parameter $\Omega_b h^2$. Figure 3.12 plots the ratio of ionization degrees $x(h = 0.5)$ and $x(h = 1)$ to $x(h = 0.65)$ as a function of redshift z. As we see from this figure, the effect of this parameter is comparable to that of Ω_b.

3.6.4   *The function $x(\Omega_m)$ in 'open' models*

In this class of models we drop the condition $\Omega_m + \Omega_b + \Omega_\Lambda = 1$, analysing differences in ionization modes in the so-called 'open' models of the Universe. In all models with $\Omega_{tot} \leq 1$ we fix the parameters $\Omega_b h^2 = 0.02$ and $h = 0.65$ and vary the parameter Ω_m from $\Omega_m = 0.3$ up to $\Omega_m = 1$. The results of the calculations of the ionization degrees and their ratios are given in Figs 3.13 and 3.14. On the basis of these numerical calculations of the dynamics of the degree of ionization given in Section 3.6.1 for various cosmological models, we can suggest the following approximation for $x(\Omega_b, \Omega, h)$ (Boschan and Biltzinger, 1998):

$$x \sim \Omega_m^{1/2} \Omega_b^{-1} h^{-1}. \tag{3.56}$$

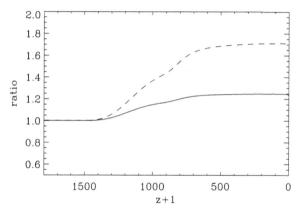

Figure 3.14 Ratio of ionization degree $\Gamma(\Omega_m) = x(\Omega_m)/x(\Omega_m = 0.3)$ in 'open' models. Solid curve: $\Omega_m = 0.5$; dashed curve: $\Omega_m = 1$.

This approximation takes into account both quantitatively and qualitatively all specific features of the function $x(\Omega_m, \Omega_b, h)$ in the framework of the standard model of recombination of cosmological hydrogen.

3.7 Spectral distortion of the CMB in the course of cosmological recombination

One of the most important features of the ionization history of the Universe was the formation of distortions of the primordial microwave background as a result of the complex dynamics of interaction between Ly-α quanta and neutral hydrogen atoms. As we mentioned in Section 3.6, this interaction results in 'delayed' recombination of hydrogen and, consequently, of the formation of excess Ly-α quanta in the Wien part of the spectrum of the primordial CMB. This specificity of recombination was emphasized in the pioneer papers by Peebles (1968), and Zeldovich *et al.* (1969), where qualitative and quantitative estimates were given of the level of distortion of the CMB spectrum in baryonic models of the Universe. Recent progress in this field has left the understanding of the physical foundations of the effect unchanged, but has led to a more detailed picture of spectral distortions (Boschan and Biltzinger, 1998; Dell'Antonio and Rybicki, 1993) and to its generalization to models that include non-baryonic fractions of matter.

Mathematically, the problem of establishing spectral distortions of the CMB by recombination quanta boils down to finding the asymptotics of the solutions of transfer equations for quanta in lines, described in Section 3.6 for $t \to \infty$. Mostly, Ly-α quanta and photons of the two-quantum decay of the 2S level of hydrogen atom constitute the sources of distortions. Following Boschan and Biltzinger (1998), we introduce variables that describe the kinetics of formation of recombination distortions in the CMB spectrum:

$$\tau = \int dt\, H(t); \quad g(v, \tau) = \frac{v}{n_b} \cdot n'(v, t); \quad n'(v, t) = n(v, t) - \tilde{n}(v, t); \quad x = \ln \frac{hv}{I},$$

(3.57)

where $H(t)$ is the Hubble parameter, $n(v, t)$ is the radiation spectrum (concentration of quanta in $cm^{-3}\, Hz^{-1}$), $\tilde{n}(v, t)$ is the equilibrium Planck distribution, $n_b(\tau)$ is the baryonic concentration and $I = 13.6\, eV$ is the hydrogen ionization potential. In terms of these variables, the equation for perturbations of the distribution function of quanta takes the form (Boschan and Biltzinger, 1998)

$$\left(\frac{\partial}{\partial \tau} - \frac{\partial}{\partial x}\right) g(x, \tau) = \frac{I \cdot e^x}{2\pi \bar{h} H(\tau) n_b(\tau)} Q(x, \tau), \tag{3.58}$$

where $Q(x, \tau)$ is the production of recombination quanta (in $\text{cm}^{-3}\ \text{Hz}^{-1}\ \text{s}^{-1}$). As we mentioned earlier, $Q(x, \tau)$ depends on the generation of $2S - 1S$ and Ly-α photons:

$$Q(x, \tau) = R_1 e^{x - x_{12}} \delta(x - x_{12}) + R_2 \Phi(x). \tag{3.59}$$

Here, $x_{12} = x_1 - x_2$ is the frequency of the transition,

$$\Phi(x) = 0.7081 e^x \Psi\left(\frac{4}{3} e^x\right), \tag{3.60}$$

where the function $\Psi(t)$ was calculated in Spitzer and Greenstein (1951) and

$$R_1 = \frac{8\pi \nu_{12}^3 H}{c^3} \left(\frac{n_{2S}}{n_{1S}} - \left(e^{\frac{h\nu_{12}}{kT}} - 1\right)^{-1}\right), \tag{3.61}$$

$$R_2 = \Lambda_{2S-1S} \cdot \left(n_{2S} - n_{1S} e^{-\frac{I}{4kT}}\right). \tag{3.62}$$

As we see from Eqs (3.59)–(3.62), detailed information on the kinetics of hydrogen recombination has to be taken into account to calculate the dynamics of formation of spectrum distortions. Following Boschan and Biltzinger (1998), we reproduce here the results of a numerical solution of the problem. Figures 3.15 and 3.16 show the behaviour of the spectra of generated quanta at different stages of cosmological recombination in models with $\Omega_{\text{tot}} = 1$, $\Omega_b = 0.1, 0.01$ and $h = 1$ ($\Omega_\Lambda = 0$).

As we see from Figs 3.15 and 3.16, the main contribution to the spectral power of the distortion source at the beginning of recombination for $z \simeq 10^3$ is provided by Ly-α quanta. However, as recombination proceeds, the gradually more and more important role shifts to the two-quantum decay of the 2S state, which dictates the shape of distortions at low frequencies. A comparison of Figs 3.15 and 3.16 shows that, amplitude-wise, the effect of the 2S decay in models with low density of the baryonic fraction is substantially suppressed in comparison with the case of $\Omega_b \sim 0.1$. However, in this case as well, the same process dominates the low-frequency asymptotics of distortions.

In Fig. 3.17 we give the results of calculating the distortions of the Wien part of today's CMB spectrum as given by a number of cosmological models (Boschan and Biltzinger, 1998, 1999). As we see from this figure, the largest distortions of the Wien part of the spectrum are expected on wavelengths $\lambda \leq 220$ μm at $\Omega_b \simeq 0.1$ and $h \simeq 1$ and on $\lambda \leq 160$ μm for all values of the parameters Ω_b, Ω_m and h. At this wavelength the corresponding intensity of quanta is 10^{-25}–10^{-24} erg $\text{cm}^{-2}\ \text{s}^{-1}\ \text{ster}^{-1}\ \text{Hz}^{-1}$ in the range $120 < \lambda \leq 160$ μ*m* (see Fig. 3.17). We see, by comparing the predicted intensity with the infrared background data (see Chapter 1), that the intensity of recombination distortions is comparable with the intensity of background ($\sim 3 \div 5 \times 10^{-25}$ erg $\text{cm}^{-2}\ \text{s}^{-1}\ \text{ster}^{-1}\ \text{Hz}^{-1}$) and, in principle, can make a contribution to the resulting IR spectrum of radiation comparable to other sources. If $\lambda \leq 120$ μm, recombination distortions begin to fall off steeply and 'disappear' under the detection threshold. Note that it would be very interesting to observe the anisotropy of the IR background distribution in the wavelength range 120–160 μm. By virtue of the cosmological nature of spectral

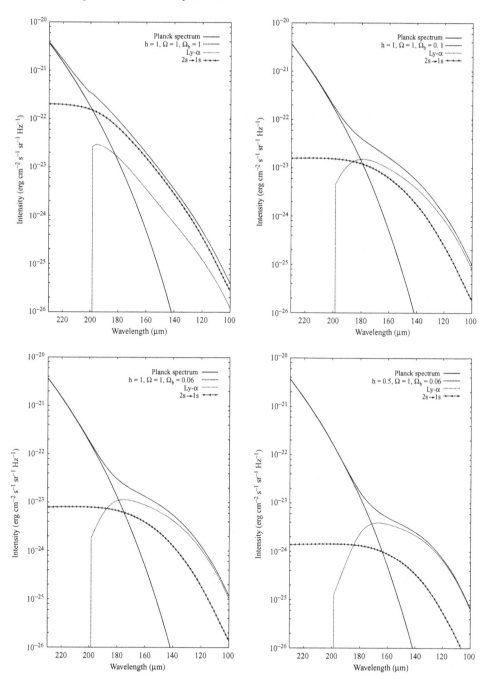

Figure 3.15 The contribution of Ly-α photons and of the two-quantum decay 2S $\rightarrow$ 1S to distortions of the CMB spectrum. Adapted from Boschan and Biltzinger (1999).

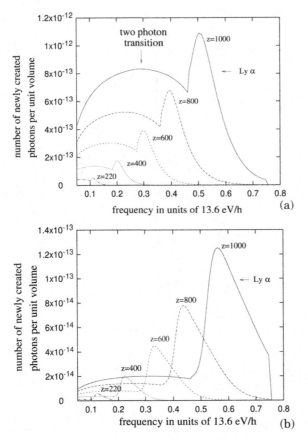

Figure 3.16 Spectra of recombination lines for a number of values of z in the models (a) $\Omega_{tot} = 1$, $\Omega_b = 0.1$, $h = 1$ and (b) $\Omega_{tot} = 1$, $\Omega_b = 0.01$, $h = 1$ for the CDM model. Adapted from Boschan and Biltzinger (1999).

distortions, the angular anisotropy must be characterized by variance comparable to that of the primordial microwave background measured by COBE in the range of radio wavelengths. It is possible that the 'Submillimetron' project would be able to measure the anisotropy of the IR background in this range of wavelengths. This result would be a most important confirmation of the theory of cosmological recombination and would, at the same time, make it possible to obtain more accurate values of the cosmological parameters Ω_b, Ω_m and h in our Universe.

3.8   The inevitability of hydrogen reionization

The standard model of hydrogen recombination presented in Section 3.3 is based on one very drastic assumption, the legitimacy of which is not necessarily obvious. We mean the hypothesis that beginning with redshifts $z \sim 3 \times 10^3$ and ending with $z = 0$, the Universe never contained any other sources of ionization of cosmic plasma in addition to the microwave background radiation. This assumption is certainly wrong for the epoch of $z < 10$, because the very fact of the existence of galaxy clusters, especially quasars with high redshifts $z_q \simeq$ 3–6, proves that after the recombination epoch at $z \sim 10^3$, gravitationally bound structures were forming in the neutral gas, constituting potential sources of gas reionization at the

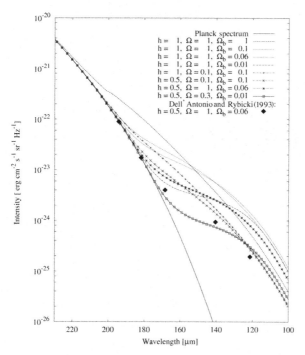

Figure 3.17 The overall pattern of distortion of the CMB spectrum by recombination quanta in various cosmological models. Adapted from Boschan and Biltzinger (1999).

expense of internal sources of energy release (explosions of stars, active galactic nuclei, etc.). The study of this epoch of reionization of cosmological hydrogen is a separate important chapter of modern cosmology that continues to be researched actively now.

The observation of Ly-α lines in the spectrum of remote quasars provides an experimental foundation for the conclusion on the inevitability of the hydrogen reionization epoch. Martin Schmidt (Schmidt, 1965) was the first to conduct the observation of the Ly-α line in the spectrum of the 3C9 quasar, which stimulated the famous work by Gunn and Peterson (1965). In this paper, the authors formulated for the first time the conclusion that the observation of Ly-α lines in quasars with redshift $z \geq 2$ signifies that, at this z, hydrogen is almost completely ionized. According to Gunn and Peterson (1965) the optical depth of neutral hydrogen is calculated on the basis of Ly-α line absorption using the following expression (see also Barkana and Loeb (2000)):

$$\tau_{\mathrm{GP}} = \frac{\pi e^2 f_\alpha \lambda_\alpha n_{\mathrm{HI}}(z_{\mathrm{S}})}{m_e C H(z_{\mathrm{S}})} \simeq 4.3 \times 10^5 x_{\mathrm{HI}} \left(\frac{\Omega_b h^2}{0.02}\right) \times \left(\frac{\Omega_m}{0.3}\right)^{-1/2} \left(\frac{1 + z_{\mathrm{S}}}{10}\right)^{3/2}.$$

$$(3.63)$$

Here, $f_\alpha = 0.4162$ is the oscillator strength for the line $\lambda_\alpha = 1216\,\text{Å}$, $H(z_{\mathrm{S}})$ is the value of the Hubble parameter for the redshift, z_{S}, of the source and $h_{\mathrm{HI}}(z_{\mathrm{S}})$ is the neutral hydrogen concentration for $z = z_{\mathrm{S}}$.

Treating the recent data of Fan *et al.* (2002) on identifying the lines of SDSS 1044-0125 quasar (shown in Fig. 3.18) in the spirit of the paper by Gunn and Peterson (1965), we can evaluate the expected degree of hydrogen ionization. Assuming in Eq. (3.63) that $z_{\mathrm{S}} = 5.8$ and $\tau_{\mathrm{GP}} \leq 0.5$, we find that the fraction of neutral hydrogen, x_{HI}, must be infinitesimally

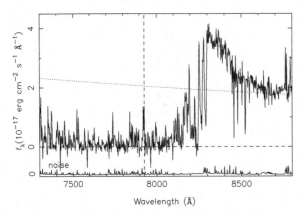

Figure 3.18 Optical spectra of the SDSS 1044-0125 quasar at $z = 54.8$. From Fan *et al.* (2000).

small: $x_{HI} \leq 10^{-6}$. Therefore, we can be absolutely certain that already at $z \sim 6$ the epoch of neutral hydrogen was replaced by the epoch of its complete ionization. We need to mention that, regardless of the specific mechanisms that produce the reionization of cosmological hydrogen at such a high redshift as $z \sim 6$, the current data on small-scale anisotropy of the CMB (see the discussion in Chapter 6) show that the optical depth relative to the Thomson scattering is unlikely to exceed $\tau_T \simeq 0.2$–0.3. A simple estimate of the maximum redshift at which secondary ionization of hydrogen could take place for this constraint follows from the definition of $\tau_T(z)$ (in Eq. (3.3)):

$$z_{max} \simeq 20 \left(\frac{\tau_T}{0.2}\right)^{2/3} \left(\frac{\Omega_b h^2}{0.02}\right)^{2/3} \left(\frac{\Omega_m h^2}{0.126}\right)^{-1/3}. \tag{3.64}$$

The whole history of the changes in the hydrogen ionization mode thus unfolds within a relatively narrow range of redshift, $6 \leq z \leq z_{max}$, at which the formation of the very first objects in the Universe occurs. We have to emphasize that the specifics of secondary ionization of hydrogen, including that at the maximum redshift, z_{max}, at which reionization actually begins, strongly depend on the type of dark matter (Barkana and Loeb, 2000) and in particular on the characteristic scale of cut-off of the density fluctuation spectrum. Because of the importance of detailed information on the inter-relation between the properties of dark matter and the ionization history of the Universe at $z \simeq 20$–30, we will now discuss this aspect in more detail.

3.9   Type of dark matter and detailed ionization balance

There is no longer any doubt that the formation of structures in the expanding Universe is driven by the evolution of small perturbations of density, velocity and gravitational potential that unfold at the early stages of cosmological expansion in the mixture of ultra-relativistic matter and the 'gas' of primordial gravitating particles – the carriers of the future dark matter of galaxies and their clusters. The theory of the origin of these fluctuations, and the description of their specifics in a multicomponent medium, constitute a separate important chapter of modern cosmology that started with the pioneering work of Lifshits (1946), I. Novikov (1964), Bonnor (1957), Zeldovich (1970), Harrison (1970) and others, and still continues to progress. Two main elements of the theory of gravitational instability in the Universe have attracted special attention in the last 10–15 years.

First, owing to the progress in inflation theory, it appears that the source of small pre-galactic irregularity in the cosmic plasma was identified for the first time. Regardless of the specific version of the theory (see Linde (1990) and Starobinsky (1979)) it is clear that small 'seed' perturbations of density, velocity and gravitational potential can be traced to quantum fluctuations of the vacuum of physical fields at the early stages of cosmological evolution. Moreover, two problems that seemed to be remote from one another, namely that of the beginning of the expansion of the Universe and that of the origin of the pre-galactic fluctuations in the framework of the inflation paradigm, appear to have a common source: the instability of the initial state of the vacuum (Linde, 1990; Starobinsky, 1979). It is beyond the scope of this book to go into a detailed description of the current theory of inflation and the origin of pre-galactic fluctuations in the Universe; they are presented in detail in Kolb and Turner (1989) and Linde (1990) (see also the original papers by Guth (1981), Mukhanov and Chibisov (1981, 1982) and Starobinsky (1979, 1980)). The problem of the early stages in cosmological evolution is in itself no less interesting and fascinating than the problem of formation of structures in the expanding Universe and the related physics of the CMB.

However, referring the reader to the above-mentioned Kolb and Turner (1989) and Linde (1990), we will limit ourselves in what follows to the role of the 'consumers' of information and assume as an unavoidable fact the existence of small inhomogeneities in the distributions in density, velocity and gravitational potential in the multicomponent medium, both before and after the cosmological hydrogen recombination; this medium included baryons, primordial electromagnetic radiation, neutrinos and any background gravitating particles – the carriers of dark matter.

Secondly, the evolution of perturbations in this multicomponent medium was accompanied by a rise of gravitationally bound structures whose mass spectrum extended formally from arbitrarily small values[1] up to the masses of clusters and superclusters of galaxies. As these structures were being formed, the baryonic fraction of matter was also evolving, accompanied by the emergence of first-generation stars. In fact, the process of the conversion of gravitational energy to the energy of electromagnetic radiation (via stars) capable of ionizing the medium begins to play an important role at this stage, along with gravitational processes.

The rate at which objects were formed in the Universe and the rate of reionization of cosmological hydrogen, especially at redshifts $z > 6$, thus appear to constitute links of the same chain. The most important factor for CMB physics is the effect of hydrogen reionization at $z \sim 10$–30. The effect of this factor on anisotropy and polarization is, in principle, detectable. The formation of galaxies with masses not very different from that of our Galaxy, M_G, occurs at a relatively late stage ($z \sim 1$–2) in CDM models, and consequently it is clear that the fact of early reionization at ($z \sim 10$–30) can only be caused by low-mass objects with $M \ll M_G$ that are formed at $z \gg 1$. Hence, by testing the observational manifestations at the early reionization of hydrogen, we could also answer the question about the extent to which the spectrum of perturbations of dark matter stretches into the range of low masses, and hence we could arrive at conclusions about the origin and nature of the dark matter.

---

[1] Provided CDM models do not contain a physical scale for fluctuation spectrum cut-off.

3.9.1   Phenomenology of reionization

In this subsection we discuss a phenomenological approach to describing the hydrogen reionization epoch, as suggested in Tegmark, Silk and Blanchard (1994). Following this paper, we define the fraction of the mass of intergalactic medium that is ionized,

$$\kappa = f_S \cdot f_{UVPP} \cdot f_{ion}, \tag{3.65}$$

where f_S is the fraction of baryons contained in non-linear (gravitationally bound) structures, f_{UVPP} is the number of UV photons emitted into the intergalactic medium per proton in non-linear structures, and f_{ion} is the number of ionizations per emitted UV photon.

As for a source of UV photons, Tegmark *et al.* (1994) suggested that they are generated in stars and quasars. Using the fact that the transformation of hydrogen into helium in the process of stellar nuclear synthesis is accompanied by the transformation of the fraction $\sim 7.3 \times 10^{-3}$ of the proton mass into UV radiation, we obtain the following estimate for the parameter f_{UVPP}:

$$f_{UVPP} \simeq 7.3 \times 10^{-3} \left(\frac{m_p c^2}{I}\right) f_H f_{burn} f_{UV} f_{esc}, \tag{3.66}$$

where I is the hydrogen ionization potential, m_p is the proton mass, $f_H = 0.76$ (76% hydrogen, 24% He[4]) is the fraction of the hydrogen mass in intergalactic medium, f_{burn} is the fraction of hydrogen mass in stars, f_{UV} is the fraction of energy contained in UV quanta, and f_{esc} is the fraction of UV quanta escaping from galaxies into the intergalactic medium. With Eq. (3.66) taken into account, we have the following value for κ from Eq. (3.65):

$$\kappa \simeq 3.8 \times 10^5 f_{net} f_S, \tag{3.67}$$

where $f_{net} = f_{burn} f_{UV} f_{esc} f_{ion}$.

As the fraction of ionized gas is definitely below or equal to unity, Eq. (3.67) implies that

$$f_{net} f_S \leq 2.6 \times 10^{-6}. \tag{3.68}$$

The presence of a small parameter on the right-hand side of Eq. (3.68) demonstrates that the condition of complete ionization of hydrogen can be achieved even if each of the co-factors on the left-hand side is extremely small. At the same time, an evaluation of f_S and f_{net} requires a more meticulous approach, with a detailed scenario of structure generation in the expanding Universe and energy transformation into ionizing radiation.

Evaluation of ionization efficiency, f_{net}

According to Eq. (3.67), the parameter $f_{net} = f_{burn} f_{UV} f_{esc} f_{ion}$ is determined by a combination of parameters that describe the transformation of the energy of baryons into ionizing radiation in the course of structure formation in the Universe. The first factor, f_{burn}, is connected to the metallicity of the gas immediately after the first stars were formed; owing to their high mass, these stars must evolve over a fairly short time scale (Miralda-Escude and Ostriker, 1990).

The value of f_{burn} was evaluated from the data in Miralda-Escude and Ostriker (1990) at the level $f_{burn} \simeq 1\%$. The upper bound of this parameter can be obtained by analysing a model in which all baryons contained in a galaxy condense into stars with a combined mass $M \simeq 30 M_\odot$, for which Woosley and Weaver (1986) evaluated the resulting metallicity at the 25% level. Consequently, the choice of $f_{burn} \simeq 1\%$ ensures a 25-fold safety margin relative to the upper bound on this parameter. For the parameters f_{esc} and f_{UV} that characterize the escape of the UV radiation from the galaxy, the following estimates were obtained in Miralda-Escude and Ostriker (1990): $f_{esc} \simeq 10\text{–}50\%$ and $f_{UV} \simeq 5\text{–}50\%$; these obviously follow from the fraction of stars in the mass spectrum of the objects formed.

In contrast to the parameters mentioned above, an estimate of the fraction of ionized matter depends on the redshift, z_{vir}, at which the first low-mass objects that include the baryonic and dark matter are virialized.[2] It was shown in Tegmark *et al.* (1994) that an approximate estimate can be used for the parameter f_{ion}:

$$f_{ion} \simeq \left[1 + 0.8\Omega_b h (1 + z_{vir})^{3/2}\right]^{-1} . \tag{3.69}$$

An estimate of $f_{ion} \simeq 0.1\text{–}0.95$ was obtained in Tegmark *et al.* (1994) for $M \sim 10^6 M_{\odot}$. Combining the parameters f_{burn}, f_{UV}, f_{esc} and f_{ion}, it is possible to obtain the upper and lower bounds on the changes in f_{net} in a model with $\Omega_{dm} = 0.3$, $h = 0.5$, $\Omega_{\Lambda} \simeq 0.7$ and $\Omega_b h^2 \simeq 0.02$:

$$10^{-6} < f_{net} < 6 \times 10^{-2}. \tag{3.70}$$

A considerable gap between the minimum and maximum values of f_{net} reflects the uncertainty in the models of formation of low-mass structures ($M \ll M_G$) in the expanding Universe in the epoch with $z < 20 \div 30$.

Evaluation of the fraction, f_S, of matter

The method of finding the fraction of matter that transferred into objects of mass M at a redshift z is based on predictions of the theory of gravitational growth of small initial fluctuations of density, velocity and gravitation potential that evolve in a multicomponent media. Substantial progress in studying the non-linear phase in the evolution of such fluctuations is predicated on the fact that the density of dark matter is greater than that of the baryonic fraction by a factor of at least 8–10 and exceeds by five orders of magnitude the CMB density. The inevitable conclusion from this is that the gravitationally bound structures in the Universe are formed by the evolution of perturbations mostly in the dark matter, especially during the epoch $z < 1.2 \times 10^4 \Omega_m h^2$ when its density dominated that of the CMB. Following Peebles (1993), we consider specific features of this process in more detail, using the framework of the hydrodynamic approximation:

$$\frac{\partial \delta_x}{\partial t} + \frac{1}{a}\vec{\nabla}[(1 + \delta_x)\vec{U}] = 0,$$
$$\frac{\partial \vec{u}}{\partial t} + H\vec{u} + \frac{1}{a}(\vec{u} \cdot \vec{\nabla}\vec{u}) = -\frac{1}{a}\vec{\nabla}\Phi, \tag{3.71}$$
$$\nabla^2 \Phi = 4\pi G \overline{\rho} a^2 \delta.$$

These equations are easily linearizable for small perturbations and reduce to a single equation for density perturbations in the dark matter 'gas' (Peebles, 1993), as follows:

$$\frac{\partial^2}{\partial t^2}\delta_x + 2H\frac{\partial \delta_x}{\partial t} - 4\pi G \overline{\rho}_x \delta_x = 0, \tag{3.72}$$

---

[2] Typically, z_{vir} corresponds to the virialization of objects with a mass $\sim 10^6 M_{\odot}$ (Tegmark *et al.*, 1994).

where $\bar{\rho}_x$ is the dark matter density. It is known (see, for example, Peebles (1993)) that Eq. (3.72) implies that δ_x has a solution that grows and then decays with time. In the general case, $\delta_x(\bar{r}, t)$ can be rewritten in the form $\delta_x(\bar{r}, t) = f(\bar{x})D(t) \ll 1$, where

$$D(t) = \frac{(\Omega_\Lambda a^3 + \Omega_k a + \Omega_x)^{1/2}}{a^{3/2}} \int^s \frac{\mathrm{d}a \cdot a^{3/2}}{(\Omega_\Lambda a^3 + \Omega_k a + \Omega_m)^{3/2}}, \tag{3.73}$$

$\Omega_k = 1 - \Omega_{\text{tot}}$, $\Omega_{\text{tot}} = \Omega_x + \Omega_\Lambda$, and the function $f(\bar{x})$ describes the spatial form of perturbation density distribution, with a value known at a certain initial moment of time, t_0. In what follows we assume that the initial adiabatic mode of perturbations is generated in the process of restructuring of the vacuum of physical fields at the inflation stage of evolution of the Universe (Kolb and Turner, 1989; Linde, 1990). This means that we can use the expansion to a Fourier integral,

$$f(\bar{x}) = \int \mathrm{d}^3 x \, f_{\bar{k}} \, \mathrm{e}^{-\mathrm{i}\bar{r}\bar{k}}, \tag{3.74}$$

for the representation of the function $f(\bar{x})$, where $\bar{k}$ is the wave vector and $f_{\bar{k}}$ are random Gaussian coefficients that satisfy the relation

$$\langle f_{\bar{k}} f_{\bar{k}'}^* \rangle = (2\pi)^3 P(k)\delta(\bar{k} - \bar{k}') \tag{3.75}$$

and $P(k)$ is the power spectrum of fluctuation density.

Since the thermal (peculiar) velocities of particles are considered negligibly small in CDM models, the spectrum of $P(k)$ fluctuations extends up to spatial scales comparable to distances between particles $n_k^{-1/3}$. At the linear stage of the evolution of fluctuations there is no interaction between various harmonics of $\bar{k}$. It is thus possible to introduce the concept of smoothed field of density fluctuations by using, for example, the filter $F(\bar{r}) = \theta(R - r)$, where the scale R is connected to the mass of dark matter within a radius R as follows: $M = \frac{4\pi}{3}\rho_{\text{dm}}R^3$. Then the density fluctuation field smoothed over the scale R retains its statistical properties, as does the field $\delta(\bar{r})$, but now it depends on the density contrast, $\delta(M)$, in the sphere of radius R of the corresponding mass. For this smooth field the variance is given by (Kolb and Turner, 1989; Peebles, 1985, 1993)

$$\sigma^2(M) = \int_0^\infty \frac{\mathrm{d}k}{2\pi^2} k^2 P(k) \left[\frac{3 j_1(kR)}{kR}\right]^2, \tag{3.76}$$

where $j_1(kR) = (\sin x - x \cos x)/x^2$.

Note that the linear evolution of fluctuations $\delta_x(\bar{r}, t)$, similar to $\delta_R(\bar{x}, t)$, follows one and the same law (Eq. (3.73)). However, at the non-linear phase, when $\sigma(M) \sim 1$, the pattern of gravitational instability becomes more complex. Following Kolb and Turner (1989), Peebles (1993), and Zeldovich and Novikov (1983) we consider the dynamics of a spherically symmetric region of radius $R \ll ct$ in terms of the Newton approach. The equation of evolution of the radius $R(t)$ in time takes the form

$$\frac{\mathrm{d}^2 R}{\mathrm{d}t^2} = H_0^2 \Omega_\Lambda R - \frac{GM}{R^2}, \tag{3.77}$$

where M is, as before, the mass of matter inside of the volume of this radius. The first term on the right-hand side of Eq. (3.77) describes the role played by the energy density of the

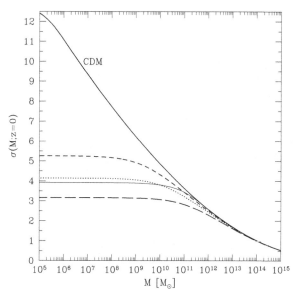

Figure 3.19 $\sigma(M, z)$ for various masses M and redshifts $z \le 30$. The thick solid curve is the cold dark mass model (CDM); the middle curve is the warm dark mass (WDM) model with corresponding cut-off masses $M \simeq 2 \times 10^{10} h^{-1} M_\odot$ (short dashed line), $M \simeq 1.5 \times 10^{11} h^{-1} M_\odot$ (dotted line and thin solid curve obtained by Sommer-Larsen and Dolgov (1999); lower curve (long dashes) is $M \simeq 1.2 \times 10^{12} h^{-1} M_\odot$. From Sommer-Larsen and Dolgov (1999).

vacuum, which is important only for perturbations on a scale not exceeding

$$R_{\rm v} = \left(\frac{GM}{H_0^2 \Omega_\Lambda}\right)^{1/3} \simeq 10 \left(\frac{\Omega_\Lambda h^2}{0.7}\right)^{-1/3} \left(\frac{M}{M_{15}}\right)^{1/3} \text{Mpc},$$

where $M_{15} = 10^{15} M_\odot$.

If $R \ll R_{\rm v}$, the expansion dynamics obeys a conventional law for spherically symmetric distribution of matter. If the total energy $E = (\dot{R}^2/z) - (GM/R) < 0$ then the spherical region, having reached the maximum radius, $R_{\rm max}$, begins to contract with the ensuing growth in density (Peebles, 1993).

Using the linear law of increase $\delta\rho_x/\rho_x$, we find the moment when collapse begins from the condition $\delta_x = \delta\rho_x/\rho_x = \delta_{\rm cr} = 1.686$ (Peebles, 1993). Normalizing to $D(z = 0) = 1$, we arrive at the criteria of formation of objects for arbitrary values of redshift z in the form $\delta_{\rm cr}(z) = 1.686 D^{-1}(z)$. Therefore, in terms of variance $\sigma(M)$ of Eq. (3.76), the condition of formation of gravitationally bound structures with mass M reduces to $\sigma(M) \simeq \delta_{\rm cr}(z)$. Figure 3.19 shows the behaviour of the function $\sigma(M, z)$ for three values of redshift z, in the case when the criteria of formation of objects with mass M is met.

As before, the background parameters of the model were chosen as follows: $\Omega_{\rm tot} = 1, \Omega_{\rm m} = 1, \Omega_x = 0.7, h \simeq 0.7$. The fraction of matter not included into objects of mass M in the course of non-linear evolution of perturbations equals, in terms of the variables $f_{\rm c}$ and $\sigma(M, z)$,

$$f_{\rm c} = \text{erfc}\left[\frac{\delta_{\rm c}}{\sqrt{2}\sigma(M, z)}\right], \tag{3.78}$$

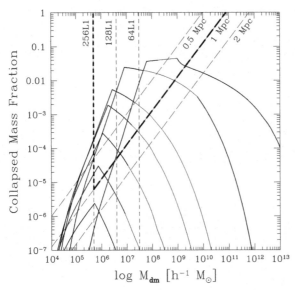

Figure 3.20 Fraction of collapsed gas as a function of dark matter halo mass at $z + 1 = 35, 30, 25, 20, 15, 10, 4$ (from upper right to lower left). The thin portions of the curves correspond to virial temperatures $10^4 \leq T_{\mathrm{vir}} \leq 10^5$ K. In these haloes, the cooling is very efficient and all the gas is collapsed; the thick portions of the curves on the left mark the objects with $T_{\mathrm{vir}} < 10^4$ K and on the right objects with $T_{\mathrm{vir}} > 10^5$ K. The vertical lines correspond to the smallest halo mass fully resolved by simulations with mass resolution $M_{\mathrm{dm}} = 4.93 \times 10^3, 3.94 \times 10^4$ and $3.15 \times 10^5 M_\odot$ (from left to right). As a quick reference they correspond to $256^3, 128^3, 64^3$ cubes with $L_{\mathrm{box}} = 1$ Mpc. The oblique lines show the largest halo mass that we can find in cubes with $L_{\mathrm{box}} = 0.5, 1, 2$ Mpc. For details of computations see Ricotti, Gnedin and Shull (2001), from which this figure is adapted.

or, using the definition for z_{vir} (Tegmark *et al.*, 1994)

$$1 + z_{\mathrm{vir}}^{(M)} \equiv \frac{\sqrt{2}}{\sigma(M, z = 0)},$$

so that we obtain

$$f_{\mathrm{c}} = \mathrm{erfc} \left[\frac{1 + z}{1 + z_{\mathrm{vir}}(M)} \right]. \tag{3.79}$$

Figure 3.20 shows the behaviour of the function $f_{\mathrm{c}}(M)$ for several values of redshifts $z + 1 = 35, 30, 25, 20, 15, 10, 4$ (upwards for solid curves) in the ΛCDM cosmological model. This dependence was obtained in the course of numerical simulation of the process of structure formation for masses from $10^4 M_\odot$ to $M \simeq 10^{13} M_\odot$ (Chiu, Gnedin and Ostriker, 2001).

We can now evaluate the efficiency of hydrogen ionization by objects of different mass using Fig. 3.20 and Eq. (3.67) and (3.70). Equations (3.67) and (3.70) imply for the lower bound $f_{\mathrm{net}} \simeq 10^{-6}$ that $\kappa \simeq 0.38 f_{\mathrm{s}}$ and that χ reaches a maximum at $f_{\mathrm{s}} \simeq (5-6) \times 10^{-2}$ in the mass range $10^8 \leq M \leq 10^9 M_\odot$ (see Fig. 3.21). The corresponding value, $\chi_{\mathrm{min}} \simeq 2 \times 10^{-2}$, indicates that in this case the efficiency of hydrogen recombination is very low. The corresponding redshifts, z_{r}, are then close to $z = 3$, which is in obvious contradiction with the observations of the Ly-α lines for the quasar SDSS 1044-0125 for $z = 5.8$. Values close to $f_{\mathrm{s}} \simeq (2-3) \times 10^{-2}$ are achieved in the mass range $8 \times 10^6 - 3 \times 10^7 M_\odot$ for $z_{\mathrm{r}} \simeq 10$.

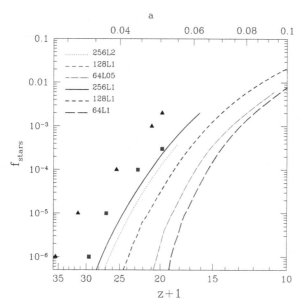

Figure 3.21 Fraction of baryons in stars versus redshift for two sets of simulations with different resolution and box sizes. The three thick lines show 256L1, 128L1 and 64L1 simulations with constant box size ($L_{box} = 1$ Mpc) and varying mass resolutions ($M_{dm} = 3.15 \times 10^5$, 3.94×10^4 and $4.93 \times 10^3 M_\odot$, respectively). The triangles show f_{star} as a function of redshift in the limit of a simulation with infinite mass resolution and $L_{box} = 1$ Mpc. The three thin lines show 64L05, 128L1 and 256L2 simulations, with constant mass resolution ($M_{dm} = 3.94 \times 10^4 M_\odot$) and varying box sizes ($L_{box} = 0.5$, 1 and 2 Mpc, respectively). The squares show f_{star} as a function of redshift in the limit of infinite box size and mass resolution $M_{dm} = 3.94 \times 10^4 M_\odot$. Note that the simulation 128L1 appears in both sets; therefore there are only five lines in the plot. Adapted from Ricotti *et al.* (2001).

However, in this case, χ_{min} is very close to 1%, which practically always signifies the absence of recombination. In the 'optimistic' limit $f_{net} \simeq 6 \times 10^{-2}$, the total ionization of the medium is achieved for $f_S \simeq 4 \times 10^{-5}$. This level, $f_{S(max)}$, is realized in the range of mass from $2 \times 10^5 M_\odot$ to $10^{13} M_\odot$ and covers the range of redshifts from $z \simeq 25$ (for $M \simeq 2 \times 10^5 M_\odot$) to $z = 3$ (for $M \simeq 10^{13} M_\odot$). Hydrogen recombination in this 'extremely' favourable model is thereby achieved in a wide range of parameters (M, z). The situation changes dramatically, however, if a 'cut-off' is observed in the spectrum of initial fluctuations in a gas of cold particles, near a mass $M \simeq 10^7 - 10^8 M_\odot$. Physically, this cut-off of the spectrum occurs if particles of dark matter possess, for instance, non-zero 'thermal' particle velocities that lead to smearing of inhomogeneities on the scale $\lambda_{dis} \sim v_T \cdot t_{eq}$, where v_T is thermal velocity and t_{eq} is the cosmological time at which the density of the dark matter becomes comparable to that of the primordial background radiation.

In this model of 'warm' dark matter, the formation of structures with $M \ll 10^7 - 10^8 M_\odot$ is hindered owing to the 'cut-off' of the spectrum (Bardeen *et al.*, 1986; Dolgov and Sommer-Larsen, 2001). However, the total recombination of hydrogen is only possible in these models for $z < 20$, with the maximum ionization efficiency ($f_{net} \simeq 6 \times 10^{-2}$) taken into account.

We need to emphasize that estimates given above and the conclusions made concerned polar values of f_{net} from Eq. (3.70), which takes into account both the lowest-efficiency and the maximum-efficiency ionization thresholds.

It is possible at the same time to try to minimize the uncertainty of the parameters f_{net} and f_S, taking into consideration the fact that quasars are known to exist at $z = 5.8$ and there are four quasars in the SLOAN review whose redshift z equals 6.28. Assuming that the fraction of ionized hydrogen $X = 1$ at such redshifts $z \simeq 6$, we immediately find from Fig. 3.21 that the maximum of f_S reached at $M \geq 10^8 - 10^9 M_\odot$ equals $(2-3) \times 10^{-2}$. Also, $\overline{f_{net}} \simeq 10^{-4}$, which approximately equals the geometric mean of $f_{net(min)}$ and $f_{net(max)}$.

We need to stress that the object of maximum interest for the CMB radiation physics is obviously not the end of the hydrogen reionization epoch (when, according to estimates of the Gunn–Peterson effect, the degree of hydrogen ionization falls to $X_H \simeq 10^{-6}$) but rather the earlier stages at which $X_H \sim 0.1$. An analysis of Ly-α lines of quasars in the SLOAN review carried out in Chiu *et al.* (2001) and Djorgovski *et al.* (2001) shows that the redshifts $z_r \simeq 6.2 \pm 0.1S \pm 0.2r$ are a good estimate at the close of the reionization period. At the same time, the beginning of this process could very well involve much larger redshifts, $z \sim 15 - 20$, and could result in distortions in the anisotropy of ΔT distribution on the celestial sphere.

This factor must at least be taken into account when processing and interpreting the data of current and future observations of anisotropy and polarization of the CMB (see the remark at the end of this chapter).

3.10   Mechanisms of distortion of hydrogen recombination kinetics

The standard model of hydrogen recombination presented in Section 3.3 predicts a rapidly decreasing concentration of free electrons at redshifts $z \leq 1400$. In a realistic model with $\Omega_{tot} = \Omega_{dm} + \Omega_b + \Omega_\Lambda = 1$ and $\Omega_m = 0.3, = \Omega_b h^2 \simeq 0.02, h = 0.7$, the degree of ionization is found to be close to $x_e \simeq 0.1$ by $z \simeq 10^3$ and reduces to $x_e \simeq 10^{-2}$ for $z \simeq 800$. The optical depth of the plasma with respect to the Thomson scattering then becomes quite low ($\tau \ll 1$), and at $z < 800$ the CMB quanta propagate freely, not being scattered on free electrons. This picture, standard for each cosmological model operating with its own set of parameters $\Omega_{tot}, \Omega_{dm}, \Omega_b, \Omega_\Lambda$ and h, is based on the assumption that it is precisely in the epoch of redshift $z \leq 1400$ that the cosmic plasma contains no sources of non-equilibrium ionization of hydrogen that would supply plasma with additional ionizing quanta not connected with the kinetics of the Ly-α part of the CMB spectrum.

It is clear that if the power at which such quanta are generated exceeds the power of the Ly-α range in the Wien part of the CMB spectrum, then the kinetics of hydrogen recombination should evolve according to a scenario that differs in principle from the standard model; therefore, the characteristics of CMB anisotropy, shaped during the period of cosmological recombination, should differ from those of fluctuations that are formed in the 'standard' model of plasma becoming transparent for radiation. A reservation is necessary here: the epoch with redshifts $z \sim 10^3$ is definitely 'peculiar' for any models explaining the origin of structures in the Universe. In Section 3.9, when we discussed possible sources of hydrogen ionization, we saw that the formation of gravitationally bound structures with masses $M \sim M_G \sim 10^{12} M_\odot$ proceeds in the framework of CDM models mostly at relatively low redshifts, $z \leq 2-3$. The low-mass part of the spectrum ($M \sim 10^5 - 10^6 M_\odot$) is responsible for the formation of objects at $z \leq 25-30$.

Note that this mass range is close to the Jeans mass, $M_{J(b)}$, in the baryonic fraction of matter precisely at the moment when the plasma becomes transparent for the CMB radiation at $z \sim 10^3$. This means that formally the spectrum of density fluctuations in the dark matter gas contains perturbations with $M \ll M_{J(b)}$ that could reach a non-linear mode at $z \sim 10^3$. However, the emerging low-mass non-linear structures would not, in fact, involve the baryonic matter, let alone form the stars. We can add that at $z \sim 10^3$ the age of the Universe was only

$$t_{\mathrm{rec}} \simeq \frac{2}{3H_0 z_{\mathrm{rec}}^{3/2} \sqrt{\Omega_{\mathrm{m}}}} \simeq 10^6 \left(\frac{\Omega_{\mathrm{dm}} h^2}{0.15} \right)^{-1/2} \text{ years,}$$

which is insufficient for transforming the rest energy of baryons into ionizing radiation, even if the primary stars were supermassive ($M \geq M_{J(b)}$) (Tegmark *et al.*, 1994). In fact, in addition to stellar energy sources, there exists another mechanism that transforms the rest energy of matter into radiation. We speak here of the electromagnetic decay of massive particles $X \to X' + \gamma$ or $X \to X' + e^+ + e$, in which the initial particle X is transformed into a new particle X' and a γ quantum or an electron–positron pair is emitted. Moreover, it is not at all necessary for the electromagnetic channel to dominate the X particle decays. It would be sufficient for the supermassive ($m \gg 10^3$) X particle to generate a quark–antiquark jet (X $\to$ q + $\bar{\mathrm{q}}$), and then the annihilation of quarks would result in both a rapid ionization of decay products and the formation of the electromagnetic component. Note that this mechanism is considered nowadays as one of the sources of generation of superhigh-energy cosmic rays ($E \geq 10^{20}$ eV) in the so-called top-down scenario (for details, see Bhattacharjee and Sigl (2000). The presented model of generation of excess ionizing quanta in the epoch with $z \sim 10^3$ practically coincides with the model of evaporation of primary black holes (PBH); the possibility of PBH formation in the early Universe was first considered by Zeldovich and Novikov (1966) then by Hawking (1971). These objects are quite special in that their formation requires only a relatively high – in comparison with galactic scales – amplitude of adiabatic inhomogeneity ($\delta\rho/\rho \simeq (3-10) \times 10^{-2}$ at the moment $t \simeq 2GM_{\mathrm{BH}}/c^3$ (where M_{BH} is the mass of matter collapsing onto a black hole on the scale of the cosmological horizon).

An important feature of this potential remnant of the very early Universe is the effect of quantum decay of PBH into particles (Hawking, 1974). The characteristic energy of particles created in this decay is related to the mass of PBH by the expression

$$E_{\mathrm{BH}} \simeq \frac{hc}{\lambda} = \frac{hc}{r_{\mathrm{g}}(M)} \propto \frac{hc^3}{GM_{\mathrm{BH}}}, \tag{3.80}$$

and the decay time is given by

$$\tau_{\mathrm{BH}} \simeq t_{\mathrm{u}} \left(\frac{M_{\mathrm{BH}}}{10^{14.5}} \right)^3. \tag{3.81}$$

A comparison of Eq. (3.81) and the characteristic time of the onset of the hydrogen recombination epoch, $t_{\mathrm{rec}} \simeq 10^6 (\Omega_{\mathrm{m}} h^2)^{-1/2}$, shows that $\tau_{\mathrm{BH}} \simeq t_{\mathrm{rec}}$ for black holes with mass $M_{\mathrm{BH}} \simeq 10^{14.5} \cdot z_{\mathrm{rec}}^{-1/2} \simeq 10^{13}$. The characteristic energy of electron–positron pairs, γ quanta and neutrinos then equals $\overline{E}_{\mathrm{BH}} \simeq 1.5$ GeV, which is close to the proton rest energy. If we assume that the transformation of the rest energy of the PBH to ionizing radiation is characterized by a factor ξ, it is easy to evaluate the ratio of the density of PBH to that of baryons at $z \simeq 10^3$ that are capable of distorting the hydrogen recombination kinetics in this period. We have already seen in Section 3.3 that to violate the recombination equilibrium it is necessary to provide additional energy release, $\varepsilon_{\mathrm{ion}} \sim I \cdot n_{\mathrm{bar}}$. If this energy release is caused by PBH of mass $M \sim 10^{13}$ g, evaporated during this period, then the energy balance yields the ratio

$$I n_{\mathrm{bar}} \simeq \xi \rho_{\mathrm{BH}} c^2, \tag{3.82}$$

and

$$\left. \frac{\rho_{\mathrm{BH}}}{\rho_{\mathrm{bar}}} \right|_{z \sim 10^3} \simeq \xi^{-1} \times \frac{I}{m_{\mathrm{p}} c^2}. \tag{3.83}$$

The factor $I/(m_{\mathrm{p}} c^2)$ characterizes the ratio of the hydrogen ionization potential $I \simeq 13$ eV to the proton rest energy, $m_{\mathrm{p}} c^2 \sim 1$, and equals 10^{-8}. Therefore, if transformation efficiencies are not too small, $\xi^{-1} \sim 10-10^2$, the PBH with $M \sim 10^{13}$ constitute an extremely small fraction of matter density ($\rho_{\mathrm{BH}}/\rho_{\mathrm{bar}} \sim 10^{-7}-10^{-6}$) (Naselsky, 1978). Interestingly, if the mass spectrum of PBH contained objects with $M_{\mathrm{BH}} \simeq 10^{13}(m_{\mathrm{p}} c^2/I) \sim 10^{24}$, then the Hawking radiation spectrum would peak exactly at the energy $E \sim I$. However, such objects evaporate completely over a time $\tau_{\mathrm{BH}} \simeq 10^6 (m_{\mathrm{p}} c^2/I)^3 \sim 10^{30}$ years, and the efficiency of their influence on ionization processes in the epoch of $z \sim 10^3$ is suppressed by a factor $t_{\mathrm{rec}}/\tau_{\mathrm{BH}} \simeq (I/m_{\mathrm{p}} c^2)^2$. Even if $\rho_{\mathrm{BH}} \simeq \rho_{\mathrm{bar}}$ and $\xi = 1$, these objects could not produce any noticeable distortion of recombination kinetics – owing to the smallness of the fraction of the mass of PBH transformed into radiation ($t_{\mathrm{rec}}/\tau_{\mathrm{BH}} \sim 10^{-24}$).

As a last step, we follow Landau, Harari and Zaladarriaga (2001) and describe another potential channel for distorting hydrogen recombination kinetics, not related directly to the injection of additional photons: time-dependent fundamental physical constants. This would inevitably result in time-dependent atomic constants which, in principle, may not be equal to their current values (Ivanchuk, Orlov and Varshalovich, 2001; Varshalovich, Ivanchuk and Potekhin, 1999). As we saw in Section 3.3, the dynamics for the plasma becoming transparent for primordial radiation is tied to the kinetics of Ly-α lines via the rate of the two-quanta decay of the metastable 2S state of the hydrogen atom; hence, a weak variation of these parameters may be accompanied by strong changes in the dynamics of the degree of ionization.

We are talking here about studying the stability of recombination kinetics, especially at its initial stages at $z \sim 10^3$, that is the actual time of formation of anisotropy of the primordial radiation background. Clearly, regardless of the specifics of mechanisms and sources of energy release, this aspect is of independent interest; we discuss it in Section 3.11.

3.11   Recombination kinetics in the presence of ionization sources

In this section we consider the possible distortion of the kinetics of cosmological hydrogen recombination if there exist sources of non-equilibrium Ly-α quanta that occur in the process of decay of hypothetical massive particles or via evaporation of PBH. In principle, the effects of distortion of ionization equilibrium depend considerably on the dynamics of the 'ionizer', on its energy density, decay energy spectrum, kinetics of transformation from the maximum energy E_{max} to the quantum energies $\sim I$ and a number of other specific features of ionization sources. However, following Peebles, Seager and Hu (2001), we can offer a sufficiently general phenomenological description of a 'non-equilibrium' hydrogen recombination by formalizing the effects of various mechanisms of pumping ionizing quanta into the plasma. Namely, by analogy to Peebles *et al.* (2001), we introduce the rate of pumping an excess of ionizing quanta into the plasma, as follows:

$$\frac{dn_i}{dt} = \chi(t) n_H H(t),$$ (3.84)

where n_H is the concentration of neutral hydrogen atoms, $H(t) = \dot{a}/a$ is the Hubble parameter, and $\chi_\alpha(t, E)$ is the efficiency of transformation of the spectrum of injected high-energy particles into ionizing radiation. Note that in contrast to Peebles *et al.* (2001), we assume that $\chi_\alpha(E, t)$ is a function of time, whereas in the model in Peebles *et al.* (2001) $\chi_\alpha(t)$ is a constant. In a general analysis of hydrogen recombination kinetics in the presence of an ionizer (Eq. (3.84)), two characteristic time intervals can be pointed out, differing in principle in the role played by Ly-α quanta in the formation of the ionization equilibrium (see Section 3.3). The first of them corresponds to redshifts $z < 1400$ when Ly-α quanta of the CMB play a decisive role in the formation of the curve $\chi_e(z)$, and the second corresponds to redshifts $z \leq 800$ when the role of Ly-α quanta becomes insignificant and recombination processes dominate over ionization processes in the absence of additional ionization sources. We need to emphasize that this important role of Ly-α quanta in the standard model of hydrogen recombination occurs entirely due to the Planckian nature of the CMB spectrum and, specifically, due to its Wien section, where the number of quanta with energy $E \sim I$ is considerably smaller than that of quanta with $E \sim \frac{3}{4}I = h\nu_{Ly-\alpha}$. For non-equilibrium ionizers (Eq. (3.84)) this condition may in general be conserved or violated (see, for example, Doroshkevich and Naselsky (2002). We make use of the fact that the initial stages of recombination – when the role of Ly-α quanta of the Planckian spectrum of the CMB is important – are limited in time to a relatively narrow interval of redshifts, $\Delta z \sim 200$ at $z \sim 10^3$.

For a qualitative analysis of the situation, we expand the function $\chi(t, E)$ into a Taylor series in the neighbourhood of a moment of time, t_{rec}, corresponding to $z = 10^3$:

$$\chi(t, E) = \chi(t_{rec}, E) + \left. \frac{\partial \chi(E_\alpha, t_{rec})}{\partial t} \right|_{t_{rec}} (t - t_{rec}).$$ (3.85)

Starting with Eq. (3.85), we introduce a characteristic time for the variation $\chi(t, E_\alpha)$:

$$\tau_\chi = \frac{\chi(t_{rec}, E_\alpha)}{\partial \chi / \partial t |_{t_{rec}, E_\alpha}}.$$

If the pumping of non-equilibrium quanta into the system is of quasi-stationary nature not connected with instantaneous energy release, then $\tau_\chi \sim t_{rec}$ and the second term in Eq. (3.85) can be dropped. Consequently, an approximate equality, $\chi(t, E) \simeq \chi(E)$, holds.

Let us consider possible types of behaviour of the function $\chi(E)$ in the neighbourhood of energy $E_\alpha = \frac{3}{4}I = h\nu_\alpha$. From the most general point of view we can simulate the dependence $\kappa(E)$ at $E \simeq E_\alpha$ as a power-law function:

$$\chi(E) = \varepsilon_\alpha \left(\frac{E}{E_\alpha} \right)^\gamma \qquad \varepsilon_\alpha = \text{const.}$$ (3.86)

If $\gamma > 0$ but is not too high ($\gamma \sim 1$), the difference in the efficiency of generation of ionizing quanta with $E \simeq I$ and of Ly-α quanta is found to be roughly given by

$$\frac{\chi(E = I)}{\chi\left(E = \frac{3}{4}I\right)} \simeq \left(\frac{4}{3} \right)^\gamma \sim 1.$$

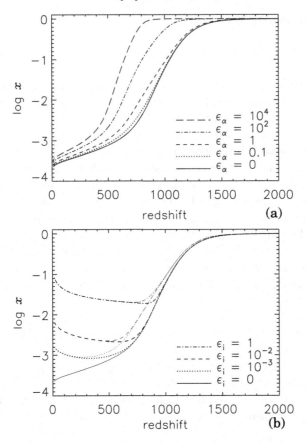

Figure 3.22 Ionization modes in models with 'delayed' recombination (a) Corresponds to Eq. (3.87); (b) corresponds to an ionizer active at small redshifts ($z < 10^3$). Adapted from Peebles, Seager and Hu (2000).

In this case, the excess Ly-α quanta do not play any important role in the kinetics of hydrogen recombination because the concentration of non-equilibrium quanta with energy $E \simeq I$ is comparable to that of quanta with $E = \frac{3}{4}I$. A similar conclusion holds also if the spectral exponent $\gamma < 0$ but still is not too large. The case when $\gamma < 0$ and $|\gamma| \gg 1$ is an exception. Then $\chi(E = I)/\chi(E = \frac{3}{4}I) \sim \left(\frac{3}{4}\right)^{|\gamma|} \ll 1$ and the role of quanta with $E = \frac{3}{4}I$ at high exponent γ becomes predominant.

First of all, these additional quanta sum up additively with Ly-α quanta in the Wien part of the CMB spectrum and start playing a significant role in the ionization balance of the medium, mostly acting as an additional source in Eq. (3.24):

$$\tilde{R} = R + \varepsilon_\alpha n_H H(t),\qquad\qquad (3.87)$$

where R is given by the expression in Eq. (3.24). Using Eq. (3.87), it is easy to take into account the renormalization of the ionization and thermal balances in the framework of the RECFAST program and calculate the function $x_e(z, \varepsilon_\alpha)$ for various cosmological models (Peebles *et al.*, 2001). Figure 3.22 plots various ionization curves in the model of 'delayed' recombination for $\Omega_b h^2 = 0.02$, $\Omega_{tot} = 1$, $\Omega_{dm} = 0.3$, $h = 0.7$ and $\Omega_\Lambda = 0.7$. We see from this figure that as the power of the ionizer (ε_α) increases, the curve $x_e(\varepsilon_\alpha, z)$ is 'flattened' in the range $700 \le z \le 1400$. Hydrogen recombination becomes increasingly delayed, even though the change in the residual degree of ionization (at $z = 0$) is not so large in comparison with a drop of 2–3 orders of magnitude in the function $x_e(z)$ for $z \simeq 800 - 10^3$. Obviously, low values of $\varepsilon_\alpha \simeq 0.1 - 1$ result in insignificant distortions of the ionization mode at $z \simeq 10^3$, which is the most important range for the formation of temperature fluctuations of the primordial background radiation.

In the opposite case of asymptotics, when the concentration of quanta with $E \sim I$ is comparable to or exceeds the concentration at $E \sim \frac{3}{4}I$, the role of non-equilibrium Ly-α quanta in the formation of the ionization equilibrium of the plasma is not so great. In the limit when the effect of excess Ly-α quanta on the recombination kinetics becomes negligible, the main mechanism of distortions is the ionization of the 1S state of the hydrogen atom. Figure 3.22 gives the results of calculations of the degree of the ionization, $x_e(z)$, in this model for various values of ionizer power, $\chi(E) = \varepsilon_i =$const. (Peebles *et al.*, 2001). As we see from this figure, non-equilibrium hydrogen ionization leads to considerable distortions of the function $x_e(z)$ at $z < 10^3$, even at relatively low values of the parameters $\varepsilon_i = 10^{-3} - 10$.

To conclude this section, we follow Peebles *et al.* (2001) and evaluate Compton distortions of the primordial radiation spectrum that are produced in the process of 'pumping' of non-equilibrium photons with $E \sim I$ into the plasma in the $z \le 10^3$ epoch. Since the rate of energy pumping is given by the relation in Eq. (3.87) that characterizes the level of distortion of the primordial radiation spectrum, we can make use of the estimate

$$y \simeq \frac{1}{4} \frac{\varepsilon_I}{\varepsilon_z} = \frac{\varepsilon_\alpha n_H I f F}{4(1+z)\sigma T_0^4} \simeq 10^{-9} \frac{\varepsilon_\alpha f F}{z/10^3}, \tag{3.88}$$

where ε_I is the density of ionizing quanta, $\varepsilon_R = \sigma T^4$ is the energy density of the CMB, $T_0 = 2.73$ K is its current temperature, σ is the Stefan–Boltzmann constant, f is the fraction of the energy ε_I contained in the distortion of the spectrum of the CMB radiation and F is the fraction of the energy of ionization sources contained in quanta with $E \sim I$. Since the observational bound on the y parameter is $y_{\text{obs}} \le 2 \times 10^{-5}$, it is clear that $\varepsilon_\alpha f F \le 2 \times 10^4$ at $z \sim 10^3$. Figure 3.22 clearly shows that even at extremely large values of the parameter $\varepsilon_\alpha \simeq 10^4$ the condition of the smallness of y distortions is always met. Therefore, spectral distortions are insensitive to the parameters of the ionizer, and the main information on its characteristics may be obtained directly from the data on the distribution of the CMB anisotropy, taking into account ionization modes more powerful than in the standard ionization model (Doroshkevich *et al.*, 2003; Naselsky and Chiang, 2004; Naselsky and Novikov, 2002; Sommer-Larsen *et al.*, 2004).

4

Primordial CMB and small perturbations of uniform cosmological model

4.1   Radiation transfer in non-uniform medium

When describing transfer processes involving quanta in a uniform and isotropic Universe in the preceding chapters, we used the Boltzmann equation in a symbolic form:

$$\frac{df}{dt} = S_t[f],$$
(4.1)

where $f(t, x^\nu, p^i)$ is the distribution function of quanta in the primordial background radiation over momenta p^i ($i = 0, \ldots, 3$) and coordinates x^ν ($\nu = 1, \ldots, 3$), and $S_t[f]$ is the collision integral. In Section 2.1 we were mostly interested in the structure of the collision integral for photons colliding with electrons, under the assumption of uniformity and isotropy, $f(t, x^\nu, p^i) = f(t, p^0)$.

At the same time, the fact that the observable (luminous) matter in today's Universe is strongly structured signifies that the hypothesis of uniformity and isotropy of matter distribution has only a limited domain of applicability and that the corresponding cosmological model needs certain modifications. For gravitating systems the role of such a source of weak non-uniformity was traditionally played, ever since the publication of Newton's famous *Philosophiae Naturalis Principia Mathematica* (*Mathematical Principles of Natural Philosophy*) and after later work by James Jeans, by the fluctuations of density, velocity and gravitational potential of the matter. It is natural that this hypothesis attracted the attention of theorists. We need to emphasize in particular E. M. Lifshits's pioneering paper (Lifshits, 1946) on the gravitational instability of matter in the expanding Universe. Nowadays the presence of primordial fluctuations in cosmological plasma is a fact, not a hypothesis. Direct observations of the anisotropy of primordial radiation background, carried out by a number of groups of experimentalists, indicated that the plasma was 'perturbed' already in the hydrogen recombination epoch. This inhomogeneity resulted in perturbed distribution functions of CMB quanta, which behaved not only as fluctuations of the collision integral $S_t[f]$, but also as perturbations of the momenta of quanta related to the perturbation of the gravitational potential.

In a uniform and on-average isotropic Universe that contains metric perturbations, the interval takes the following form:

$$ds^2 = \left(\bar{g}_{ik} + h_{ik}\right) dx' \, dx^k = -dt^2[1 + 2\Psi(\vec{x}, t)] + a^2(t)[1 + 2\Phi(\vec{x}, t)] + \gamma_{\alpha\beta} \, dx^\alpha \, dx^\beta,$$
(4.2)

where $\Psi(\vec{x}, t)$ and $\Phi(\vec{x}, t)$ are conformal perturbations of the metric given in the Newtonian reference form, $\gamma_{\alpha\beta}$ is the metric tensor of the three-dimensional space, $a(t)$ is the scale factor, and i and j run through the values from 0 to 3, and α and β run from 1 to 3.

Metric perturbations given by Eq. (4.2) are one of the sources of distortion of the distribution function of quanta as a result of the gravitational shift in their frequency. Taking this factor into account, we can rewrite the left-hand side of Eq. (4.1) as follows (Peebles and Yu, 1970):

$$\frac{df}{dt} \equiv \frac{\partial f}{\partial t} + \frac{\partial f}{\partial x^\alpha}\frac{dx^\alpha}{dt} + \frac{\partial f}{\partial p_0}\frac{dp^0}{dt} + \frac{\partial f}{\partial \gamma^\alpha}\frac{d\gamma^\alpha}{dt}.$$ (4.3)

The following notation was used in this equation: $dx^\alpha/dt = p^\alpha/dp^0$ is the coordinate velocity of photons, $g_{ik}p^i p^k = 0$ is the relation between the energy and momentum of the photon, $dp_i/dt = \frac{1}{2}g_{jk,i}(p^i p^k/p^0)$ is the change in the photon's momentum in its motion along the geodesic, and $\gamma^\alpha = -(p^\alpha/p^0)(1/a)(1 + \Phi)$ are the direction cosines.

As before, the distribution function of quanta, without taking into account the collision term, conserves the number of particles; i.e.

$$N = \int d^3x\, d^3p f(x, p),$$ (4.4)

regardless of the presence of metric perturbations h_{ik}.

We need to specify immediately that as far as the problem of generation of CMB anisotropy is concerned, the term on the right-hand side of Eq. (4.3), proportional to $d\gamma/dt$, is in itself a small quantity $\sim \dot{\Phi}$ and $\partial f/\partial\gamma^\alpha$ is non-zero only when the geodesic deviates from a straight line. This condition is met because of non-zero spatial curvature, whose role in the epoch of formation of small-scale anisotropy of primordial background radiation is not important. In any case, the last term on the right-hand side of Eq. (4.3) is of a higher order of smallness in comparison with other terms and can be neglected in the first approximation in the amplitude of metric perturbations. Let us turn now to analysing the gravitational frequency shift of quanta dp^0/dt in Eq. (4.3). First of all, p^0 changes with time because of the expansion of the Universe (see Section 2.1); this is simulated by the metric perturbation $h_{\alpha\beta}$. The following equation is obtained from Eq. (4.3) for the rate of change of the energy of a quantum:

$$\frac{1}{p_0}\frac{dp_0}{dt} = \left(-H(t) + \frac{1}{2}\dot{h}_{\alpha\beta}\gamma^\alpha\gamma^\beta\right),$$ (4.5)

where $h_{\alpha\beta}$ corresponds to a spatial part of the perturbations for $\alpha, \beta = 1, 2, 3$.

We turn now to an analysis of perturbations of the collision integral in Eq. (4.1). We have already discussed in Section 2.1 the basic processes of interaction between quanta and plasma electrons characterized by the corresponding expressions for $S_t[f]$. In the general case the collision integral has the following form in the approximation linear in velocity:

$$S_t[f] = \frac{3}{4}\sigma_T n_e \int dp' \frac{p'}{p} \int \frac{d\Omega}{4\pi}[C_0[f] + C_{v_b}[f]].$$ (4.6)

Here, $C_0[f]$ follows from Eq. (2.14) and $C_{v_b}[f]$ follows from Eq. (2.15). Substituting $C_0[f]$ from Eq. (2.14) into Eq. (4.6) and integrating in photon momentum, we obtain

$$S_{t_0}[f] = \sigma_T n_e[f_0 - f + \gamma_\alpha\gamma_\beta f^{\alpha\beta}],$$ (4.7)

where

$$f^{\alpha\beta}(t, \vec{x}, \vec{p}) = \frac{3}{4}\int \frac{d\Omega}{4\pi}\left(\gamma^\alpha\gamma^\beta - \frac{1}{3}\delta^{\alpha\beta}\right) f.$$

Likewise, we find the following expression for the term $C_{v_b}[f]$ corresponding to the linear Doppler effect,

$$St_v[f] = -\sigma_T n_e (\gamma_\alpha v^\alpha) p \frac{\partial f_0}{\partial p}, \qquad (4.8)$$

where f_0 corresponds to the non-perturbed (Planckian) distribution function.

The last step in deriving an equation for the perturbation of the distribution function of CMB quanta in the presence of fluctuations of density, velocity and space-time metric consists in the transition from perturbations $f(t, \vec{x}, \vec{p})$ to temperature fluctuations. To achieve this, we multiply Eq. (4.3) by $(p^0)^3$ and integrate it in dp^0, taking into account the definition

$$\Theta = \frac{\Delta T}{T} = \frac{1}{4} \frac{\delta \rho_\gamma}{\rho_\gamma} = \frac{1}{4} \left[\frac{1}{\pi^2 \rho_\gamma} \int dp^0 (p^0)^3 f - 1 \right]. \qquad (4.9)$$

The resultant equation is as follows:

$$\dot{\Theta} + \frac{\gamma_\alpha}{a} \frac{\partial \Theta}{\partial x^\alpha} - \frac{1}{2} \gamma^\alpha \gamma^{-\beta} \dot{h}_{\alpha\beta} = \sigma_T n_e \left(\frac{1}{4} \sigma_\gamma - \gamma'_\alpha v^\alpha - \Theta + \frac{1}{16} \gamma_\alpha \gamma_\beta \Pi^{\alpha\beta} \right), \qquad (4.10)$$

where

$$\Pi^{\alpha\beta} = \frac{3}{\pi} \int d\Omega \; \Theta(t, \vec{x}, \vec{\gamma}) \left(\gamma^\alpha \gamma^\beta - \frac{1}{3} \delta^{\alpha\beta} \right).$$

Note that the equation for temperature fluctuations Θ up to the quadrupole term $\Pi^{\alpha\beta}$ was first obtained in Peebles and Yu (1970). The Doppler part of the collision integral was studied in Doroshkevich, Zeldovich and Sunyaev (1978) and Zeldovich and Sunyaev (1970). In fact, the significant difference between the predictions of observable manifestations of primary Doppler perturbations in the CMB anisotropy and predictions made in the 1970s lies in taking into account the cosmological hidden mass and its effect on the formation of gravitationally bound structures. At first glance, the inclusion of an additional component of matter that manifests itself only gravitationally cannot greatly change the regime of generating the anisotropy of ΔT. However, there exist a number of principal differences that show that this hypothesis is incorrect. We have already mentioned that the presence of dark matter in the Universe changes the kinetics of the hydrogen recombination. Formally, this effect manifests itself through the time dependence of the Thomson optical depth on the right-hand side of Eq. (4.10). Another important difference lies in the specific features of perturbation evolution in the gas of carriers of the cosmological hidden mass and, as a result, in an absolutely different sequence of structure formation stages compared with models that were popular at the beginning of the 1970s. These 'details' gave rise to specific quantitative factors that sometimes reach 10^2–10^3 for the characteristic levels of fluctuations ΔT. Section 4.2 describes additional details of this aspect of evolution of perturbations and classification of their types.

4.2   Classification of types of initial perturbations

To describe the dynamics of small perturbations in an expanding Universe we first need to introduce an orthonormalized set of eigenfunctions that make it possible to monitor time changes in their amplitude. The assumption of a uniform and isotropic expansion of the Universe forms the background against which perturbations of density, velocity and the metric of space-time are defined; it is thus an important starting point for this classification. In contrast to the non-perturbed ground state, these perturbations are functions of spatial coordinates and time. The idea of classifying the possible lengths of time of perturbations in the expanding Universe was first formulated in Lifshits (1946) (see also Lifshits and Khalatnikov (1960)). The main results of that 1946 paper constitute one of the most important chapters in modern cosmology and withstood with flying colours more than half a century of testing.[1] Following this paper, we define three types of eigenfunctions of the Laplace operator defined on the hypersurface $t = $const.: scalar modes $Q(\vec{x})$, vector modes $Q^{\alpha}(\vec{x})$ and tensor modes $Q^{\alpha\beta}(\vec{x})$; when the coordinate reference frame transforms, these three types transform according to the laws of transformational of scalars, vectors and tensors, respectively (Landau and Lifshits, 1962) as follows:

$$Q = Q',$$
$$Q^{\alpha} = \frac{\partial x^{\alpha}}{\partial x^{\gamma}} Q'^{\gamma}, \tag{4.11}$$
$$Q^{\alpha\beta} = \frac{\partial x^{\alpha}}{\partial x'^{\gamma}} \frac{\partial x^{\beta}}{\partial x'^{\delta}} Q'^{\gamma\delta}(x').$$

Taking into account Eqs (4.11), we consider the problem of eigenfunctions of the Laplace operator on the hypersurface $t = $ const.

4.2.1 Scalar modes

Assume now that the metric properties of the hypersurface $t = $ const. are given by the metric tensor $\gamma_{\alpha\beta}$, where α and $\beta = 1, \ldots, 3$. The problem of finding the principal values of the Laplace operator in the metric $\gamma_{\alpha\beta}$ reduces to finding the solutions of the equation

$$\nabla^2 Q(\vec{x}) \equiv \gamma^{\alpha\beta} Q_{;\alpha;\beta} = -k^2 Q, \tag{4.12}$$

where k are the eigenvalues and the semicolon indicates the operation of covariant differentiation in the metric $\gamma_{\alpha\beta}$, corresponding to constant curvature $E = -H_0^2(1 - \Omega_0 - \Omega_\lambda)$. For a flat space ($\Omega_0 + \Omega_\lambda = 1$) the space curvature $E \equiv 0$ and $\gamma_{\alpha\beta} \equiv \delta_{\alpha\beta}$, where $\delta_{\alpha\beta}$ is the Kronecker delta. In this model the spectrum of eigenvalues is continuous and the natural representation for the function $Q(\vec{x})$ is a set in plane waves of the type $e^{i\vec{k}\vec{x}}$. If the space curvature is non-zero, the problem of finding the spectrum and the Laplace operator eigenfunctions depends on the sign of E. Note that E vanishes in 'close' models, $E > 0$, while in 'open' models $E < 0$; this signifies that the total density of matter may exceed ($E > 0$) or be less than ($E < 0$) the critical density of matter $\rho_{cr} = 3H_0^2/8\pi G$. Let us consider a specific model for $\gamma_{\alpha\beta}$ that corresponds to a three-dimensional space of constant (positive or negative) curvature E:

$$\gamma_{\alpha\beta} \, dx^{\alpha} \, dx^{\beta} = -E^{-1} \left[d\chi^2 + \sinh^2 \chi (d\theta^2 + \sin^2 \theta \, d\varphi^2) \right], \tag{4.13}$$

where $\chi = \sqrt{-E} \cdot \xi$ are spatial distances and θ and φ are the corresponding angles in the spherical reference frame of the metric, Eq. (4.13). Then the problem of finding the Laplace operator eigenfunctions reduces to finding the solutions of the following equations:

$$\nabla^2 Q(\chi, \theta, \varphi) = -E \sinh^{-2} \chi$$
$$\times \left[\frac{\partial}{\partial \chi} \left(\sinh^2 \chi \frac{\partial Q}{\partial \chi} \right) + \sin^{-1} \theta \frac{\partial}{\partial \theta} \left(\sin \theta \frac{\partial Q}{\partial \theta} \right) + \sin^{-2} \theta \frac{\partial^2 Q}{\partial \varphi^2} \right] = -k^2 Q,$$
$$\tag{4.14}$$

---

[1] Additional types of possible initial perturbations, not connected with metric perturbations, will be discussed later.

where, by virtue of the linearity of the equation for the description of the angular velocity $Q(\chi, \theta, \varphi)$, we can use spherical harmonics $Y_l^m(\theta, \varphi)$. Then

$$Q(\chi, \theta, \varphi) = G_\nu^l(\chi) Y_l^m(\theta, \varphi) \tag{4.15}$$

and Eq. (4.14) immediately implies

$$G_\nu^l(\chi) = (-1)^{l+1} M_l^{-1/2} \nu^{-2} (\nu^2 + 1)^{-1/2} \sinh^l \chi \frac{d^{l+1}(\cos \nu \chi)}{d(\cosh \chi)^{l+1}}, \tag{4.16}$$

where

$$\nu^2 = -\left(\frac{K^2}{E} + 1\right); \quad M_l \equiv \prod_{i=0}^{l} k_i; \quad k_0 = 1; \quad k_i = 1 - \frac{(l^2 - 1)E}{k^2}; \quad l \geq 1.$$

Since the spherical harmonics $Y_l^m(\theta, \varphi)$ are orthonormal, the condition of orthonormalization for $Q(\chi, \theta, \varphi)$ in the form (4.15) requires that the functions $G_\nu^l(\chi)$ be orthogonal and normalizable:

$$\int G_\nu^l(\chi) G_{\nu'}^{l'}(\chi) \sin h^2 \chi \, d\chi = \frac{\pi}{2\nu^2} \delta(\nu - \nu') \delta(l - l'). \tag{4.17}$$

Note that the functions G_ν^l satisfy the following recursive relation:

$$\frac{d}{d\xi} G_\nu^l = \frac{l}{2l+1} k K_l^{1/2} G_\nu^{l-1} + \frac{l+1}{2l+1} k K_{l+1}^{1/2} G_\nu^{l+1}. \tag{4.18}$$

This last relation is especially important for finding the 'response' of the CMB anisotropy to the metric perturbation in models with $E \neq 0$, since Eq. (4.15) implies that Θ from Eq. (4.10) satisfies Eq. (4.18).

Let us have a good look at the model with $E < 0$, which is of special interest as one of the possible alternatives to the ΛCDM cosmological model, with the data on the age of the Universe taken into account. Following Wilson (1983), the spatial part of the metric tensor $\gamma_{\alpha\beta}$ in this model for $k \geq \sqrt{-E}$ can be chosen in the Cartesian coordinate system in the form $\gamma_{\alpha\beta} = \delta_{\alpha\beta}/z^2(-E)$; for each fixed z the metric of this model corresponds to the metric of flat space with $|x| < \infty$, $|y| < \infty$ and $z \geq 0$. Then the Laplace equation in the $\gamma_{\alpha\beta}$ metric has the simple form

$$\nabla^2 Q = -Ez^2 \left(\frac{\partial^2 Q}{\partial x^2} + \frac{\partial^2 Q}{\partial y^2} + \frac{\partial^2 Q}{\partial z^2}\right) + Ez \frac{\partial Q}{\partial z} = -k^2 Q, \tag{4.19}$$

and the eigenfunctions are the generalized functions of the type

$$Q(x, y, z) = z K_{i\nu}((k_\perp z) e^{i(k_1 x + k_2 y)}, \tag{4.20}$$

where $K_{i\nu}(x)$ is the modified Bessel function of the second kind of index $i\nu$, $i = \sqrt{-1}$, $k_\perp^2 = k_1^2 + k_2^2$, and k_1 and k_2 are the components of a two-dimensional wave vector. In view of the Kantorovich–Lebedev relations

$$g(y) = \int_0^\infty f(x) K_{ix}(y) \, dx; \quad f(x) = \frac{2}{\pi^2} x \sinh(\pi x) \int_0^\infty g(y) K_{ix}(y) \frac{dy}{y}, \tag{4.21}$$

we can show that

$$\int_0^\infty \nu \sinh(\pi \nu) K_{i\nu}(k_\perp z) K_{i\nu}(k_\perp z') \, d\nu = \frac{\pi^2}{2} z \delta(z - z').$$

Then, any function $F(\vec{x})$ integrable on a square can be presented as an expansion in generalized plane waves (Wilson, 1983):

$$F(\vec{x}) = \int_0^\infty \nu \sinh(\pi \nu) \, d\nu \int_{-\infty}^\infty dk_1 \int_{-\infty}^\infty F(\vec{k}) Q(\vec{x}, \vec{k}) \, dk_2,$$

$$F(\vec{k}) = \frac{1}{2\pi^4} \int_0^\infty \frac{dz}{z^3} \int_0^\infty dx \int_0^\infty F(\vec{x}), \, Q(\vec{x}, \vec{k}) \, dy, \tag{4.22}$$

where $Q(\vec{x}, \vec{k})$ corresponds to the definition (4.20), and $\vec{x} = (x, y, z)$ and $\vec{k} = (k_1, k_2, \nu)$. In this way the representation (4.22) makes it possible to define the concept of the perturbation spectrum of the open Universe following similar definitions for the 'plane' case with $E = 0$.

In conclusion it should be remarked that the transition from an 'open' model to a closed one is done by a formal substitution $\chi \to i\chi$.

4.2.2 Vector and tensor functions

Following Lifshits' classification, we introduce vector eigenfunctions of the Laplace operator on the hypersurface $t = \text{const}$:

$$\nabla^2 Q_\alpha = -(k^2) Q_\alpha. \tag{4.23}$$

Note that $Q^\alpha_{;\alpha} = 0$.

The zero-divergence condition signifies that it will not be possible to construct a Q_α-based scalar in the first order in amplitude Q_α. At the same time, it is possible to use covariant differentiation operations to define a tensor as follows:

$$S^\beta_\alpha = Q^{;\beta}_\alpha + Q^\beta_{;\alpha}. \tag{4.24}$$

Under the condition $Q_\alpha Q^\alpha = 1$, the corresponding normalized tensor for flat space is defined by $\bar{S}^\beta_\alpha = (1/|k|)(k^\beta Q_\alpha + k_\alpha Q^\beta)$.

Finally, following Lifshits (1946), we introduce tensor eigenfunctions of the Laplace operator, G^β_α, that satisfy the conditions $G^\beta_{\alpha,\beta} = 0$ and $G^\alpha_\alpha = 0$, where, as usual, repeated indexes imply summation. In view of the definitions given above, any perturbation of the metric, density and velocity of the matter can be presented as an expansion in scalar, vector and tensor modes. In a flat Universe this expansion takes an especially simple form because conventional plane waves of the type $e^{i\vec{k}\vec{x}}$ are used as base functions,

$$h^\beta_\alpha = \left\{ \frac{1}{3}\delta^\beta_\alpha \mu_1(t) Q_{\vec{k}} + \left(\frac{1}{3}\delta^\beta_\alpha - \frac{k_\alpha k^\beta}{k^2} \right) Q_{\vec{k}} \mu_2(t) + \mu_3(t) \bar{S}^\beta_\alpha + G^\beta_{\alpha(k)} \mu_4(t) \right\} e^{i\vec{k}\vec{x}}, \tag{4.25}$$

where $\mu_1(t)$ and $\mu_2(t)$ describe the temporal dependence of the potential modes, while $\mu_3(t)$ and $\mu_4(t)$ characterize the vector and tensor perturbations, respectively. It is clear from Eq. (4.25) that $\text{Spur}(h^\beta_{\alpha(k)}) = h = \mu_1(t) Q_k$ is determined only by the function $\mu_1(t)$ from the expansion of h^β_α in scalar functions.

Likewise for the 'vector' part of perturbations of the metric tensor h^0_α the corresponding expansion in mode types will contain only the gradient part of the scalar mode $\sim Q_{;\alpha}$ and explicitly the vector mode S_α,

$$h^0_\alpha(\vec{x}, t) = \{v_1(t) S_\alpha + ik_\alpha v_2(t) Q\} e^{i\vec{k}\vec{x}}, \tag{4.26}$$

where $v_1(t)$ and $v_2(t) - h_\alpha^0$ are the corresponding functions taking into account the evolution of h_α^0 in time.

We need to emphasize especially that in what follows we are mostly interested in the observational manifestations of scalar and tensor modes in the anisotropy and polarization of the primordial CMB. The vector modes, whose role was discussed so widely at the end of the 1970s and the beginning of the 1980s (see Zeldovich and Novikov (1983) do not comply with the hypothesis of uniform and isotropic expansion of the Universe, at least during the epoch of cosmological nuclear synthesis (see Zeldovich and Novikov (1983) and references therein). However, the role of vortex (vector) perturbations is important at the non-linear phase of the evolution of the perturbations. In fact, in view of the extreme smallness of space scales for which the generation of the vortex component could be significant, they leave the characteristics of anisotropy of the CMB radiation practically unaffected.

4.3   Gauge invariance

The choice of the class of reference frame plays a very important role in describing the dynamics of evolution and spatial distribution of perturbations of metric, density and velocity of matter (Landau and Lifshits, 1962). This problem of 'choice' is not anything specific and inherent in perturbations *per se* in the expanding Universe. The root of the problem may be traced back to the covariant formulation of the equations of general relativity in which the form of equations is independent of the choice of reference frame whereas the form of solutions does depend on it. This factor is of principal importance in analysing the behaviour of perturbations in the expanding Universe because without a special analysis the effects of coordinates can be easily mistaken for true physical effects. Following Hu (1995), we will consider this aspect in more detail for a scalar mode of perturbations. The most general form of presenting the perturbation metric for this type of perturbations has the following form:

$$
\begin{aligned}
g_{oo} &= -a^2 \big[1 + 2\varphi_{(\xi)}^{\mathrm{G}} Q\big], \\
g_{0j} &= -a^2 \psi_{(\xi)}^{\mathrm{G}} Q_j, \\
g_{\alpha\beta} &= a^2 \big[\gamma_{\alpha\beta} + 2H_{L(\xi)}^{\mathrm{G}} \gamma_{\alpha\beta} Q + 2H_T^{\mathrm{G}}(\xi) Q_{\alpha\beta}\big],
\end{aligned}
\tag{4.27}
$$

where $Q_j \equiv -k^{-1} Q_{;j}$ and $Q_{ij} \equiv k^{-2} Q_{;i;j} + \frac{1}{3}\gamma_{ij} Q$, the index G is a mark that the corresponding functions belong to a specific reference frame and $\xi = \int dt/a$ is conformal time. Let us consider a shift in spatial (x^α) and temporal ξ variables in response to scalar perturbations Q that also automatically perturb the reference frames:

$$
\tilde{\xi} = \xi + T\dot{Q}, \qquad \tilde{X}^\alpha = X^\alpha + L\dot{Q}^\alpha.
\tag{4.28}
$$

Here the tilde corresponds to the perturbed reference frame, and T and L are the corresponding functions of time ξ. We now make use of the rule of transformation of the metric tensor g_{ik} from one reference frame to another:

$$
g_{ik}(\xi, x^\alpha) = \frac{\partial x^m}{\partial \tilde{x}^i} \frac{\partial x^n}{\partial \tilde{x}^k} g_{mn}(\xi - TQ; x^\alpha - LQ^\alpha).
\tag{4.29}
$$

Then the metric perturbations $h_{ik} = g_{ik} - \tilde{g}_{ik}$ are related to perturbations of coordinates by the following relation:

$$
h_{ik} = \tilde{g}_{ik}(\xi, x^\alpha) - g_{ik}(\xi, x^\alpha) = g_{nk}(\delta x)_{;i}^n + g_{ni}(\delta x^n)_{,k} - g_{ik,n}(\delta x)^n.
\tag{4.30}
$$

Substituting Eqs (4.28) and (4.29) into Eq. (4.30), we arrive at the following law for function transformation:

$$\widetilde{\varphi}^{\mathrm{G}} = \varphi^{\mathrm{G}} - T' - \frac{a'}{a}T, \tag{4.31a}$$

$$\widetilde{\psi}^{\mathrm{G}} = \psi^{\mathrm{G}} + L' + kT, \tag{4.31b}$$

$$\widetilde{H}_L^{\mathrm{G}} = H_L^{\mathrm{G}} - \frac{k}{3}L - \frac{a'}{a}T, \tag{4.31c}$$

$$\widetilde{H}_T^{\mathrm{G}} = H_T^{\mathrm{G}} + kL, \tag{4.31d}$$

where the prime (') stands for the derivative with respect to time t.

4.3.1   The Newtonian gauge

Following Hu (1995), we define the Newtonian gauge by the condition $\widetilde{\psi} = \widetilde{H}_T^{\mathrm{G}}|_{\mathrm{G=N}} = 0$. The reference frame with this gauge will be referred to as Newtonian. From Eq. (4.31b) we obtain

$$T = -\frac{\Psi^{\mathrm{G}}}{k} - \frac{L'}{k}, \tag{4.32}$$

and Eq. (4.31d) implies that

$$L = -\frac{1}{k}H_T^{\mathrm{G}}. \tag{4.33}$$

Substituting Eq. (4.33) into Eq. (4.32), we see that the transformation of perturbations from an arbitrary reference frame to a Newtonian reference frame follows the law

$$T = -\frac{\Psi^{\mathrm{G}}}{k} + \frac{H_T'^{\mathrm{G}}}{k^2}; \qquad L = -\frac{1}{k}H_T^{\mathrm{G}}. \tag{4.34}$$

The components of the perturbation metric in the Newtonian reference frame are then given by

$$\Psi \equiv \varphi^{\mathrm{N}} = \varphi^{\mathrm{G}} + \frac{1}{a^k}\left[\frac{a\xi^{\mathrm{G}}}{k} - \frac{aH'^{\mathrm{G}}}{k^2}\right]',$$

$$\Phi \equiv H_L^{\mathrm{N}} = H_L^{\mathrm{G}} + \frac{1}{3}H_T^{\mathrm{G}} + \frac{a'}{a}\left(\frac{\xi^{\mathrm{G}}}{k} - \frac{H_T'^{\mathrm{G}}}{k^2}\right). \tag{4.35}$$

For the sake of completeness, we give the expressions for perturbations of density δx, pressure δP_x and velocity of matter v_x in the Newtonian gauge (Hu, 1995):

$$\delta_x^{\mathrm{N}} = \delta_x^{\mathrm{G}} + 3(1 + \omega_x)\frac{a'}{a}\left[-\frac{B^{\mathrm{G}}}{k} + \frac{H_T'^{\mathrm{G}}}{k^2}\right],$$

$$\delta P_x^{\mathrm{N}} = \delta P_x^{\mathrm{G}} + 3(1 + \omega_x)C_x^2\frac{a'}{a}\left[-\frac{B^{\mathrm{G}}}{k} + \frac{H_T'^{\mathrm{G}}}{k^2}\right], \tag{4.36}$$

$$V_x^{\mathrm{N}} = V - x^{\mathrm{G}} - \frac{H_T'^{\mathrm{G}}}{k},$$

where the index x stands for the matter component and $\omega_x = P_x/\rho_x$. The relation between the Newtonian gravitational potential and density perturbations is given by the following relation that follows from the Poisson equation:

$$\Phi = 4\pi G(a^2)\rho_T \left[\delta_T^G + 3\frac{a'}{a}(1 + \omega_T)(v_T^G - B^G)/k \right]. \tag{4.37}$$

4.3.2   Synchronous gauge

The choice of a synchronous reference frame is one of the most widespread in analysing the behaviour of perturbations in the expanding Universe. Formally it corresponds to the conditions $\widetilde{\varphi}^S = \widetilde{\xi}^S = 0$. As we see from Eqs (4.27), the condition $\varphi^S = 0$ signifies that perturbations of the g_{00} component of the metric tensor are absent and $\xi^S = 0$ automatically turn the g_{0j} component to zero. Equations (4.31d) immediately imply that

$$T = \frac{1}{a}\int a\varphi^G \, d\xi + \frac{C_1}{a}d, \qquad L = -\int (\xi^G + kT) \, d\xi + C_2, \tag{4.38}$$

where C_1 and C_2 are arbitrary constants determined by the initial conditions. It is then straightforward to find from Eqs (4.38) the corresponding relations between $\widetilde{H}_L^S$, H_L^G and H_T^S, H_N^G, by substituting Eq. (4.38) into Eqs (4.31d). To conclude this section we give the relation between perturbations of the metric Φ and Ψ given in the Newtonian reference frame, and perturbations $h_S \equiv 6H_L^S$ and $\xi_S = -H_L^S - \frac{1}{3}H_T^S$ determined in the synchronous gauge:

$$\Psi = \frac{1}{2k^2 a}\left[a(h_S' + 6\sigma\xi_S') \right]', \qquad \Phi = -\xi_S + \frac{a'}{2k^2 a}(h_S' + \sigma\xi_S'). \tag{4.39}$$

4.4   Multicomponent medium: classification of the types of scalar perturbations

An important feature of the modern theory of the evolution of scalar perturbations in the expanding Universe is that the theory takes into account the multicomponent nature of the medium; it includes electromagnetic radiation, the baryonic fraction of matter, electrons, muons, τ neutrinos and massive particles – carriers of cosmological hidden mass. Each of the components listed above has its own history of interaction with other components which inevitably modify the dynamics of evolution of scalar-type perturbations, both within each subsystem and on the whole for matter density perturbations (including all its components) and for metric perturbations related to them. The situation becomes even more complicated if we assume that the 'cold' hidden mass that today constitutes practically 30% of the critical matter density may itself have a more complicated composition; that is, it may include several components, each of which is identified with its own massive carrier (particle, black hole, etc.) that had its own history of evolution in the course of the expansion of the Universe.

In this section we concentrate mainly on what types of perturbations are possible in such multicomponent media, by detailing the classification of perturbations types that typically belong under the scalar mode (see Section 4.2).

4.4.1   Adiabatic (isentropic) modes

We begin by specifying the definition of the total density of matter in the Universe ε_{tot} and the total pressure in the framework of the hydrodynamic approach:

$$\varepsilon_{\text{tot}} = \sum_i \varepsilon_i, \qquad P_{\text{tot}} = \sum_i P_i, \tag{4.40}$$

where i stands for the type of the corresponding component. We write ε_i in the form

$$\varepsilon_i = \varepsilon_i(t)(1 + \delta_i(t, \vec{r})), \tag{4.41}$$

where $\delta_i(t, \vec{r})$ is the corresponding density perturbation in each component:

$$\delta_i(\vec{r}, t) = \frac{\varepsilon_i - \varepsilon_i(t)}{\varepsilon_i(t)}. \tag{4.42}$$

By analogy to Eq. (4.42) we define the perturbations of partial pressures

$$\delta_i^{(P)}(\vec{r}, t) = \frac{P_i - P_i(t)}{P_i(t)} \tag{4.43}$$

and partial entropy densities

$$\delta S_i = \frac{S_i - S_i(t)}{S_i(t)}. \tag{4.44}$$

One of the more important assumptions that we use to classify scalar types of perturbations in a multicomponent medium is that the temperature of each component was the same at the very earliest stages of cosmological expansion and equal to radiation temperature, and, as a consequence, its perturbations can be expressed in terms of fluctuations of this temperature:

$$\frac{\delta S}{S} = 3\frac{\delta T}{T}. \tag{4.45}$$

Following Ma and Bertschinger (1995) and Zeldovich and Novikov (1983), we introduce the specific entropy of radiation normalized to the density of the baryonic charge n_{bar}, such that $S_{\gamma b} = S_\gamma/n_{\text{bar}}$. The fluctuations of this quantity are given by

$$\frac{\delta S_{\gamma b}}{S_{\gamma b}} = \frac{\delta S_\gamma}{S_\gamma} - \frac{\delta n_{\text{bar}}}{n_{\text{bar}}}. \tag{4.46}$$

First of all we define the standard adiabatic mode of perturbations by the condition $\delta S_{\gamma b}/S_{\gamma b} = 0$ and the standard entropic mode by the condition $\delta S_{\gamma b}/S_{\gamma b} = \text{const.}$ (Zeldovich and Novikov, 1983). For the adiabatic mode we obtain, from Eqs (4.45) and (4.46),

$$3\frac{\delta T}{T} = \frac{3}{4}\frac{\delta \varepsilon_\gamma}{\varepsilon_\gamma} = \frac{\delta \varepsilon_\gamma}{\varepsilon_\gamma + P_\gamma} = \frac{\delta n_{\text{bar}}}{n_{\text{bar}}}. \tag{4.47}$$

Following Ma and Bertschinger (1995), we generalize the definition of Eq. (4.46) to the case of multicomponent media, as follows:

$$\frac{\delta n_j}{n_j} = \frac{\delta \varepsilon_\gamma}{\varepsilon_\gamma + P_\gamma}, \tag{4.48}$$

where the index j refers to all massive components, including baryons. Taking into account that $\delta n_j/n_j \equiv \delta_j$ and $\delta_\gamma = \delta \varepsilon_\gamma/\varepsilon_\gamma = 4(\delta T/T)$, we obtain from Eq. (4.48) the following equation that relates perturbations in each component to perturbations in electromagnetic radiation:

$$\delta_b = \delta_x = \ldots \delta_y = \frac{3}{4}\delta_\gamma = \frac{3}{4}\delta_\nu = \frac{3}{4}\delta_z. \tag{4.49}$$

Here δ_b represents the baryonic fraction of matter, $\delta_x, \ldots, \delta_y$ corresponds to various fractions of cold hidden mass, and δ_ν and δ_z refer to massless neutrinos and hypothetical light particles, respectively.

4.4.2   'Isopotential' modes

In the West, the term 'isocurvature perturbations' is usually used instead of 'isopotential modes', indicating that this sort of perturbation does not perturb the spatial curvature. In its turn, this means that the total density perturbation of the multicomponent medium is zero – perturbations in each component have such phases that the combined perturbation $\delta\varepsilon_{\text{tot}} = 0$. The condition $\delta\varepsilon_{\text{tot}} = 0$ leads to the following equation that relates perturbations in each component:

$$\sum_i m_i \delta n_i + 4\varepsilon_\gamma (1 + R_{\nu\gamma})\frac{\delta T}{T} = 0. \tag{4.50}$$

Here m_i is the mass of the ith non-relativistic fraction of matter (baryons, various types of dark matter), $R_{\nu\gamma} = \varepsilon_\nu/\varepsilon_\gamma$ and ε_ν is the sum of energy densities of massless neutrinos.

Assume now that the adiabaticity condition, $\delta_i = \delta n_j/n_j = \frac{3}{4}\delta_\gamma$, is violated for one of the components of hidden mass, while it holds for all other components including the baryonic fraction. In this case we have that for $i \neq j$ $\delta_i = \frac{3}{4}\delta_\gamma$ and from Eq. (4.50) we obtain

$$\varepsilon_i \delta_i + \left[3\sum_{i \neq j} \varepsilon_i + 4\varepsilon_\gamma (1 + R_{\nu\gamma}) \right]\frac{\delta T}{T} = 0. \tag{4.51}$$

This result implies that the relation between temperature perturbations $\delta T/T$ and δ_j is as follows:

$$\frac{\delta T}{T} = -\frac{\varepsilon_j}{3\sum_{i \neq j} \varepsilon_i + 4\varepsilon_\gamma (1 + R_{\nu\gamma})}\delta_j. \tag{4.52}$$

As we see from Eq. (4.52), the condition of isopotentiality $\delta\varepsilon_{\text{tot}} = 0$ in a multicomponent medium, including partial adiabaticity and non-adiabaticity of at least one non-relativistic component, results in certain phasing of perturbations in electromagnetic radiation and massless neutrinos in this 'peculiar' massive component (see Eq. (4.52)). The relation between the perturbations $\delta S_{\gamma j}/S_{\gamma j}$ and $\delta T/T$,

$$\frac{\delta S_{\gamma j}}{S_{\gamma j}} = 3\frac{\delta T}{T} - \delta_j = 3\frac{\delta T}{T}\left[1 + \frac{\sum_{i \neq j} \varepsilon_i + \frac{4}{3}\varepsilon_\gamma (1 + R_{\nu\gamma})}{\varepsilon_j} \right], \tag{4.53}$$

is easily obtained by using the expressions for the perturbations of specific entropy of radiation for each jth kind of particle (see Eq. (4.46)). Now, Eq. (4.53) shows that isopotentiality of the perturbation signifies at the same time the existence of non-zero specific perturbations of entropy. However, in the general case this type of perturbation cannot be reduced to the purely entropic mode, owing to the adiabaticity of perturbations in all massive components with the exception of the jth component. This effect reflects the specificity of the multicomponent media in addition to the properties of the baryon–photon 'fluid' discussed in Zeldovich and Novikov (1983).

4.4.3   Entropic–isopotential mode

Let us turn again to Eq. (4.50) and consider a situation in which perturbations in all massive components i do not satisfy the adiabatic relation $\delta_j = \frac{3}{4}\delta_\gamma$. We introduce the specific entropy of radiation normalized to the concentration of particles of the species i, assuming as before that neutrinos are massless. Consider the perturbations

$$\sigma_i = \frac{\delta S_{\gamma(\text{tot})}}{S_{\gamma(\text{tot})}} = \frac{\delta S_\gamma}{S_\gamma} - \frac{\delta n_i}{n_i} = 3\frac{\delta T}{T} - \delta_i. \tag{4.54}$$

We again assume that the condition of isopotentiality of perturbations in the form (4.40) continues to hold. Substituting Eq. (4.54) into Eq. (4.50), we obtain

$$\frac{\delta T}{T} = \frac{\sum_i \varepsilon_i \sigma_i}{3\varepsilon_{\text{tot}} + 4\varepsilon_\gamma(1 + R_{\nu\gamma})}, \tag{4.55}$$

where $\varepsilon_{\text{tot}} = \sum_i \varepsilon_i$. It is clear from Eq. (4.55) that for this mode of perturbations, fluctuations of specific entropy σ_i are the source of temperature fluctuations of CMB and massless neutrinos. The perturbations of the radiation energy density and radiation temperature at the early stages of expansion, when $\varepsilon_\gamma \gg \varepsilon_{\text{tot}}$, are small but finite: $\delta T/T = (\sum_i \varepsilon_i \sigma_i)/4\varepsilon_\gamma(1 + R_{\nu\gamma})$. As the condition $\varepsilon_i \propto a^{-3}$ is satisfied for each non-relativistic component, and $\varepsilon_\gamma \sim a^{-4}$, then $\delta T/T \propto a$ and tends to zero as $a \to 0$. As the scale factor increases, the densities ε_{tot} and ε_γ reach equality at $a = a_{\text{eq}}$, and the asymptotic behaviour (for $a \gg a_{\text{eq}}$) $\delta T/T$ reaches the maximum $\delta T/T \simeq \frac{1}{3}\sum x_i \sigma_i$, where $x_i = \varepsilon_i/\varepsilon_{\text{tot}}$.

4.4.4 Isothermal mode

The multicomponent nature of the medium allows the emergence of new modes that have no analogues in a simple baryon–phonon model of perturbation evolution in the expanding Universe that was widely discussed in the middle of the 1960s to the end of the 1970s (Zeldovich and Novikov, 1983). The most impressive example of these sorts of differences is the isothermal mode in which $\delta T/T \equiv 0$ and the isopotentiality condition, Eq. (4.11), is satisfied. The absence of perturbations in the relativistic component signifies that the distribution of baryons and hidden mass is phased in some special manner as follows:

$$\sum_i \rho_i \delta_i = 0. \tag{4.56}$$

As follows from Eq. (4.56), the positive density contrast in baryons is compensated for by the negative contrast in all components of dark matter,

$$\delta_b = -\frac{1}{\rho_b} \sum_i \rho_i \delta_i, \tag{4.57}$$

where ρ_b is the baryonic density and the sum is taken only over the sorts of particles comprising the hidden mass. For $i = 1$ (i.e. in the model with only one species of particle that carry the hidden mass) Eq. (4.57) is found to be especially simple: $\delta_b = -(\rho_x/\rho_b)\delta_x$, where ρ_x is the density of these sorts of hidden mass carriers. Taking into account that $\rho_x/\rho_b \gg 1$, we obtain $|\delta_b| \gg |\delta_x|$. Formally, at the limit of applicability of the linear theory, perturbations in the 'gas' of hidden mass cannot be less than -1. This corresponds to the density contrast in the baryonic gas:

$$\delta_b \simeq \frac{\rho_x}{\rho_b} \simeq \frac{\Omega_x h^2}{\Omega_b h^2}, \tag{4.58}$$

where $\Omega_x h^2$ and $\Omega_b h^2$ are the current values of the hidden mass and baryon densities in units of critical density of matter. Assuming $\Omega_x h^2 \simeq 0.127$ ($\Omega_x \sim 0.3$; $h = 0.65$) and $\Omega_b h^2 \simeq 0.02$, we obtain the upper bound on the density contrast in the baryonic component to be $\delta_b \leq 6.3$. It should be emphasized again that this estimate goes beyond the linear approximation and is used here only as an example. However, using the idea of isothermal isopotential modes as a launching pad, we can pose a much more important problem: what are the properties of the ground state of multicomponent matter that we previously assumed to be uniform and isotropic? Section 4.4.5 deals with this aspect in more detail.

4.4.5   *The origin of potential functions*

This subsection deals with possible modifications of the model of the on-average uniform isotropic Universe in the presence of various types of scalar perturbations, undoubtedly including non-linear configurations. In contrast to the preceding subsection where we made no assumptions concerning the existence of characteristic perturbation scales, in what follows we assume that a special scale λ_{max} exists, such that isothermal perturbations can develop in a non-linear mode if $\lambda < \lambda_{max}$ while the amplitude of these perturbations is zero if $\lambda > \lambda_{max}$. At first glance, assumptions made about some putative peculiar linear sizes smaller than the physical scale of the current galaxies may appear artificial. A hypothesis that no such peculiar scales exist, first formulated by Zeldovich (1970) and Harrison (1970) in relation to the adiabatic mode appears more appealing. The corresponding spatial spectrum of perturbations is known as the Harrison–Zeldovich spectrum. When discussing possible types and characteristic scales of perturbations in a multicomponent cosmological plasma, we should not forget that without specific implementation of the models of the origin of the perturbations, it would be just as 'dangerous' to arrive at conclusions about the presence of specific parameters as to suggest that no such parameters exist. It is widely recognized that the most promising approach to explaining the source of the initial irregularity in the Universe is through progress in the theory of the early inflationary stages of the expansion of the Universe. Therefore, all possible characteristics of perturbations must be treated within the framework of the inflation theory.

A considerable amount of work has covered the analysis of this approach, for example Kolb and Turner (1989) and Linde (1990). We need to point out here that the first correct derivation of the shape of the spectrum of small perturbations in terms of inflation theory was obtained for tensor perturbations by Starobinsky (1979) and for scalar perturbations by Guth (1981), Hawking (1982) and Starobinsky (1982). Important contributions to this problem were made by Lukash (1980), Mukhanov and Chibisov (1981), Starobinsky (1980) and Kompaneets, Lukash and Novikov (1982). We cannot for obvious reasons go into a detailed discussion of these impressive achievements of current inflation theories in this brief subsection, and therefore we refer the reader to the original papers cited above. As in the preceding chapter, we act as consumers of information and consider as adequate for the purposes of this subsection to make use of the fact that, in the framework of the inflation scenario, the current irregularities on galactic and larger scales originate with the quantum noise of the physical fields – precisely in the period of the inflational expansion of the Universe. Furthermore, it is found to be sufficient for the generation of non-adiabatic perturbation modes that there exist a mixture of various fields, for example scalar φ and χ fields, whose evolution triggers the generation of both adiabatic and isopotential modes – under a certain combination of parameters of the φ–χ interaction (Novikov, Schmalzing and Mukhanov, 2000; Polarski and Starobinsky, 1994; Turok, 1996).

The level of predicted fluctuations of both the adiabatic and isopotential types may be different, depending on the specific choice of parameters in the inflationary mode, and may reach extreme values, $\delta\rho/\rho \sim 1$. It is then natural that the spatial scales of such non-linear perturbations cannot reach the scale of clusters and superclusters of galaxies because they would inevitably cause strong anisotropy of the primordial background radiation in the hydrogen recombination epoch. To illustrate the possibility of the existence of adiabatic and isopotential types of characteristic scales permitting the existence of well developed ($\delta\rho/\rho \sim 1$) isopotential modes in the spectrum of perturbations, we turn to an analysis of the specific models of early inflationary stages.

4.4.6 Peculiarities of the inflaton potential

In accordance with the general ideology of inflationary scenarios we assume that the dynamics of the expansion of the Universe at the earliest stages of cosmological evolution ($t \to 0$) is dictated by the scalar field $\varphi(\vec{x}, t)$ that simulates the vacuum state of matter. The scalar field $\varphi(\vec{x}, t)$ contains a classical part, $\varphi(t)$, and quantum fluctuations, $\delta\varphi(\vec{x}, t)$, which are functions of the spatial and temporal coordinates (Linde, 1990). The classical component of the field φ in the uniform and isotropic Universe evolves in conformity with the standard equation

$$\frac{\partial^2 \varphi_c}{\partial t^2} + 3H\frac{\partial \varphi_c}{\partial t} = -\frac{\partial V}{\partial \varphi_c}, \tag{4.59}$$

$$3H^2 = 8\pi \left(\frac{\partial \varphi_c^2}{\partial t} + V(\varphi_c) \right), \tag{4.60}$$

where $V(\varphi)$ is the potential of the scalar field φ, $\dot{\varphi} \equiv d\varphi/dt$, and we use the system of units $\hbar = C = G = 1$. For the adiabatic mode we use the gauge-invariant v_k related to the quantum component of the scalar field $\hat{\varphi}_q$ (Linde, 1990):

$$\hat{\varphi}_q(\vec{x}, t) = \frac{1}{2\pi^{3/2}} \int \left(\hat{a}_k v_k(t) e^{-i\vec{k}\vec{x}} + \hat{a}_k^+ v_k^*(t) e^{-i\vec{k}\vec{x}} \right) d^3k, \tag{4.61}$$

where $\hat{a}_k$ and $\hat{a}_k^*$ are the creation and annihilation operators, respectively. The dynamics of $v_k(t)$ is described by the following equation:

$$v_k'' + \left[k^2 - \frac{2}{\xi^2} + \frac{\partial^2 V}{\partial \varphi^2} a^2 \right] v_k = 0, \tag{4.62}$$

where $v_k'' \equiv \partial^2 v_k/\partial\xi^2$, $\xi = \int(1/a)\,dt$. We need to stress that the set of equations (4.59)–(4.62), a basis for any inflationary model, relates the dynamics of evolution of the classical component of the field φ and quantum noise $\hat{\varphi}_q(\vec{x}, t)$. In the absence of singularities of the potential $V(\varphi)$, as for example in the model $V(\varphi) \simeq (\lambda/4)(\varphi)^4$ or a simpler model $V(\varphi) \simeq m^2\varphi^2/2$, the most important condition that allows us to calculate the spectrum of adiabatic perturbations generated from the function $v_k(t)$ is the approximation of slow evolution of the field φ at the very start of inflation:

$$\left| \frac{\partial^2 \varphi}{\partial t^2} \right| \ll 3H|\dot{\varphi}|, \qquad \frac{\partial \varphi^2}{\partial t} \ll 2V(\varphi). \tag{4.63}$$

In this approximation the theory predicts that the spectrum of fluctuations in the adiabatic-type metric is given by (Linde, 1990)

$$P(k) \sim k \frac{V^3(\varphi)}{(\partial V / \partial \varphi)^2} \Big|_{k=aH(\varphi)},$$
(4.64)

where $H(\varphi)$ is the value assumed by the Hubble parameter when the field φ reaches a certain value φ_{min} that corresponds to the termination of the inflationary mode. For monotone potentials $V(\varphi)$ that possess no singularities of the type $\partial V / \partial \varphi \to 0$ on some range $(\varphi_1; \varphi_2)$, the predictions of the inflation theory are quite specific. The spectrum of adiabatic perturbations of density becomes

$$P_\delta(k) \simeq Ak,$$
(4.65)

where A is the spectrum amplitude, which coincides in form with the Harrison–Zeldovich spectrum. However, the situation changes radically if the inflaton potential contains an area $(\varphi_1 \leq \varphi_1 \leq \varphi_2)$ where $\frac{\partial V}{\partial \varphi} \simeq 0$ (Ivanov *et al.*, 1994; Starobinsky, 1992a,b). In this case, Eq. (4.64) formally implies that the amplitude of fluctuations becomes infinite. Obviously, this singularity is non-physical. Its formal emergence indicates that the slow rolloff approximation at the plateau of the potential $V(\varphi)$ is inapplicable. In view of this, Ivanov *et al.* (1994) arrived at the following general form for the spectrum of adiabatic density perturbations in the simple model of potential $V(\varphi)$ discussed above:

$$P(k) = A^2 k D(k),$$
(4.66)

where $D(k) \simeq 1$ for $k < k_2$ and

$$D(k) \simeq \left[1 + \frac{A_1^-}{A_2^+} \gamma^3\right]^2 \left[1 + 3 \frac{\sin(2kR_2)}{kR_2}\right] \quad \text{for} \quad k_2 < k < k_1;$$
(4.67)

R_2 is the characteristic scale of perturbations that intersect the horizon at $\gamma = k_1/k_2 \gg 1, k_i = a(t_i)H(t_i), i = 1, 2, t_i$ is the moment of time when $\varphi(t_2) = \varphi_i$, and A_1^- and A_+^2 characterize the step in the derivatives of the potential at $\varphi = \varphi_1$ and $\varphi = \varphi_2$.

As follows from Eq. (4.67), a region of sharp increase in fluctuation amplitude, reaching the level $D^{1/2}(k) \sim (A_1^-/A_2^+)\gamma^3$, will form in the spectrum of adiabatic perturbations at $\gamma \gg 1$. Assuming, for the sake of evaluation, $A_1^-/A_2^+ \sim 1$ and $\gamma \simeq 20$, we obtain that the 'enhancement' factor $D^{1/2}$ may reach $\sim 10^4$ or higher values, remaining nevertheless localized in the zone of wave factors $k_2 < k < k_1$. Note that the position of the 'plateau' in the spectrum of adiabatic perturbations is a naturally free parameter that depends on the positions of φ_1 and φ_2 on the potential $V(\varphi)$. Note also that the effect of formation of the 'plateau' on the spectrum of primary perturbations was used by Ivanov *et al.* (1994) to evaluate the fraction of mass of the matter that was transferred to primary black holes at early, post-inflation, stages of cosmological expansion. The parameters φ_1 and φ_2 were chosen in such a way that the typical mass of the primary black hole was close to $1M_\odot$. The density of such objects today was chosen to be equal to the density of MACHOs in our Galaxy.

It is clear therefore that the specifics of the epoch when primary fluctuations are formed may have very important cosmological consequences. One of them, the formation of massive primary black holes, is directly related to the excess of power in the short-wavelength part of the spectrum of adiabatic perturbation modes.

4.4.7 Multicomponent inflation and generation of isopotential modes

One of the non-trivial ideas of modern inflation theory is the explanation of the possible causes of generation of isopotential modes by perturbations of isopotential type, in addition to adiabatic perturbations. We know from Section 4.4.6 that if inflation is controlled by a single scalar field, the emerging perturbations are always adiabatic. The quantum part of the field always produces perturbations of the metric, and hence the initial irregularity of the Universe is purely adiabatic. In this scenario, isopotential modes can only arise at the decay phase of the field φ_0 into particles at the end of inflation when reheating of the Universe takes place. We discuss this aspect in detail in the following. Together with this process, however, the isopotential mode can also be generated during the inflation itself if it is sustained by another field χ (assumed to be scalar for simplification), in addition to the scalar field φ (Kofman and Linde, 1987; Linde, 1984; Mukhanov and Chibisov, 1981; Mukhanov and Steinhard, 1998). Let us consider the simplest model of the isopotential mode following the ideas of Mukhanov and Steinhard (1998). We assume that the energy–momentum tensor for the fields φ and χ is diagonal and we choose the Newtonian gauge $\Phi = -\Psi$ in Eq. (4.35). Then the equations describing the evolution in time of the non-perturbed fields φ and χ have the following form (Mukhanov and Steinhard, 1998):

$$3H\dot{v} = -2\Phi; \qquad 3H\dot{\mu} = -2\Phi; \qquad \Phi = \frac{1}{H}(\dot{V}_1 \cdot v + \dot{V}_2\mu), \tag{4.68}$$

where H is the Hubble parameter, $V_1(\varphi)$ and $V_2(\chi)$ are the two components of the potential $V(\varphi, \chi) = V_1(\varphi) + V_2(\chi)$, $V_1' \equiv \partial V_1(\varphi)/\partial\varphi$, and $V_2' \equiv \partial V_2(\chi)/\partial\chi$.

As in Section 4.4.6, we use the approximation of slow variation of the fields φ and χ in the form (4.59). By analogy with the model with a single scalar field we introduce perturbations $\delta\varphi$ and $\delta\chi$, assuming formally the wave factor to be zero, $|\vec{k}| = 0$. This means that we have selected the long-wavelength approximation. Then, making the replacements $\delta\varphi = V_1'v$ and $\delta\chi = V_2'\mu$ we obtain the set of equations for the perturbations v and μ as follows:

$$3H\dot{v} = -2\Phi; \qquad 3H\dot{\mu} = -2\Phi; \qquad \Phi = \frac{1}{H}(\dot{V}_1 \cdot v + \dot{V}_2\mu), \tag{4.69}$$

where $\dot{V} \equiv dV/dt$. It is not difficult to see that in this set the difference between the perturbations v and μ reduces to a single constant,

$$\mu = v + g, \tag{4.70}$$

where $g = \text{const}$. Then we have, for metric perturbations,

$$\Phi = \frac{1}{H}\left[(\dot{V}_1 + \dot{V}_2)v + g\dot{V}_2\right]. \tag{4.71}$$

Returning to Eqs (4.69), we immediately note that

$$\dot{v} + \frac{\dot{V}_1 + \dot{V}_2}{V_1 + V_2}v = -g\frac{\dot{V}_2}{V_1 + V_2}. \tag{4.72}$$

The solution of this equation has the following form (Mukhanov and Steinhard, 1998):

$$v = \frac{C - gV_2}{V_1 + V_2}; \qquad \mu = \frac{C + gV_1}{V_1 + V_2}, \tag{4.73}$$

where the integration constants C and g are found from the conditions

$$\mu \simeq \alpha H|_{K=aH}, \qquad \nu = \beta H|_{K=aH}, \tag{4.74}$$

and α and β are coefficients ~ 1:

$$g = \frac{1}{3H}\left(\frac{\nu}{\dot{\varphi}} - \frac{\mu}{\dot{\chi}}\right)\Bigg|_{K=aH}, \tag{4.75}$$

$$C = -\frac{1}{2}H\left(\frac{V_1}{V_1+V_2}\frac{\nu}{\dot{\varphi}} + \frac{V_2}{V_1+V_2}\frac{\mu}{\dot{\chi}}\right)\Bigg|_{K=aH}.$$

With this normalization of constants, the corresponding perturbation of the gravitational potential is given by

$$\Phi = 2C\frac{\dot{H}}{H^2} + g\frac{1}{H}\frac{V_1\dot{V}_2 - V_2\dot{V}_1}{V_1+V_2}. \tag{4.76}$$

The first term in Eq. (5.18) (see Section 5.2) corresponds to the standard adiabatic mode that is characteristic of the single-component model of inflation. The second term describes the entropy mode (using the terminology of Mukhanov and Steinhard (1998)).

Let us consider the possibility of formation of the isothermal mode with $\Phi = 0$ in the framework of the two-field model of inflation. In view of Eqs (4.70) and (4.71), we obtain for this mode of perturbations the following relations:

$$\nu = -g\frac{\dot{V}_2}{\dot{V}_1+\dot{V}_2}, \qquad \mu = g\frac{\dot{V}_1}{\dot{V}_1+\dot{V}_2}. \tag{4.77}$$

It is not difficult to see that this solution is a particular case of the general solution of Eqs (4.73). A comparison of Eqs (5.19) and (4.69) immediately yields that $\dot{V}_2/(\dot{V}_1+\dot{V}_2) = \text{const.} = A$ and $\dot{V}_1/(\dot{V}_1+\dot{V}_2) = \text{const.} = B$; that is, for the isothermal mode with zero perturbations of the metric to exist, it is necessary that the potentials $V_1(\varphi)$ and $V_2(\chi)$ be linearly dependent functions in at least one interval $\varphi \in (\varphi_1, \varphi_2)$ and $\chi \in (\chi_1, \chi_2)$. Indeed, in view of the relations $\dot{V}_1 = V_1' \cdot \dot{\varphi}$ and $\dot{V}_2 = V_1' \cdot \dot{\chi}$ and Eqs (4.68), we arrive at the following relations:

$$\frac{(V_2')^2}{(V_1')^2 + (V_2')^2} = A, \qquad \frac{(V_1')^1}{(V_1')^2 + (V_2')^2} = B, \qquad A + B = 1. \tag{4.78}$$

Equation (5.20) immediately implies an equation that relates $V_1(\varphi)$ and $V_2(\chi)$ in the region $[\varphi_1, \varphi_2] \bigcup [\chi_1, \chi_2] = \Omega$:

$$\frac{\partial V_1}{\partial \varphi} = \pm\sqrt{\frac{B}{A}}\frac{\partial V_2}{\partial \chi}. \tag{4.79}$$

The simplest model of the potentials V_1 and V_2 in the domain Ω is the linear dependence $V_1(\varphi) \sim C_1\varphi$ and $V_2(\chi) \sim C_2\chi$. Then, owing to Eq. (5.21), C_1 and C_2 must be related by

$$C_1 = \pm\sqrt{\frac{B}{A}}C_2. \tag{4.80}$$

In this case the total potential of the system, $V_{\text{tot}}(\varphi, \chi)$, is also a linear function in the fields φ and χ:

$$V_{\text{tot}}(\varphi, \chi) = \pm\sqrt{\frac{B}{A}}C_2\varphi + C_2\chi = C_2\left(\chi \pm \sqrt{\frac{B}{A}}\varphi\right). \tag{4.81}$$

Note that, as in the preceding section, the localizations of singularity of the potential result in a violation of the adiabaticity of fluctuations and in the emergence of new isopotential-type modes localized within a specific interval of the perturbation spectrum. We need to emphasize especially that the simplest model of two-scalar non-interacting fields could have a much more complex generalization. It is important to know, however, that the conclusion on the relation of these anomalies to the peculiarities of perturbations in the cosmological plasma will obviously remain unchanged. This is the reason why the role of observational data in the anisotropy and polarization of primordial radiation background becomes enormously important when we finally try to answer the question of how was the Big Bang unfolding in the framework of the inflation paradigm?

4.5 Newtonian theory of evolution of small perturbations

Beginning here and until the end of the chapter we will discuss various modes of evolution of perturbations, both of adiabatic and isopotential type in a uniform and, on average, isotropic Universe. Following the classical tradition, we begin this analysis with a discussion of the model of the standard gravitating multicomponent medium, neglecting expansion. The importance of, and necessity in, this analysis stem first of all from its simplicity, which makes it possible to define the main parameters that dictate the pace of the evolution of perturbations, and from the ease with which the main results can be generalized to the model of the expanding Universe.

Note that the basics of the theory of gravitational instability of matter were developed in J. Jeans' classical paper (Jeans, 1902). A description of Jeans' approach is given in Zeldovich and Novikov (1983). We will follow the ideology of Jeans' pioneering paper and the description in the Zeldovich–Novikov monograph practically without any changes. An important assumption used in this section, one that allows us to reproduce correctly the general properties of the cosmological plasma, is the assumption of the existence of N mutually independent components that evolve in a self-consistent gravitational potential. We also assume, in the spirit of the Newtonian approach to describing gravitation, that the characteristic velocities v of processes, $v \ll c$, and that any created gravitational fields are weak: $\varphi \ll c^2$, where c is the velocity of light in vacuum.[2] Then the dynamics of the multicomponent medium is described by hydrodynamic equations of the type (see Peebles (1993) and Zeldovich and Novikov (1983))

$$\frac{\partial \rho_i}{\partial t} + \nabla(\rho_i \vec{V}_i) = 0, \tag{4.82}$$

$$\frac{\partial \vec{V}_i}{\partial t} + (\vec{V}_i \nabla)\vec{V}_i + \frac{\nabla P_i}{\rho_i} + \nabla\varphi = 0, \tag{4.83}$$

$$\frac{\partial S_i}{\partial t} + (\vec{V}_i \nabla)S_i = 0, \tag{4.84}$$

$$\Delta\varphi = 4\pi G \sum_i \rho_i, \tag{4.85}$$

where ρ_i is the density of the ith component of the medium, $\vec{V}_i$ is its vector of velocity, P_i is pressure and S_i is entropy. Equations (4.82)–(4.85) show that there are no external sources

---

[2] Further on in this section we assume that $c = 1$.

of mass, momentum or entropy. Moreover, components with different 'i' interact with one another only through gravitation. Later, when applying this description to specific situations, we usually assume that pressure is significant in only one component and negligible in the others. Following Jeans (1902), we assume the non-perturbed state of the medium to be at rest ($\vec{v}_i = 0$) and that the density and pressure of each component are constant:

$$\rho_i^{(0)} = \text{const.}, \qquad S_i^{(0)} = \text{const.}, \qquad P_i^{(0)} = P\left(\rho_i^{(0)}\right) = \text{const.} \tag{4.86}$$

Note, however, that this prescription of the initial non-perturbed state of the medium demands that $\nabla\varphi$ vanish, which automatically contradicts the Poisson equation, Eq. (4.85). Nevertheless, the assumption of stationarity of the non-perturbed solution will allow us to find fairly simply (and correctly!) the perturbation parameters that separate stable solutions from unstable ones, and to classify perturbations. The reasons why this is possible are analysed in great detail in Zeldovich and Novikov (1975, 1983), so we need not go into that here. We therefore assume

$$\varphi^{(0)} = 0. \tag{4.87}$$

As we mentioned before, this assumption leads to correct values of the parameters that determine the stability or instability. However, the law describing the evolution of perturbations with time will have to be improved. The exact relativistic theory will be treated in Section 4.6.

Let us begin our analysis of the stability of small perturbations of uniform distribution assuming that condition (4.87) holds. Following the standard method of investigating the stability of a multicomponent scheme, we write the density, velocity, pressure, entropy and gravitational potential in the following form:

$$\{X\} = \{\vec{X}\}_0 + \delta X(t)e^{i\vec{k}\vec{x}}, \tag{4.88}$$

where $\{\vec{X}\}_0$ is the ground state and $\delta X(t)e^{i\vec{k}\vec{x}}$ is the perturbed state, and we take into account that pressure perturbations are related to density and entropy perturbations in each component:

$$\delta P_i = \left(\frac{\partial P_i}{\partial \rho_i}\right)_S \delta\rho_i + \left(\frac{\partial P_i}{\partial S_i}\right)_{\rho_i} \delta S_i. \tag{4.89}$$

Introducing the notation $C_S^2 = (\partial P_i/\partial \rho_i)_S$ and $C_\rho^2 = (\partial P_i/\partial S_i)_{\rho_i}$ for the adiabatic and isothermal speed of sound, respectively, we obtain a set of linear equations, Eqs (4.82)–(4.89), for perturbations of thermodynamic quantities, velocities and gravitational potential:

$$\frac{d\delta_i}{dt} + i\vec{k}\vec{u}_i = 0, \tag{4.90a}$$

$$\frac{d\vec{u}_i}{dt} + i\vec{k}\widetilde{\varphi} + i\vec{k}C_S^2\delta_i + i\vec{k}C_\rho^2\sigma_i\left(\rho_i^{(0)}\right)^{(-1)} = 0, \tag{4.90b}$$

$$\frac{d\sigma_i}{dt} = 0, \tag{4.90c}$$

$$k^2\widetilde{\varphi} = -4\pi G \sum_i \rho_i\delta_i, \tag{4.90d}$$

where $\delta \equiv \delta\rho_i/\rho_i^{(0)}$ is the relative density perturbation in the ith component, σ_i are perturbations of entropy, $\widetilde{\varphi}$ are perturbations of potential, and C_ρ and C_S are constants within the linear approximation.

The study of various types of behaviour of perturbations will be conducted in accordance with the classifications of their types that we gave in the preceding sections of this chapter.

4.5.1 Adiabatic perturbations

For this type of perturbation we need to recall that $\sigma_i = 0$ and $\vec{u}_i = u_i(\vec{k}/|\vec{k}|)$ (Zeldovich and Novikov, 1983). As the coefficients in Eq. (4.90d) are time-independent, we make use of the representation δ_i, u_i and $\widetilde{\varphi}$ as a Fourier integral in the variable t:

$$\begin{pmatrix} \delta_i \\ u_i \\ \widetilde{\varphi} \end{pmatrix} \propto \int \begin{pmatrix} \delta_{i,\omega} \\ u_{i,\omega} \\ \widetilde{\varphi}_{i,\omega} \end{pmatrix} e^{-i\omega t}\, d\omega,$$

whence

$$-i\omega\delta_{i,\omega} + iku_{i,\omega} = 0,$$

$$-i\omega\frac{\vec{k}}{k}u_{i,\omega} + i\vec{k}\widetilde{\varphi}_{i,\omega} + i\vec{k}C_{S(i)}^2\delta_{i,\omega} = 0, \tag{4.91}$$

$$k^2\widetilde{\varphi}_\omega = -4\pi G \sum_i \rho_i^{(0)}\delta_{i,\omega}.$$

A non-trivial solution for the perturbations in Eqs (4.91) is only possible if the determinant of the system vanishes, resulting in the following dispersion equation:

$$4\pi G \sum_i \frac{\rho_i^{(0)}}{\omega^2 - k^2 C_{S(i)}^2} = -1. \tag{4.92}$$

For a single-component medium this equation has a standard form,

$$\omega^2 - k^2 C_S^2 = -4\pi G\rho_0, \tag{4.93}$$

and its solutions depend on the relation between the terms $k^2 C_S^2$ and $4\pi G\rho_0$. Assuming $\omega = 0$ in Eqs (4.91), we obtain an expression for the critical scale $k_j = C_S^{-1}(4\pi G\rho^{(0)})^{1/2}$ and the corresponding wavelength $\lambda_j = 2\pi/k_j = C_S\left(\pi/G\rho^{(0)}\right)^{1/2}$ known as the Jeans wave vector and the Jeans wavelength, respectively. As we see from the definition of k_j, for $k > k_j$ the term with ω^2 is positive and the expression for frequency ω takes the following form:

$$\omega = \pm kC_S\left(1 - \frac{k_j^2}{k^2}\right)^{1/2}. \tag{4.94}$$

We immediately see that this branch of the dispersion equation is stable and describes acoustic oscillations of the medium. If $k < k_j$ the solution for ω has only an imaginary part:

$$\omega = \pm i4\pi G\rho\left(1 - \frac{k^2}{k_j^2}\right)^{1/2}. \tag{4.95}$$

Note that the mode with $\omega = -i4\pi G\rho^{(0)}[1 - (k^2/k_j^2)]$ corresponds to the growing (unstable) branch for which an increase in amplitude is exponential, $\sim e^{|\omega|t}$.

We turn now to analysing particular cases of Eq. (4.95). In a medium with index 2, we use, by analogy to a realistic model, the term 'dark matter'; medium 1 will simulate a mixture of radiation and baryons. Let us consider the dispersion equation, Eq. (4.92). It is not difficult to show that this equation has a solution

$$\omega_{1,2}^2 = \frac{A \pm \sqrt{D}}{2},$$ (4.96)

where

$$A(k) \left[k^2 \left(C_1^2 + C_2^2 \right) - 4\pi G \rho_{\text{tot}} \right] / 2,$$

$$D(k) = \left[k^2 \left(C_1^2 - C_2^2 \right) + 4\pi G \rho_{\text{tot}} \right]^2 = 16\pi G k^2 \rho_2 \left(C_1^2 + C_2^2 \right),$$ (4.97)

$$\rho_{\text{tot}} = \rho_1 + \rho_2.$$

The set of equations (4.97) shows that $|A^2| < D$ for $\rho_2 \neq 0$ and $D(k) > 0$ for any value of the modules of the vector k. This means that for $\omega(k)$ there exist an increasing (ω_+) branch and a decaying (ω_-) branch and two isolated branches ω_1 and ω_2:

$$\omega_+ = -\mathrm{i} \left(\frac{\sqrt{D(k)} - A(k)}{2} \right)^{1/2}, \qquad \omega_- = \mathrm{i} \left(\frac{\sqrt{D(k)} - A(k)}{2} \right)^{1/2},$$

$$\omega_{1,2} = \pm \left(\frac{\sqrt{D(k)} + A(k)}{2} \right)^{1/2}.$$ (4.98)

When simulating the realistic situation that arose in the cosmological plasma long before the recombination epoch, we assume the component $i = 2$ to be cold, $C_2 = 0$. Therefore we consider a special case of $C_2 = 0$, $k^2 C_1^2 = 4\pi G \rho_1$, which corresponds in the single-component approximation to the condition $k = k_i$ for this medium. Furthermore we assume that $\rho_2 \ll \rho_1$. We are going to be interested in the following question: what is the effect of an insignificant-density admixture of cold $(C_2 = 0)$ matter on the behaviour of perturbations on the scale $k = k_j^{(i)}$? In the limit $\rho_2 \ll \rho_1$ we obtain from Eqs (4.97) and (4.98)

$$\omega_+ = -\mathrm{i} \sqrt{4\pi G \rho_1} \left(\frac{\rho_2}{2\rho_1} \right)^{1/4}.$$ (4.99)

We see from Eq. (4.99) that the presence of a dark matter component results in a biased Jeans criterion for one component, which triggers an instability of the system.

Let us consider now the behaviour of the ω_+ branch under the approximation $k \to \infty$. In this approximation,

$$\omega_+ = -\mathrm{i} \left(\sqrt{4\pi G \rho_2} \right)$$ (4.100)

and, as previously, a mixture of 'cold' and hot subsystems manifests an unstable branch of perturbations for $k \to \infty$.

4.5.2   Isopotential perturbations

Entropy-type perturbations in the set of equations (4.90d) will not evolve with time as the condition $\mathrm{d}\sigma_i / \mathrm{d}t = 0$ would dictate. Note that Eq. (4.90da) implies that the peculiar velocity u_i automatically vanishes. The relation between density and gravitational potential perturbations implied by Eq. (4.90dc) has the form

$$\widetilde{\varphi} = C_{S(i)}^2 \widetilde{\delta}_i + C_\rho^2 \sigma_i \rho_{0,i}^{-1} = 0.$$ (4.101)

Expressing δ_i from Eq. (4.101) and substituting the result into Eq. (4.90dd), we obtain

$$\widetilde{\varphi} = \frac{4\pi G \sum_i \dfrac{C^2_{\rho(i)}}{C^2_{S(i)}} \sigma_i}{k^2 - 4\pi G \sum_i \dfrac{\rho_{0,i}}{C^2_{S(i)}}}, \tag{4.102}$$

$$\widetilde{\delta}_i = -\frac{C^2_{\rho(i)}}{C^2_S(i)} \frac{\sigma_i}{\rho_{0,2}} - \frac{4\pi G \sigma_j \dfrac{C^2_{\rho(i)}}{C^2_{S(i)}} \sigma_i}{C^2_{S(i)} \left[k^2 - 4\pi G \sum_j \dfrac{\rho_{0,j}}{C^2_{S(j)}} \right]}. \tag{4.103}$$

Equations (4.102)–(4.104) imply for the single-component mode that

$$\widetilde{\delta} = \frac{k^2 C^2_\rho \sigma}{\rho_0 \left(C^{2k}_S 2 - 4\pi G \rho_0 \right)}. \tag{4.104}$$

We remarked in the previous sections of this chapter that the isothermal mode, in which there are no fluctuations of the gravitational potential, is a particular case of isopotential perturbations. Assuming $\widetilde{\varphi} = 0$ in Eq. (4.101) we obtain for this mode

$$\widetilde{\delta}_i^* = -\frac{C^2_{\rho(i)}}{C^2_{S(i)}} \frac{\sigma_i}{\rho_{0,i}}, \tag{4.105}$$

and the condition $\widetilde{\varphi} = 0$ immediately implies a relation with entropy fluctuations in each component:

$$\sum_i \frac{C^2_{\rho(i)}}{C^2_{S(i)}} \sigma_i = 0. \tag{4.106}$$

This signifies that the phases of entropy perturbations in each subsystem are such that they completely cancel out the perturbations of the total matter density and the corresponding perturbations of the gravitational potential. Note that, according to Eq. (4.106), this compensation is stable and can be caused by the initial conditions under which fluctuations emerge (see Section 4.4).

4.6 Relativistic theory of the evolution of perturbations in the expanding Universe

The theory of gravitational instability in the Newtonian approximation that we discussed in Section 4.5 does not take into account two very important factors that drastically change the general concept of the evolution of perturbations. First of all, we should discuss the incorporation of the effect of expansion of the Universe into the general scheme of analysing the evolution of perturbations of metric, velocity and density of the multicomponent medium. Moreover, ultrarelativistic matter, for which the speed of sound is close to the speed of light in a vacuum, plays a most important role in the thermal history of the Universe. Therefore, relativistic effects must be taken into account along with that of expansion. A second important factor is the response of gravitation (metric perturbations) to density and velocity fluctuations in the multicomponent medium. We have already discussed this factor in Section 4.5 in the

Newtonian approximation. Since we need a relativistic generalization of the equations of motion to find metric perturbations, it is clear that on the whole the theory of gravitational instability must be based on the equations of general relativity.

We have already mentioned that the problem of the evolution of perturbations in single-component matter characterized by the equation of state $P = C_{S,\rho}^2$, where C_S is the adiabatic velocity of sound, was first formulated and solved by E. M. Lifshits (1946) practically at the same time that work began on the creation of the theory of the 'hot' Universe. Note that the title of Lifshits' paper, 'On gravitational stability...' seemed to indicate that the primordial plasma must be stable relative to small perturbations of density, velocity of motion and gravitational potential. The fact that the conclusion on the gravitational instability of matter in specific cosmological conditions follows from Lifshits' equations was first pointed out by Novikov (1964). For details about the history of this idea, see Peebles (1971). In fact this work offered a new promise that it would be possible to predict a scenario of transition of cosmic matter from a structureless state with a very low fluctuation level to a highly structured mass distribution within regions of $\delta\rho/\rho \gg 1$ identifiable with galaxies and their clusters. The sequence of events in such a transition, and the rate at which perturbations grow, depends on the composition of the matter. Let us consider this process in more detail using the Newtonian gauge for metric perturbations. For simplicity we restrict the analysis to a flat space Universe, for which the metric tensor is given by

$$g_{ik} = g_{ik}^{(0)} + h_{ik}, \tag{4.107}$$

(Zeldovich and Novikov, 1975), where $g_{00}^{(0)} = -a^2$; $g_{0\alpha}^{(0)} = 0$; $g_{\alpha\beta}^{(0)} = a^2\gamma_{\alpha\beta}$; $\gamma_{\alpha\beta}$ is the Minkowski space metric,

$$h_{00} = -2a^2\Psi_{\vec{k}}Q; \qquad h_{0\alpha} = 0; \qquad h_{\alpha\beta} = 2a^2\Phi_{\vec{k}}Q\gamma_{\alpha\beta}.$$

Note that $\Psi_{\vec{k}}^{(\xi)}$ and $\Phi_{\vec{k}}^{(\xi)}$ are metric perturbations in the Newtonian reference frame, Q denotes scalar eigenfunctions of the Laplace operator (see Section 4.2), the index α runs from 1 to 3 and $\xi = \int dt/a$ is the conformal time.

Following Peebles and Yu (1970) we describe perturbations of the CMB using the kinetic approach (see Section 4.1); we use hydrodynamic equations for perturbations in the baryonic fractions of matter, hidden mass and in other possible components.

When deriving equations describing the evolution of perturbations in a multicomponent medium, we use a relativistic analogue of the set of equations (4.82)–(4.85) in which the equations of motion for hydrodynamic components follow from the conservation condition,

$$T_{i;k}^k = 0, \tag{4.108}$$

and equations describing perturbations for the Einstein tensor play the role of the Poisson equation,

$$G_i^k = R_i^k - \frac{1}{2}\delta_i^k R, \tag{4.109}$$

where T_i^k is the energy tensor of matter, R_i^k is the Ricci tensor, $R = R_i^i, i = 0, \ldots, 3$ and δ_i^k are the Kronecker deltas. The components of the Einstein tensor's perturbations in the chosen reference frame, Eq. (4.107), are given in terms of the potentials $\Phi_{\tilde{k}}$ and $\Psi_{\tilde{k}}$ (Hu, 1995):

$$\delta G_0^0 = \frac{2}{a^2}\left[3\left(\frac{a'}{a}\right)^2\Psi_{\tilde{k}} - 3\frac{a'}{a}\Phi_{\tilde{k}} - (k^2 - 3E)\Phi_{\tilde{k}}\right]Q,$$

$$\delta G_\alpha^0 = \frac{2}{a^2}\left[\frac{a'}{a}k\Psi_{\tilde{k}} - k\Phi_{\tilde{k}}'\right]Q_\alpha,$$

$$\delta G_\alpha^\beta = \frac{2}{a^2}\left\{\left[2\frac{a''}{a} - \left(\frac{a'}{a}\right)^2\right]\Psi_{\tilde{k}} + \left(\frac{a'}{a}\right)\left[\Psi_{\tilde{k}}' - \Phi_{\tilde{k}}'\right] - \frac{k^2}{3}\Psi_{\tilde{k}}\right.$$
$$\left.- \Phi_{\tilde{k}}' - \frac{a'}{a}\Phi_{\tilde{k}} - \frac{1}{3}(k^2 - 3E)\Phi_{\tilde{k}}\right\}\delta_\alpha^\beta Q - \frac{1}{a^2}k^2(\Psi_{\tilde{k}} + \Phi_{\tilde{k}})Q_\alpha^\beta. \tag{4.110}$$

The following notation was introduced in the set of equations (4.110): Q denotes the eigenfunctions of the Laplace operator in the metric, Eq. (4.107) (see also Section 4.2), and

$$Q_\alpha \equiv -\frac{1}{k}Q_{;\alpha}, \qquad Q_{\alpha\beta} \equiv k^{-2}Q_{;\alpha;\beta} + \frac{1}{3}\gamma_{\alpha\beta}Q.$$

The following relativistic analogue of the continuity equation for each component is obtained from Eq. (4.108):

$$\delta_j' = 3\frac{a'}{a}(\delta\omega_j) = -(1 + \omega_j)(kV_{j,\tilde{k}} + 3\Phi_{\tilde{k}}'). \tag{4.111}$$

The index j in Eq. (4.111) enumerates the components, and

$$\omega_j = \frac{P_j}{\rho_j}; \qquad (\delta\omega_j)_{\tilde{k}} = \left(\frac{\delta P_j}{\delta\rho_j} - \omega_j\right)\delta_{j(\tilde{k})}. \tag{4.112}$$

Substituting Eqs (4.112) into Eq. (4.111) and introducing the notation $\Gamma_j = ((\delta P_j/\delta\rho_j - C_{S(j)}^2)\delta_{j(\tilde{k})}$, we obtain from Eq. (4.111)

$$\frac{d}{d\xi}\left[\frac{\delta_{j(\tilde{k})}}{1 + \omega_j}\right] = -(kV_{j(\tilde{k})} + 3\Phi_{(\tilde{k})}') - 3\frac{a'}{a}\omega_j\Gamma_j. \tag{4.113}$$

As we see from the definition, Γ_j characterizes the isopotential mode of perturbations for which $\Gamma_j \neq 0$. In the adiabatic mode, $\delta P_j/\delta\rho_j = C_{S(j)}^2$ and $\Gamma_j = 0$.

We turn now to analysing perturbations of hydrodynamic velocity for each component of matter. To do this, we take into account Eq. (4.108) and define a perturbation for the vector

$$\delta\left(\tilde{T}_{i;k}^k - u_i u^l \tilde{T}_{l;p}^p\right) = 0, \tag{4.114}$$

where $\tilde{T}_i^k$ is the energy–momentum tensor of each component and u_i is the four-dimensional velocity. Applying standard techniques of perturbations theory in linear approximation we find (Hu, 1995)

$$v_{j(\tilde{k})}' + \frac{a'}{a}(1 - 3\omega_j)v_{j(\tilde{k})} + \frac{\omega_j'}{1 + \omega_j}v_{j(\tilde{k})}$$
$$= \frac{(\delta P_j/\delta\rho_j)}{1 + \omega_j}k\delta_{j(\tilde{k})} - \frac{2}{3}\frac{\omega_j}{1 + \omega_j}\left(1 - \frac{E}{k^2}\right)k\Pi_{j(\tilde{k})} + k\Psi_{j(\tilde{k})}, \tag{4.115}$$

where $\Pi_{j(\vec{k})}$ is the anisotropic part of the energy–momentum tensor perturbations,

$$\delta T_\alpha^\beta = P_j \left(\frac{\delta P_j}{P_j} \delta_\alpha^\beta Q + \Pi_{j(\vec{k})} Q_\alpha^\beta \right), \tag{4.116}$$

and all other notation remains the same.

In a multicomponent medium, the relation between metric perturbations and perturbations of the energy–momentum tensor is realized through the perturbed Einstein equations

$$\delta G_i^k = \frac{8\pi G}{C^4} \delta T_i^k, \tag{4.117}$$

where the tensor components δT_i^k are obtained by summing over all types of matter using the generalized density, pressure and momentum of matter:

$$\begin{aligned}
\rho_{tot} \delta_{tot(\vec{k})} &= \sum_i \rho_i(\xi) \delta_{i(\vec{k})}; \\
\delta P_{tot(\vec{k})} &= \sum_i \delta P_{i(\vec{k})}; \\
(\rho_{tot} + P_{tot}) v_{tot(\vec{k})} &= \sum_i [\rho_i(\xi) + P_i(\xi)] v_{i(\vec{k})}; \\
P_{tot} \Pi_{tot(\vec{k})} &= \sum_i P_i(\xi) \Pi_{i(\vec{k})}; \\
\rho_{tot}' C_{tot}^2 &= \sum_i \rho_i''(\xi) C_i^2(\xi),
\end{aligned} \tag{4.118}$$

where the variables $\{x_i(\xi)\}$ stand for non-perturbed quantities. The set of equations (4.110), (4.111), (4.115) and (4.118) thus exhausts the mathematical aspect of the relativistic formulation of the problem of perturbations evolution in the expanding Universe in its most general form.

The general formulation of the problem implies not only an analysis of complex systems that involve several species of matter, including that of collisionless particles (for example neutrinos), but also yields a number of important conclusions on the dynamics of the evolution of perturbations in simple systems. Thus the optical depth of the plasma relative to the Thomson scattering was very high ($\tau \gg 10^2$) long before the hydrogen recombination epoch, that is at the radiation-dominated phase of the evolution of matter.

Ignoring the role of collisionless neutrinos and the massive gravitational component implied by the non-relativistic dark mass, we can treat radiation and baryons as an ideal liquid characterized by its adiabatic speed of sound, $C_S^2 = \frac{1}{3}$. We assume for simplicity that in this kind of medium there are no isopotential modes and that perturbations in the radiation and baryons mixture are adiabatic. If $\rho_\gamma \gg \rho_b$, the presence of 'strong binding' between electrons, protons and radiation results in the suppression of the anisotropic part of the energy–momentum tensor of this compound liquid and automatically generates a relation between the potentials, $\Psi = -\Phi$, in the Newtonian reference frame. Then we immediately obtain from the component δ_0^0 of Eq. (4.117) that

$$k^2 \Phi_{\vec{k}} = \frac{16}{3} \pi G a^2 \rho_T \Delta_{T(\vec{k})}, \tag{4.119}$$

where $\Delta_{T(\vec{k})} = \frac{3}{4}(\delta_{\vec{k}} + 4\frac{a'}{a}v_{\vec{k}})$ and the relation between Δ_T and $v_{\vec{k}}$ is given by the following equations (see Eqs (4.111), (4.115) and (4.117)):

$$\Delta'_{T(\vec{k})} - \frac{a'}{a}\Delta_{T(\vec{k})} = -kv_{\vec{k}},$$

$$v'_{\vec{k}} + \frac{a'}{a}v_{\vec{k}} = \left(C_S^2 k - \frac{16}{3}\pi G\rho_T a^2\right)\Delta_{T(\vec{k})}. \tag{4.120}$$

Non-perturbed Einstein equations yield the following relation for the early phases of expansion of the Universe where matter obeys the equation of state $P = \varepsilon/3$ ($C_S^2 = \frac{1}{3}$):

$$\left(\frac{a'}{a}\right)^2 = \frac{8}{3}\pi G\rho a^2, \qquad a'' = 0. \tag{4.121}$$

Then Eqs (4.119–4.121) yield a simple equation for $\Delta_{T(\vec{k})}$,

$$\Delta''_{T(\vec{k})} + \left(\frac{k^2}{3} - \frac{2}{\xi^2}\right)\Delta_{T(\vec{k})} = 0, \tag{4.122}$$

whose solution is given in the form

$$\Delta_{T\vec{k}} = \left[C_1 J_{3/2}\left(\frac{k\xi}{\sqrt{3}}\right) + C_2 Y_{3/2}\left(\frac{k\xi}{\sqrt{3}}\right)\right]\left(\frac{k\xi}{\sqrt{3}}\right)^{1/2}, \tag{4.123}$$

where $J_\nu(x)$ and $Y_\nu(x)$ are Bessel functions of the first and second kind, respectively, and C_1 and C_2 are integration constants.

Let us consider asymptotes of this solution in the long- and short-wavelength limits. In the low-wavelength approximation, $k\xi/\sqrt{3} \ll 1$, the growing mode of perturbations is given by the expression

$$\Delta_{T(\vec{k})} \simeq C_1 \left(\frac{k\xi}{2\sqrt{3}}\right)^{3/2}\left(\frac{k\xi}{\sqrt{3}}\right)^{1/2} \propto \left(\frac{k\xi}{\sqrt{3}}\right)^2. \tag{4.124}$$

For short-wavelength waves, $k\xi/\sqrt{3} \gg 1$, the corresponding asymptotes of the Bessel functions yield the following result:

$$\Delta_{T(\vec{k})} \propto C_1 \cos\frac{k\xi}{\sqrt{3}} + C_2 \sin\frac{k\xi}{\sqrt{3}}. \tag{4.125}$$

We need to stress that the oscillating solution of the form in Eq. (4.125) was investigated in a number of papers (see, for example, Bisnovatiy-Kogan, Lukash and Novikov (1983)) and that it undoubtedly played a major cosmological role for the physics of the cosmic microwave background. In the next chapter we deal in more detail with improving the solution that was used to describe the medium composed of the cosmological hydrogen–helium plasma and radiation at a time very close to the moment of recombination. In what follows here we consider an absolutely different approximation, when the speed of sound in the system vanishes, $C_S = 0$; that is, perturbations of density, velocity and gravitational potential evolve in a 'dusty' medium. This approximation arises inevitably in the 'baryonic' model, which is the basis for creating more complicated cosmological models, including various types of primordial massive and massless particles. After hydrogen recombination, perturbations evolve independently in baryons and in radiation, and since $\rho_b \gg \rho_\gamma$ (for $\Omega_b h^2 > 0.1$), the main dynamic component is the non-relativistic gas of baryons. Indeed, Eqs (4.111), (4.115) and (4.117) imply that the evolution of perturbations in dust is described by equations of the type

$$\Delta'_{T(\vec{k})} = -kV_{\vec{k}}; \qquad V'_{\vec{k}} + \frac{a'}{a}V_{\vec{k}} = k\Phi; \qquad k^2\Phi_{\vec{k}} = 4\pi G\rho_b\Delta_{T(\vec{k})}a^2, \quad (4.126)$$

whence an equation for $\Delta'_{T(\vec{k})} = \delta b + 3(a'/a)V_{\vec{k}}$:

$$\Delta''_{T(\vec{k})} + \frac{a'}{a}\Delta'_{T(\vec{k})} - 4\pi G\rho_b a^2\Delta_{T(\vec{k})} = 0. \qquad (4.127)$$

In view of the $\begin{pmatrix} 0 \\ 0 \end{pmatrix}$ component of non-perturbed Einstein equations,

$$3\left(\frac{a'}{a}\right)^2 = 8\pi G\rho_b a^2, \qquad (4.128)$$

and $a \propto \xi^2$, we obtain for the growing and decay modes of density perturbations in baryons

$$\Delta'_{T(\vec{k})} = -kV_{\vec{k}}; \qquad V'_{\vec{k}} + \frac{a'}{a}V_{\vec{k}} = k\Phi, \qquad (4.129)$$

where C_1^+ and C_2^+ are integration constants. Therefore, the rate of growth in perturbations in the baryonic (dusty) matter after the hydrogen recombination epoch differs in principle from the Newtonian exponential mode discussed in Section 4.5. Nevertheless, the power-law growth of the amplitude of density fluctuations is found to be of principal importance for understanding the possible mechanism leading to formation of structures in the expanding Universe (Novikov, 1964).

We need to emphasize that from the standpoint of modern cosmology based on the concept of baryonic hidden mass, the formulation of the problem of gravitational instability in the baryonic medium appears to be somewhat academic. We know that density perturbations in dark matter evolve independently of whether hydrogen recombination has occurred or not. At the same time, we are aiming at a very specific goal in reviving the classical baryonic model.

An analysis of this model will allow us to describe in the simplest possible manner the creation of systematic oscillations in the spectrum of density perturbations emerging after hydrogen recombination. The mechanism of generation of these oscillations was discovered by A. D. Sakharov (1965). For a description, see Zeldovich and Novikov (1983), in which they were given the name 'Sakharov oscillations'. Sakharov's discovery proved to be of immense importance for cosmology. This phenomenon determines the nature of modulations of the angular spectrum of anisotropy and the polarization of primordial radiation background that we are to describe in the subsequent sections. These modulations are also known as the 'Sakharov modulations' (Jørgensen *et al.*, 1995; Naselsky and Novikov, 1993). They are related to the existence of acoustic modes in small perturbations of the cosmological medium prior to the recombination epoch. Obviously, acoustic modes of perturbations exist not only in the baryonic model, but also in all currently considered cosmological models.

4.7   Sakharov modulations of the spectrum of density perturbations in the baryonic Universe

'We can choose a hypothesis that the matter at the early stages of the expansion of the Universe was very nearly uniform and that the "primordial" astronomical objects appeared as a result of gravitational instability. Even though this point of view meets with objections from a number of astronomers and astrophysicists, an investigation of this aspect appears to be necessary. For investigation of this hypothesis, it is very important to study the laws governing the growth of small density inhomogeneities and to find statistical characteristics of the initial inhomogeneities.'

We began this section with a quotation from the introduction to A. D. Sakharov's paper 'The initial stage in the expansion of the Universe and generation of inhomogeneities in the distribution of matter' (Sakharov, 1965). Incidentally, this publication is known to experts in quantum gravitation because Andrei Dmitrievich attempted there to justify a relation between classical pregalactic perturbations of matter and quantum noise; he used the framework of the 'cold' model of the Universe, widely discussed at the time, and low initial temperatures of the plasma and low specific entropy of radiation. Dmitrievich recalled his impressions of this paper: 'Using the "cold" model substantially undervalued my first work in cosmology. The results relevant to the theory of gravitational instability are not devoid of certain interest, among them (and especially) quantum instability, as well as the hypothesis of the equation of state at superhigh densities. I treated the quantum case of instability using the exact self-similar solution for the wave function of a harmonic oscillator with variable parameters: a point of great difficulty there was to take into account the effects of pressure, but I was able to overcome them (I refer the reader to this paper for details of how I achieved this; I still remember on what day I was able to find the solution – on 22 April 1964).' (see Sakharov (1999).)

Three important moments in Sakharov's paper and comments to it should be highlighted.

(1)   Sakharov considered for the first time the evolution of perturbations in a medium with a complicated equation of state, $\varepsilon = A(n) \cdot n^{4/3}$, that also includes possible zones of negative pressure.

(2)   For this medium he developed a method of analysing perturbations both for the classical and the quantum approaches.

(3)   He was the first to demonstrate that as the equation of state is changed from a more 'rigid' to a 'dust'-type one ($P \ll \varepsilon$), the perturbation amplitudes at the final phase 'inherit' the modulations due to the effect of a pressure gradient at the preceding 'rigid' phase (see formula (38a) of Sakharov (1965)).[3]

We need to point out that Sakharov (1965) has one important specific feature: it uses calculation techniques that sharply differ from the conventional ones. Moreover, the transition of the plasma from the ionized to the neutral state in the course of hydrogen recombination is, for formal reasons, included in the model of the equation of state; however, nowhere does Sakharov speak specifically about this type of 'phase transition'. It appears that the first to give a clear and detailed interpretation of the Sakharov effect of perturbation amplitude

---

[3] In today's notation the function used by Sakharov (1965), namely $F(N) = \Delta N/N$, where ΔN is the mean square deviation of the number of particles N in the selected volume V, is simply the density contrast $\delta\rho/\rho = v\Delta N/mN$, where m is the particle mass.

modulations in the course of hydrogen recombination were Ya. B. Zeldovich and I. D. Novikov (1983), which was well known in Russia following the publication of the first Russian edition in 1975. Later, in 1981, Peebles conducted a detailed investigation of the transition effects in the baryon–photon gas in the hydrogen recombination epoch and was able to show how the Sakharov oscillations can manifest themselves in the correlation function of galaxies (Peebles, 1981). Later on we follow the general ideology of describing the Sakharov oscillations in the spectrum of adiabatic perturbations in the 'baryonic' Universe as suggested in Zeldovich and Novikov (1983) (for an historical aspect, see Novikov (2001)).

First of all, we take into account that the oscillating mode of density and velocity fluctuations in the photon–baryon 'liquid' is typical of high-frequency perturbation modes whose size is much smaller than the cosmological horizon $\sim ct$ in the pre-hydrogen recombination epoch. Obviously, the effects of gravitational interaction between particles are not significant in this case, and perturbations evolve in the acoustic-wave regime:[4]

$$\delta'_{b(\vec{k})} \simeq C_1 \cos \frac{k\xi}{\sqrt{3}} + C_2 \sin \frac{k\xi}{\sqrt{3}}. \tag{4.130}$$

The continuity equation for the mixture of baryons and photons immediately implies that the velocity field is related to density perturbations by a simple expression, as follows:

$$v_{b(\vec{k})} = +ik^{-1}\delta'_{b(\vec{k})}, \tag{4.131}$$

and is also oscillating in time:

$$v_k(\xi) = \frac{i}{\sqrt{3}}\left[C_2 \cos \frac{k\xi}{\sqrt{3}} - C_1 \sin \frac{k\xi}{\sqrt{3}} \right]. \tag{4.132}$$

Hydrogen undergoes recombination at the moment $\xi = \xi_{rec}$ after which perturbations in baryons and radiation evolve independently. We need to qualify this statement by saying that perturbations in the dust component begin to grow after the moment of recombination. The rate of growth, being a function of conformal time ξ, is a function of the ratio of baryon and radiation densities. In deriving Eq. (4.132) we assume that $\rho_\gamma \gg \rho_m$ and the speed of sound in the mixture is practically identical to $C_\xi^2 = 1/3$. If this ratio of speeds is maintained until the moment $\xi = \xi_{rec}$, then immediately after recombination the rate of growth of perturbations among baryons will be somewhat lower than given by Eq. (4.129), owing to the effect of radiation density. In the following we are interested in the qualitative side of the effect and therefore neglect the influence of radiation on the behaviour of $\delta_b(\xi > \xi_{rec})$, referring the reader to Peebles (1981), where these effects are described in detail. We thus simulate the behaviour of perturbations in the baryonic fractions of matter immediately after the moment of recombination using a solution of the type in Eq. (4.129). The corresponding expression for the velocity field is given in the continuity equation (4.131), whence we obtain

$$v_k(\xi)^+ = \frac{i}{k}\left[2C_1^+\xi - \frac{3C_2^+}{\xi^4} \right]. \tag{4.133}$$

Equation (4.129) continues to hold for $\delta_b^+(\xi)$. The '+' index signifies that the corresponding quantities refer to the moment $\xi \geq \xi_{rec}$. Our task is to find the coefficient C_1^+ that characterizes

---

[4] Note that owing to viscosity the primary acoustic waves damp out over short scales. This is the so-called Silk damping (Silk, 1968). This effect is of no major significance for the present discussion.

the amplitude of the growing mode of perturbations δ_b^+ as a function of the coefficients C_1 and C_2 assigned to the acoustic phase of perturbation evolution. The principal idea of the 'matching' of solutions lies in that the field of velocities does not suffer 'jumps' in the baryonic fraction at the moment $\xi = \xi_{rec}$ of hydrogen recombination. This means that

$$\frac{i}{k}\left[2C_1^+\xi_{rec} - \frac{3C_2^+}{\xi_{rec}^4}\right] = \frac{i}{\sqrt{3}}\left[C_2\cos\frac{k\xi_{rec}}{\sqrt{3}} - C_1\sin\frac{k\xi_{rec}}{\sqrt{3}}\right]. \tag{4.134}$$

Likewise, for the field of density perturbations at the moment $\xi = \xi_{rec}$ we have

$$\left(C_1\cos\frac{k\xi_{rec}}{\sqrt{3}} + C_2\sin\frac{k\xi_{rec}}{\sqrt{3}}\right) = C_1^+\xi_{rec}^2 + C_2^+\xi_{rec}^{-3}. \tag{4.135}$$

For perturbations with $k\xi_{rec}/\sqrt{3} \gg 1$ we find from Eqs (4.134) and (4.135)

$$C_1^+ \simeq \frac{k\xi_{rec}}{5\sqrt{3}}\left(C_2\cos\frac{k\xi_{rec}}{\sqrt{3}} - C_1\sin\frac{k\xi_{rec}}{\sqrt{3}}\right)\left(\frac{\xi}{\xi_{rec}}\right)^2, \tag{4.136}$$

and therefore, the growing perturbations mode at $\xi > \xi_{rec}$ has the form

$$\delta_b^+(\xi) \simeq \frac{k\xi_{rec}}{5\sqrt{3}}\left(C_2\cos\frac{k\xi_{rec}}{\sqrt{3}} - C_1\sin\frac{k\xi_{rec}}{\sqrt{3}}\right)\left(\frac{\xi}{\xi_{rec}}\right)^2. \tag{4.137}$$

Two very important conclusions follow from Eq. (4.137). First, the amplitude of perturbations increases with decreasing wavelength (or, similarly, with increasing k as $k\xi_{rec}/\sqrt{3} \gg 1$. This 'jumping' effect was first predicted by Sakharov (1965). Secondly, oscillations due to the acoustic modulation of perturbations amplitude at the transition moment at $\xi = \xi_{rec}$ are formed both in the amplitude C^+ and in the spectrum of perturbations:

$$\langle|\delta_{b(\vec{k})}^{(\xi)}|^2\rangle \simeq \frac{k^2\xi_{rec}^2}{75}\left(\frac{\xi}{\xi_{rec}}\right)^4\left\{\langle|C_{2(\vec{k})}|^2\rangle\cos^2\frac{k\xi_{rec}}{\sqrt{3}}\right.$$

$$\left. + \langle|C_{1(\vec{k})}|^2\rangle\sin^2\frac{k\xi_{rec}}{\sqrt{3}} - \langle|C_{1(\vec{k})}C_{2(\vec{k})}|\rangle\sin\frac{2k\xi_{rec}}{\sqrt{3}}\right\}. \tag{4.138}$$

In fact, this effect was also predicted by (1965). We need to emphasize that in the most general case the modulational part in the spectrum (4.139) contains harmonic functions with the phase doubled, $\varphi = 2k\xi_{rec}/\sqrt{3}$,

$$V(k) = \frac{1}{A(k)} + \frac{B^{(k)}}{A(k)}\cos\frac{2k\xi_{rec}}{\sqrt{3}} + \frac{C^{(k)}}{A(k)}\sin\frac{2k\xi_{rec}}{\sqrt{3}}, \tag{4.139}$$

where

$$A^{(k)} = \frac{1}{2}\left[\langle C_{1(\vec{k})}^2\rangle + \langle C_{2(\vec{k})}^2\rangle\right]; \quad B^{(k)} = \frac{1}{2}\left[\langle C_{2(\vec{k})}^2\rangle - \langle C_{1(\vec{k})}^2\rangle\right]; \quad C^{(k)} = -\langle C_{1(\vec{k})}^2 C_{2(\vec{k})}^2\rangle.$$

As we mentioned above, a detailed analysis of the evolution of adiabatic perturbations in the epoch of recombination of cosmological hydrogen was given in Peebles (1981), in which Peebles approximated the results of numerical calculations using the modulation function

$$M_p(k) = \cos^2(\Phi + Wk + vk^2), \tag{4.140}$$

where Φ, W and v are numerical constants that are functions of the parameter $\Omega_b h^2$. Thus Peebles (1981) gives the following values for these constants in a model with $\Omega_b h^2 = 0.03$: $\Phi = 5.75$ rad, $W = 182.9$ Mpc^{-1} and $v = -18.3$ Mpc^{-2}. In order to make use of the results of this paper, we specify the relation between the modulation functions $M(k)$ and $M_p(k)$. We denote $\xi_{\text{rec}}/\sqrt{3}$ in Eq. (4.139) by r_{rec}, which is the acoustic horizon of the recombination epoch recalculated to the current moment $z = 0$. A comparison of Eqs (4.139) and (4.140) shows that $r_{\text{rec}} = W$. Now we drop the quadratic term in the modulus of the wave vector k in the phase (4.140), which results in a corresponding constraint on non-uniformity scales:

$$1 \ll k r_{\text{rec}} \ll \left| \frac{W^2}{v} \right| \simeq 1.8 \times 10^3.$$

Now a comparison of Eqs (4.139) and (4.140) yields the following normalization of these functions:

$$g(k) = \frac{B(k)}{A(k)} = \cos 2\Phi = 0.98 \simeq 1,$$

$$f(k) = f = \frac{C(k)}{A(k)} = -\sin 2\Phi \simeq -0.2. \tag{4.141}$$

Our question now is: how would the Sakharov oscillations manifest themselves, and, most importantly, for what characteristics of baryonic matter distribution? At first glance, the formulation of the question looks strange. We have already emphasized an academic nature of this baryonic model which ignores the cosmological hidden mass, and therefore cannot even pretend to provide predictions that could be seriously considered for comparing with modern observational data. However, the specificity of manifestations of the Sakharov modulations could offer an additional argument to support the putative existence of the cosmological hidden mass, and even suggest that it is indeed of non-baryonic nature. Therefore, choosing *reductio ad absurdum* we assume that there is no non-baryonic hidden mass in the Universe. Then, as we saw a little earlier, the spectrum of density perturbations in matter immediately after recombination must contain Sakharov modulations of the type in Eq. (4.140). The temporal evolution of perturbations at the $\xi \gg \xi_{\text{rec}}$ phase is readily identifiable: $\delta_b(\xi) \propto \xi^2$ for the growing mode of perturbations in the flat-space model Universe. Following Peebles (1981) we introduce a two-point correlation function

$$\xi(r, \xi) = \langle \rho(\vec{r} + \vec{s})\rho(\vec{s}) \rangle / \langle \rho \rangle^2 - 1 \tag{4.142}$$

for density perturbations and rewrite its spatial part in the form

$$\xi(r) = \int k^2 \langle |\delta_k^2| \rangle \frac{\sin kr}{kr} \, dk, \qquad r = |\vec{r}|, \tag{4.143}$$

where $\langle |\delta_k|^2 \rangle$ is the spatial power spectrum of fluctuations. Equations (4.139) and (4.140) imply that the spectrum $\langle |\delta_k|^2 \rangle$ can be written in the form of a product of a monotonic and an oscillating factor,

$$\langle |\delta_k|^2 \rangle \simeq C(k)[1 + g(k)\cos 2kr_{\text{rec}} + f(k)\sin 2kr_k], \tag{4.144}$$

where $C(k)$ corresponds to the non-modulated part of the spectrum. Substituting Eq. (4.144) into Eq. (4.143) we obtain

$$\xi(r) = \xi_0(r) + \xi_1(r) + \xi_2(r), \tag{4.145}$$

where

$$\xi_0(r) = \int k^2 C(k) \frac{\sin kr}{kr} \, dk,$$

$$\xi_1(r) = \int k^2 C(k) \cdot g(k) \frac{\sin kr}{kr} \cos 2k r_{\text{rec}} \, dk,$$

$$\xi_2(r) = \int k^2 C(k) \cdot f(k) \frac{\sin kr}{kr} \sin 2k r_{\text{rec}} \, dk.$$

We make use of the fact that $g(k) \simeq 1$; then, after elementary trigonometric manipulations, we rewrite $\xi_1(r)$ in terms of the function $\xi_0(r)$:

$$\xi(r) = \frac{1}{3} \left[\frac{2r_{\text{rec}} - r}{r} \xi_0(2r_{\text{rec}} - r) - \frac{2r_{\text{rec}} + r}{r} \xi_0(2r_{\text{rec}} + r) \right]. \tag{4.146}$$

We see from this formula that the behaviour of the function $\xi_1(r)$ at $r \to 2r_{\text{rec}}$ reveals a number of specific features. First, as $r \to 2r_{\text{rec}}$, we can drop the second term on the right-hand side of the Eq. (4.146) because $\xi_0(4r_{\text{rec}}) \to 0$ at $r_{\text{rec}} \gg r_c$, where r_c is the correlation radius for the function $\xi_0(r)$:

$$r_c^2 = -\frac{\xi_0(0)}{\xi''(0)}. \tag{4.147}$$

The correlation function $\xi_0(r)$ can be expanded in the vicinity of $r \to 2r_{\text{rec}}$ into a Taylor series in the small parameter $r - 2r_{\text{rec}} \ll |r_c|$. We now introduce a function $y = (r - 2r_{\text{rec}})/r_c$. The behaviour $\xi_0(r \to 2r_{\text{rec}})$ in the vicinity $y \ll 1$ in terms of y is then given by the following expression:

$$\xi_1(y) = \frac{1}{6} \frac{r_c}{r_{\text{rec}}} \sigma^2 \cdot y \left(1 - \frac{y^2}{2} + \beta \frac{y^4}{4!} \right), \tag{4.148}$$

where $\sigma^2 \equiv \xi_0(0)$, $\beta = \xi(0) \xi_0^{\text{IV}}(0)/|\xi''_0(0)|^2$, $\xi_{(0)}^{\text{IV}}$ is the fourth derivative of $\xi_0(y)$ with respect to y, taken at a point $y = 0$. We turn now to analysing the behaviour of the function $\xi_2(r)$ in the vicinity of the point $r = 2r_{\text{rec}}$,

$$\xi_2(r \to 2r_{\text{rec}}) \simeq \frac{f}{2r_{\text{rec}}} \int k G(k) \left[1 - \frac{k^2 r_c^2}{2} y^2 + \frac{k^4 r_c^4}{24} y^4 \right] dk. \tag{4.149}$$

An important feature of the function $\xi_2(r)$ is that its behaviour cannot be expressed in terms of the moments ξ_0, ξ''_0 and $\xi_{(0)}^{\text{IV}}$. Thus the value $\xi_2(y = 0)$ is expressed via the integral of the regular part of the spectrum,

$$\xi_0(y = 0) = \frac{f}{2r_{\text{rec}}} \int k C(k) \, dk, \tag{4.150}$$

that differs from $\xi_0(0)$ by a factor k^{-1} in the integrand:

$$\xi_0(k) \simeq \int k^2 C(k) \, dk. \tag{4.151}$$

We define now the mean value of the wave vector k^n, weighed over the spectrum $C(k)$ for positive and negative values of n, as

$$\langle k^n \rangle = \frac{\int k^2 k^n C(k) \, dk}{\int k^2 C(k) \, dk}. \tag{4.152}$$

We assume that the monotonic part of the spectrum has the form

$$G(k) = Ak^m T_b(k), \qquad (4.153)$$

where k^m corresponds to the primary perturbation spectrum (for example $P(k)Ak$, the Harrison–Zeldovich spectrum) and $T_b(k)$ is the transfer function taking into account, in addition to the modulation part, the difference in behaviour of perturbations in the $k\xi \gg 1$ and $k\xi \ll 1$ modes. For a qualitative description of the effect, we simulate $T_b(k)$ by an exponential function of the form

$$T_b(k) = e^{-k^2 r_\alpha^2}, \qquad (4.154)$$

where r_α is the characteristic scale of the high-frequency cut-off of the spectrum. Then Eq. (4.152) yields a simple estimate for $\langle k^n \rangle$ in a class of spectra of type (4.153):

$$\langle k^n \rangle \simeq r_\alpha^{-n} \frac{\Gamma\left(\frac{n+m+1}{2} + 1\right)}{\Gamma\left(\frac{m+1}{2} + 1\right)}, \qquad (4.155)$$

where $\Gamma(x)$ is the gamma function. In terms of $\langle k^n \rangle$, the expressions for the correlation scale r_c^2, the parameters β and $\xi_2(y \to 0)$ are as follows:

$$r_c^2 = \frac{3}{2}\langle k^2 \rangle^{-1}; \quad \beta = \frac{g}{20}\frac{\langle k^4 \rangle}{\langle k^2 \rangle^2};$$

$$\xi_2(y) = \frac{f}{2r_{\text{rec}}}\langle k^{-1}\rangle \sigma^2 \left[1 - \frac{1}{2}\frac{r_c^2 \langle k \rangle}{\langle k^{-1}\rangle}y^2 + \frac{r_c^4}{24}y^4\frac{\langle k^3 \rangle}{\langle k^{-1}\rangle}\right]. \qquad (4.156)$$

Since $\xi_0(r = 2r_{\text{rec}})$ is negligibly small, the behaviour of the function $\xi(r)$ as $r \to 2r_{\text{rec}}$ is determined by a superposition of the functions $\xi_1(r)$ and $\xi_2(r)$. In the most general form, Eqs (4.148) and (4.156) yield that $\xi(r) \simeq \xi_1(y) + \xi_2(y)$ is a fourth-order curve in parameter y. In the most general case, the equation for determining the extrema of the function $\xi(y)$ for the corresponding y_* will be of third order. Therefore, the structure of $\xi(y)$ in the vicinity of these points will either have two minima and one maximum between them, or two maxima and a minimum in the case of three real roots. Peebles' calculations show that the former of these versions is implemented (Peebles, 1981). Therefore, the Sakharov modulations of the spectrum are best pronounced on spatial scales $r \simeq 2r_{\text{rec}}$ and not on small r, where high-frequency modulations of the spectrum are averaged out and greatly smoothed. The characteristic scale of peculiarities for the 'baryonic' model of the Universe is found to be extremely large. When $\Omega_b h^2 \simeq 0.03$ it is very close to 360 Mpc, which is far beyond the technical possibilities of the current surveys of galaxies and clusters. At the same time, detecting such an anomaly in the correlation function would be a powerful argument in favour of the baryonic model of the Universe with primordial adiabatic perturbations.

Note another important specific feature of manifestation of the Sakharov modulations in the correlation function of density perturbations, connected with isopotential models of perturbations. Such baryonic models were treated carefully in Kotok, Naselsky and Novikov (1995), and the conclusions obtained can be regarded as complementary to the so-called PIB model of the baryonic Universe (Peebles, 1983). The main difference between isopotential (isocurvature) perturbations and adiabatic ones is that as the contribution of the baryonic component to the rate of expansion of the Universe increases ($\rho_\gamma/\rho_b \to 1$), the initial perturbations of entropic type generate secondary perturbations of adiabatic type. This aspect was investigated in detail in Chernin (2001). The emergence of a mixture of modes in an epoch close to the moment of hydrogen recombination affects the form of the transfer function $T_b(k)$ that has the following form (Shandarin, Doroshkevich and Zeldovich, 1983):

$$T(k) \simeq \frac{k^4 R^4}{(1+k^2 R^2)^2} \left[1 + \exp\left(-\frac{kr_\alpha}{2}\right) \sin kR \right]^2, \tag{4.157}$$

where $R \equiv r_{\text{rec}}$ and r_α is the characteristic scale of damping of secondary adiabatic perturbations that results from friction of matter against the isotropic radiation background.

Along with modulations with phase $\Phi = 2kR$, the specificity of the transfer function is the emergence of the spectrum modulations $\tilde{\Phi} = kR$. These modulations reflect directly the specifics of generation of the adiabatic mode from the primordial entropic one. This means, however, that in contrast to the 'pure' adiabatic mode the peculiarities in the correlation function, $\xi(r)$, should arise both as $r \to 2R$ and as $r \to R$, which is even more remarkable (Jørgensen *et al.*, 1993). The behaviour of $\xi(r)$ in the vicinity of the point $r \to R$ was studied in detail in Kotok *et al.* (1995). It is worthy of note that the manifestation of the Sakharov oscillations in the correlation function of density perturbations of matter for the initial entropy fluctuations and in the related correlation function of galaxies produced a much stronger effect than it does for the adiabatic mode. The corresponding amplitude of anomalies reaches $\sim 10\%$ for the normalization $\xi_0(r = 0) = 1$. The reason for such a considerable modulation amplitude $\xi(r)$ is crystal clear. In contrast to the adiabatic mode for which the characteristic scale of anomaly $\xi(r)$ is very nearly twice the scale of the acoustic horizon r_{rec} at the recombination moment, in the entropic mode the modulations $\xi(r)$ arise very close to $r = r_{\text{rec}}$. Since the scale $r = r_{\text{rec}}$ is closer to the correlation scale r_c, it is clear that the correlation level, including the effect of the Sakharov modulations, is found to be higher for the entropic mode than for the adiabatic one (Jorgensen *et al.*, 1993).

4.8 Sakharov oscillations: observation of correlations

We have seen that the presence of the Sakharov modulations in the spectrum results in anomalously high correlations in a 'baryonic' Universe in the distribution of density perturbations on the scales $r = r_{\text{rec}}$ and $r = 2r_{\text{rec}}$ for the entropic and adiabatic fluctuations, respectively. However, at some level they may be observed even in the CDM cosmological models, including the dark energy (see Eisenstein *et al.* (2005) for references). The recombination horizon r_{rec} depends on the parameter $\Omega_b h^2$ and varies between 100 and 200 Mpc as recalculated to the moment $z = 0$. It is natural to assume that there must be a linear relation between the correlation function of density perturbations and the observed correlation function of galactic distribution density at least on very large spatial scales.

The reader will recall that we are now trying to find in modern observations an answer to a fundamental question: is it possible to detect Sakharov modulations in the galaxy distribution and thereby gain another independent argument in favour of the existence of the cosmological (non-baryonic) hidden mass and the dark energy? The key role in the implementation of this approach is played by choosing the objects which, in our opinion, delineate the distribution of correlations in the field of matter density fluctuations. Galaxies are most often chosen. Let us turn to the analysis of the observational data for galaxy distribution catalogues that fix the position of these objects in the sky and their stellar magnitudes. Recently, Eisenstein *et al.* (2005) presented the large-scale correlation function measured from a spectroscopic sample of about 47 000 luminous red galaxies from the Sloan Digital Sky Survey (SDSS). This

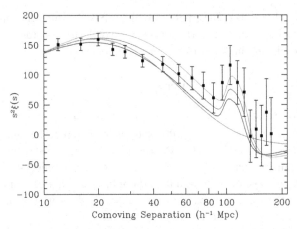

Figure 4.1 The large-scale redshift–space correlation function $s^2\xi(s)$ of the SDSS LRG sample. The lines from top to the bottom correspond to $\Omega_m h^2 = 0.12$, $\Omega_m h^2 = 0.13$ and $\Omega_m h^2 = 0.14$. The pure CDM model ($\Lambda = 0$) is shown as the lowest line. Adapted from Eisenstein *et al.* (2005).

survey covers $0.72 h^{-3} \text{Gpc}^3$ over 3816 square degrees and corresponds to redshifts of galaxies $0.16 < z < 0.47$. The result of the analysis is presented using the Landy–Szalay estimator (Landy and Szalay, 1993) for the redshift–space correlation function $\xi(s)$. In Fig. 4.1 we plot the correlation function $\xi(s)$ from Eisenstein *et al.* (2005) and corresponding theoretical curves for different ΛCDM cosmological models and adiabatic initial perturbations.

As we see from Fig. 4.1, the peculiarity $\xi(s)$ related to the Sakharov modulations of the spectrum of adiabatic perturbations falls within the error zone of determining $\xi(s)$, so that it is extremely difficult to make any conclusions about the presence or absence of this effect. We can only state in the most general case that the theoretical behaviour of ξ does not contradict the data of the SDSS survey. However, it is practically very difficult to exclude or confirm the presence of the Sakharov oscillations in the spectrum of perturbations.

There is hope, however, that progress in the observational tools of astronomers will allow us within several years to move significantly closer to reliable measurements of correlations on spatial scales of $r \sim 100\text{–}200\,\text{Mpc}$ and thereby confirm or reject in an independent way the hypothesis of the baryonic nature of hidden mass; the Sakharov modulations of the spectrum of primordial density perturbations are a part of this hypothesis.

5

Primary anisotropy of the cosmic microwave background

5.1   Introduction

The subject of this chapter is an analysis of one of the most impressive predictions of the modern theory of the structure and evolution of the Universe: the prediction of angular anisotropy in the temperature distribution of the cosmic microwave background on the celestial sphere; we will also analyse observations of this phenomenon. This anisotropy results from the interaction of the primordial radiation background with perturbations of density and velocity of the baryonic matter and with perturbations of the metric; these make an inseparable part of any scenario of structure formation in the expanding Universe. We also use this introductory section to share our personal recollections of the period when the theory of anisotropy of the CMB was being created and the first attempts were being made to detect it by observations. These notes are quite subjective and do not attempt to provide historical analysis. We hope nevertheless that the reader will find them interesting. We wish to point out first of all that while the microwave background itself was discovered by accident, its anisotropy was discovered as a result of well-planned observational searches based on a carefully developed theory.

One of the present authors (I. Novikov) was there at the moment of inception of this field of astrophysics – nowadays a fully-fledged field – and remembers well the history of conception of the theory and the drama of the experimental attempts to detect the anisotropy of primordial radiation at the end of the 1960s and the beginning of the 1970s. We should recall that the cosmology of this period was securely based on the 'baryonic' model of the hot universe (the CMB had already been discovered by Penzias and Wilson) and that the pioneer paper of Sachs and Wolfe had already been published (Sachs and Wolfe, 1967). Furthermore, the theory of cosmological recombination of hydrogen had already been created (Peebles, 1968; Zeldovich, Kurt, Sunyaev, 1969) and the role of this process in the thermal history of cosmic plasma and in the kinetics of electromagnetic radiation was already clear. To make the picture complete, we also mention the pioneering work on the gravitational instability of matter (Lifshits, 1946; Lifshits and Khalatnikov, 1960; Novikov, 1964) which was further extended to the 'pancake' theory of structure formation in the universe (Zeldovich, 1970).

In fact, the theoretical basis 'around' the prediction of the anticipated anisotropy of the microwave background radiation was quite ready, and hot Universe theory was 'pregnant' with this anticipation. The 'birth of the baby' did not disappoint. In 1968, Silk published a paper on the evolution of perturbations in the hydrogen recombination epoch and predicted the effect of remnant fluctuations $\Delta T / T$ due to the adiabatic compression of matter (Silk, 1968). Just two years later, Peebles and Yu (1970) and Sunyaev and Zeldovich (1970a,b) published their papers in which they predicted the characteristics of remnant fluctuations

$\Delta T/T$ generated by macroscopic movements of matter that take place during the period of hydrogen recombination. It immediately became clear that the variance of $\Delta T/T$ must be extremely small ($\simeq 10^{-2}$–10^{-3}), even though it looks enormous from our standpoint today. However, any attempt to measure the anisotropy of the CMB at this low level seemed at best a fantasy. To pay history its due, it must be said that precisely at this period (in the 1970s), unique conditions arose in the Soviet Union (as it was then known) both for theoretical (Ya. B. Zeldovich's group) and experimental (Yu. N. Pariisky's group, RATAN-600) research on the anisotropy of the CMB. It seemed that everything was ready for a successful experiment. The theory has provided excellent predictions, the 600 m radio telescope (still the largest in the world) began operation and the first observations were accumulated, but there was still no signal that would indicate anisotropy of primordial radiation background. It was absent in intensity recordings of the radio sky fluctuations on the RATAN-600, and was just as absent from recordings on the most powerful radio telescopes abroad.

This situation was reminiscent of a game of hide-and-seek. Theoreticians tried to explain why the anisotropy of the CMB was unobservable at the level achieved in observations (it was at that moment that the possibility of reheating and reionization of hydrogen at large redshifts $z > 10$ was first seriously discussed) while experimenters continued to reduce methodically the observational threshold of signal fluctuation detection down to 10^{-3} and, by the beginning of the 1980s, down to 10^{-4} (Berlin *et al.*, 1983).

We believe that a qualitative step in the minds of both theoreticians and experimenters occurred at the beginning of the 1980s when Yu. N. Lyubimov's group at the Institute of Theoretical and Experimental Physics in Moscow reported a putative discovery of a non-zero (at the level of 30 eV!) mass of the electron neutrino (Lyubimov *et al.*, 1980). The bomb went off! A hidden mass – possibly in the form of neutrinos and later in the form of other hypothetical particles – announced its existence. Discussions of the cosmological consequences of this phenomenon and the term 'neutrino' could be heard not only at the famous Colloquium of Vitaly L. Ginzburg (at the P. N. Lebedev Institute of Physics) and the Joint Astrophysics Colloquium at the Shternberg Astronomical Institute, but also on buses or metro trains. Each generation of scientists retains their own reminiscences of the fascination of this era.

While Igor Novikov was living through this period as a mature scientist (see Bisnovatiy-Kogan and Novikov (1980) and the earlier works by Doroshkevich, Lukash and Novikov (1974), Doroshkevish, Novikov and Polnarev (1977) and Novirov (1968)), another author of this book, Pavel Naselsky, had just submitted and defended his Ph.D. thesis, and the third author, Dmitri Novikov, was still in the sixth form of high school and could hardly think that fate would bring him into astrophysics or, even more specifically, the physics of primordial radiation. It is natural therefore that our impressions of the beginning of the 1980s, which heralded the arrival of massive neutrinos, are quite different. One of us (Naselsky) remembers as a curiosity that immediately after Yu. Lubimov's experiment he had, together with N. Zabotin, calculated the anticipated anisotropy of the primordial radiation in a model with massive neutrinos, and was very surprised to find it to be on the same ($\simeq 10^{-4}$) level as in a number of baryonic models (Zabotin and Naselsky, 1982b). It was therefore too large and in contradiction with the experimental data. These results were then discussed – in a sauna – among the then young V. Lukash, D. Kompaneets, P. Naselsky and (also still young) A. Melott who was then visiting Moscow; Melott realised that the sauna was not the best place for jotting notes on paper, so he ordered for the occasion a t-shirt printed with a number of formulas. In a spectacular way, the early and mid 1980s everywhere smelled of an inevitable discovery. It was clear that theoretical models had reached saturation point and that the CMB anisotropy would be discovered in the very near future. Indeed, a special satellite 'Relikt' was launched in the USSR in 1983 in order to detect the large-scale anisotropy ΔT on the angular scale $\theta \gg 1°$ in which the effect of plasma recombination became insignificant and the distribution of intensity fluctuations of this radiation on the sky retained its pristine state, regardless of the possible peculiarity of the ionization history of the Universe (Klypin *et al.*, 1987; Strukov and Skulachev, 1984). Using the metaphor of the times, the 'entire scientific community' participated in discussing the results of this experiment. No anisotropy was detected at the $(2–3) \times 10^{-5}$ level, and the only chance of somehow bringing together the predictions of the theory and the observations was to suggest that the 'hidden mass' was not 'hot' as were, for instance, massive neutrinos, but 'cold'! Or, if one believed in the hot hidden mass, then the spectrum of density perturbations should differ from the Harrison–Zeldovich spectrum and decrease with increasing spatial scale of fluctuations; this would also be a bad result. Matching the observations of the 'Relikt' project with theoretical predictions meant that the fall-off of the spectrum towards the region of $k \to 0$ automatically announced its increase in the short-wavelength range, the closest to the scales of galaxies and galaxy clusters. However, the level of fluctuations ΔT on the angular scale $\theta \propto k^{-1} \sim 5–10'$ must be even higher than for the Harrison–Zeldovich spectrum, and the problem with observational constraints on ΔT would be even more complicated at these angles.

The cold hidden mass (CDM) was indeed a lifeline for the theory, even though its emergence on the podium of CMB physics did not augur well for the future. It seemed that not only details of the large-scale structure distribution in the Universe, but even the fact of its existence could not survive in the framework of CDM models. Metaphorically speaking, the formation sequence in the 'hot' hidden mass model (a 'downwards' structure formation, from massive pancakes to less massive galaxies) in CDM models was reversed: from smaller to larger scales. And it was absolutely unclear how the flat conurations and filaments, so typical of the large-scale distribution of matter in the Universe, could arise in this clustering. Fortunately, this problem was successfully resolved in the early and mid 1980s, even though we are still not completely sure how cold the cosmological hidden mass is (see Dolgov and Sommer-Larsen (2001)).

After this very brief exposition of the history of the discovery of the cosmological role of CDM models, we find ourselves facing one of the most important chapters in the physics of primordial radiation, dealing with predicting the main mechanisms of generating ΔT during the epoch of the cosmological hydrogen recombination. To be fair to the history of our subject, we begin the discussion of the mechanisms of generation of primordial radiation anisotropy with the analysis of the Sachs–Wolfe effect (Sachs and Wolfe, 1967) which played the key role in raising the current status of observational cosmology – indeed, it was the anisotropy involved in this effect that was the first to be detected.

5.2 The Sachs–Wolfe effect

Let us consider the process of anisotropy formation on the scale $\lambda \gg r_{\text{rec}}$ where r_{rec} is the scale of the recombination horizon, neglecting the scattering of quanta by electrons and taking into account only the metric perturbation as a source of ΔT. In this approximation, Eq. (4.10) immediately implies that temperature fluctuations $\theta = \Delta T / T$ are related to metric perturbations $h_{\alpha\beta}$ in a simple manner:

$$\frac{\partial \theta}{\partial t} + \frac{\gamma_\alpha}{a} \frac{\partial \theta}{\partial x^\alpha} + \dot{\gamma}_\alpha \frac{\partial \theta}{\partial \gamma^\alpha} - \frac{1}{2} \gamma^\alpha \gamma^\beta \dot{h}_{\alpha\beta} = 0, \qquad (5.1)$$

where γ_α is the direction of arrival of the quanta.

In the Newton gauge, metric perturbations $h_{\alpha\beta}$ are given in terms of the functions $\Psi(\vec{x}, t)$ and $\Phi(\vec{x}, t)$ (Hu, 1995; Hu and Sugiyama, 1994). Transforming Eq. (5.1) to a variable $\xi = \int dt/a$ and taking into account Φ and Ψ, we obtain an equation for temperature perturbations as follows:

$$\theta' + \gamma^\alpha \frac{\partial(\theta + \Psi)}{\partial x^\alpha} + \gamma'_\alpha \frac{\partial \theta}{\partial \gamma^\alpha} + \Phi = 0. \tag{5.2}$$

For the sake of simplicity, we limit the analysis to a flat-space-model Universe. Despite the simple form, searching for solutions of Eq. (5.2) while taking into account the spatial curvature of the universe $E \neq 0$ remains a fairly complicated problem. In the general case, we choose a system of coordinates in an arbitrary manner, with the origin at the observation point, and we can rewrite the perturbation distribution $\theta(\xi, \vec{x}, \vec{\gamma})$ as a series (Wilson, 1983):

$$\theta(\xi, \vec{x}, \vec{\gamma}) = \sum_{l=0}^{\infty} \theta_l(\xi) M_l^{-1/2} G_l(\vec{x}, \vec{\gamma}), \tag{5.3}$$

where M_l was defined in Section 4.2, and the functions $G_l(\vec{x}, \vec{\gamma})$ are a set of eigenfunctions that satisfy a certain set of conditions. We assume that the CMB temperature and perturbations of the metric originate with the potential modes. A vector Q_α and a tensor Q_{ij} can be constructed from the potential function Q according to the following definitions:

$$Q_\alpha \equiv -\frac{1}{k} Q_{|\alpha},$$
$$Q_{\alpha\beta} \equiv k^{-2} Q_{|\alpha\beta} + \frac{1}{3} \gamma_{\alpha\beta} Q, \tag{5.4}$$

and $Q_{|\alpha}$ stands for the covariant derivative in the spatial metric $\gamma_{\alpha\beta}$. The following scalars can be constructed using the vectors Q_α and γ^α: $G_1 = \gamma^\alpha Q_\alpha$ and $G_2 = \frac{3}{2} \gamma^\alpha \gamma^\beta Q_{\alpha\beta}$. Obviously, the sequence G_l can be readily extended by increasing the order of differentiation of the scalar Q and by projecting the result to γ^α. Wilson (1983) suggested using this method to generalize the expansion of an arbitrary function of coordinates $\vec{x}$ and direction $\vec{\gamma}$ in the metric $\gamma_{\alpha\beta}$,

$$f(\vec{x}, \vec{\gamma}) = \sum_{k} \sum_{l=0}^{\infty} f_l(\vec{k}) g_l(\vec{x}, \vec{\gamma}, \vec{k}), \tag{5.5}$$

where

$$g_l(\vec{x}, \vec{\gamma}, \vec{k}) = (-k)^{-l} Q_{|\alpha_1\alpha_2 \ldots \alpha_l}(\vec{x}, \vec{k}) \times P_l^{\alpha_1\alpha_2\ldots\alpha_l}(\vec{x}, \vec{\gamma}) \tag{5.6}$$

and

$$P_0 = 1; \qquad P_1^\alpha = \gamma^\alpha; \qquad P_2^{\alpha\beta} = \frac{1}{2}(3\gamma^\alpha \gamma^\beta - \gamma^{\alpha\beta}) \ldots \tag{5.7}$$

$$P_{l+1}^{\alpha_1\alpha_2\ldots\alpha_{l+1}} = \frac{2l+1}{l+1} \gamma^{(\alpha_1} P_l^{\alpha_2\ldots\alpha_{l+1})} - \frac{l}{l+1} \gamma^{(\alpha_1\alpha_2} P_{l-1}^{\alpha_3\ldots\alpha_{l+1})}.$$

The round brackets used in Eqs (5.7) indicate a symmetry operation with respect to the corresponding indices. In a flat-space Universe with $E = 0$, Eqs (5.6) and (5.7) transform into a simple combination of a plane wave and Legendre polynomials of index l, as follows:

$$g_l(\vec{x}, \vec{\gamma}, \vec{k}) = (-i)^l e^{i\vec{k}\vec{x}} P_l(\vec{k} \cdot \vec{\gamma}) \tag{5.8}$$

and are a recognizable combination of the Fourier expansion in spatial coordinates $\vec{x}$ and an expansion in Legendre polynomials in the angular coordinate $\vec{k} \cdot \vec{\gamma} = \cos \theta_k$, where θ_k is the angle between the direction of arrival of quanta and the wave vector of perturbations $\vec{k}$. For models with an arbitrary topology of spatial cross-section the transition from the functions $G_l(\vec{x}, \vec{\gamma})$ to the functions $g_l(\vec{x}, \vec{\gamma}, \vec{k})$ is carried out directly from Eq. (5.5):

$$G_l(\vec{x}, \vec{\gamma}) = \sum_{\vec{k}} a_l(\vec{k}) g_l(\vec{x}, \vec{\gamma}, \vec{k}), \tag{5.9}$$

where $a_l(k) = M_l^{-1/2}$. An important feature of the functions $G_l(\vec{x}, \vec{\gamma})$ is the recurrence relation

$$\gamma^\alpha G_{l|_\alpha} = \frac{d}{d\xi} G[\vec{x}(\xi), \vec{\gamma}(\xi)] \equiv \dot{x}^\alpha \frac{\partial G_l}{\partial x^\alpha} + \dot{\gamma}^\alpha \frac{\partial G_l}{\partial x^\alpha}$$

$$= k \left\{ \frac{l}{2l+1} k_l G_{l-1} - \frac{l+1}{2l+1} G_{l+1} \right\}, \tag{5.10}$$

where $k_l = 1 - (l^2 - 1)(E/k^2)$, $l \geq 1$, $k_0 = 1$ and $dG/d\xi$ stands for the complete (Lagrangian) derivative of the function G_l. With these qualifications we can return to Eq. (5.2). In the most general case, ($E > 0$ and $E < 0$), the equations for the components $\theta_l(\xi)$ have the following form (Hu, 1995):

$$\theta_0' = -\frac{k}{3} \theta_1 - \Phi', \tag{5.11a}$$

$$\theta_1' = -k \left[\theta_0 + \Psi - \frac{2}{5} k_2^{1/2} \theta_2 \right], \tag{5.11b}$$

$$\theta_l' = -k \left[\frac{l}{2l-1} k_l^{1/2} \theta_{l-1} - \frac{l+1}{2l+3} k_{l+1}^{1/2} \theta_{l+1} \right]. \tag{5.11c}$$

It is clear from these equations that all k_l in the flat-space model of the Universe ($E = 0$) for $l \geq 0$ are identically equal to unity. In the approximation $k\xi \ll 1$, the main components in Eq. (5.11ca) are θ_0 and Φ, while the dipole term with θ_1 on the right-hand side of Eq. (5.11ca) can be ignored. Consequently,

$$\theta_0' = -\Phi'. \tag{5.12}$$

In the long-wavelength limit we can ignore the effect of pressure on the gravitational potential and assume $\Phi = -\Psi$. Then Eq. (5.12) yields

$$(\theta_0 + \Psi)(\xi) = \theta_0(0) + 2\Psi(\xi) - \Psi(0). \tag{5.13}$$

Equation (5.13) implies that the temperature fluctuations at the moment of the last scattering $\xi = \xi_{rec}$ are given by

$$\Delta = (\theta_0 + \Psi)|_{\xi = \xi_{rec}} = \theta_0(0) + \Psi_0(0) + 2[\Psi(\xi_{rec}) - \Psi(0)]. \tag{5.14}$$

The relation between $\theta_0(0)$ and $\Psi(0)$ for adiabatic perturbations has the following form: $\theta_0(0) = -\frac{1}{2}\Psi(0)$ (Hu, 1995). Taking then into account how Ψ changes from the radiation-dominated epoch to the epoch when 'hidden mass' dominates, we obtain

$$[\theta_0 + \Psi] = \begin{cases} \frac{1}{2}\Psi & \text{for } z > z_{eq} \\ \frac{1}{3}\Psi & \text{for } z < z_{eq}, \end{cases} \tag{5.15}$$

where z_{eq} is the value of the redshift that corresponds to the moment when the densities of the relativistic and non-relativistic subsystems become equal. The reader will recall that the solution for θ_0 was obtained in the long-wavelength approximation with $k\xi \ll 1$ and one could neglect in Eqs (5.11ca–c) all multipoles with $l > 0$. However, for an arbitrary Fourier harmonic with scale $k\xi_{now} \gg 1$, where ξ_{now} is today's horizon, there is always a moment of time ξ_k such that the corresponding wavelength of inhomogeneity first becomes equal to the horizon of particles ($k\xi_k \simeq 1$), and then at $\xi \gg \xi_k$ diminishes below the horizon ($k\xi \gg 1$ for $\xi \gg \xi_k$). The solution of Eqs (5.11ca–c) for such Fourier modes has the form

$$\frac{\theta_l(\xi, k)}{2l + 1} = [\theta_0 + \Psi](\xi_{rec}, k) j_l [k(\xi - \xi_{rec})], \tag{5.16}$$

where $j_l(x)$ are the spherical Bessel functions. By virtue of the relation between θ_0 and $\theta_l(\xi, k)$ and the perturbations of the gravitational potential it is clear that the statistical properties of the distribution of fluctuation temperature on the celestial sphere are dictated by the properties of the potential perturbations distribution and will reflect specific features of its formation.

Assume now that the spatial distribution of potentials $\Phi(\vec{x}, \xi)$ and $\Psi(\vec{x}, \xi)$ is a realization of a random Gaussian process. Owing to the linearity of the transfer equation (5.3), the angular distribution of temperature fluctuations is also a realization of a random Gaussian process with the following correlation function (Bond and Efstathiou, 1987):

$$\langle \theta^*(\xi, \vec{x}, \vec{\gamma})\theta(\xi, \vec{x}, \vec{\gamma})\rangle = \frac{V}{2\pi^2} \int_0^\infty \sum_{l=0}^\infty \frac{k^3}{2l + 1} |\theta_l(\xi, k)|^2 \times P_l(\vec{\gamma}' \cdot \vec{\gamma}) \, dk, \tag{5.17}$$

where $P_l(\cos\theta)$ are Legendre polynomials, $\cos\theta = \vec{\gamma}' \cdot \vec{\gamma}$, and θ is the angle between the directions of arrival of the quanta $\vec{\gamma}'$ and $\vec{\gamma}$. One specific feature of a Gaussian random process is that its statistical characteristics are dictated by its spectrum (see, for example, Adler (1981). Then the power spectrum has the form

$$\frac{2l + 1}{4\pi} C(l) = \frac{V}{2\pi^2} \int_0^\infty k^2 \frac{|\theta_l(\xi, k)|^2}{2l + 1} dk. \tag{5.18}$$

Substituting Eq. (5.15) into Eq. (5.17), we finally obtain

$$C(l)^{SW} = \frac{9}{200\sqrt{\pi}} B\xi_0^{1-n} \frac{\Gamma\left(\frac{3-n}{2}\right) \Gamma\left(l + \frac{n-1}{2}\right)}{\Gamma\left(\frac{4-n}{2}\right) \Gamma\left(l + \frac{5-n}{2}\right)}, \tag{5.19}$$

where $\xi_0 = 2(\Omega_{tot} H_0^2)^{-1/2}(1 + \ln \Omega_{tot}^{0.085})$ is today's horizon, Ω_{tot} is the total density of all forms of matter in units of critical density, n is the exponent of the spectrum of primary density perturbations and B is the spectrum amplitude. Note that $n = 1$ for the Harrison–Zeldovich spectrum, which is the most attractive one from the theory standpoint. Equation (5.19) then immediately implies

$$C(l)^{SW} \propto \frac{1}{l(l + 1)}, \tag{5.20}$$

and most of the power ΔT of fluctuations due to the Sachs–Wolfe effect concentrates on small multipoles $l \simeq 0, 1, 2$. The mode with $l = 0$ corresponds to uniform distribution of ΔT, independent of angle, and must therefore be eliminated by an ordinary change of the datum line for ΔT. The mode with $l = 1$ corresponds to the mean value of the dipole component and may in principle be observable. However, a similar anisotropy is also created by the local motion of our Galaxy relative to the CMB. Therefore, the cosmological dipole anisotropy is considerably distorted by this effect. Finally, the mode with $l = 2$ corresponds to the contribution from the quadrupole anisotropy to the variance of the CMB temperature perturbations. In fact, this mode is powerwise one of the largest contributors to the angular distribution of ΔT. It was for these reasons that the first experiments on searching for large-scale anisotropy ΔT were specifically focused on achieving the maximum possible precision in measurements of the quadrupole component; this is also true for the Cosmic Background Explorer (COBE) project.

Concluding this subsection we need to remark that the Sachs–Wolfe effect is sometimes additionally split in two: into the 'local' and the 'global' effects. The details can be found in, for example, Hu, Sugiyama and Silk (1997) and Hu *et al.* (1995a,b).

5.2.1 Dipole anisotropy

Let us look in greater detail at the properties of the dipole anisotropy of the CMB revealed in careful measurements made by the COBE satellite. We shall assume that the observer moves at a speed $\vec{v}$ in a certain direction relative to the CMB. The angular distribution of the intensity of quanta that is observed, recalculated to the blackbody temperature, is given by the following well known formula (Landau and Lifshits, 1984):

$$T_{\text{obs}}(\theta) = T_0 \frac{(1 - \beta^2)^{1/2}}{(1 - \beta \cos \theta)}, \tag{5.21}$$

where θ is the angle between the vector $\vec{v}$ and the direction of observation. We assume that $\beta = |\vec{v}|/c| \ll 1$; that is, that the movement proceeds at a non-relativistic speed. We expand Eq. (5.21) in a Taylor series in a small perimeter $\beta \ll 1$. The temperature distortions are then equal, up to terms $\sim \beta^2$, to

$$\frac{\Delta T(\theta)}{T_0} \simeq \beta \cos \theta + \frac{\beta^2}{2} \cos 2\theta + O(\beta^3). \tag{5.22}$$

As we see from Eq. (5.22), the motion of an observer relative to the CMB generates a dipole component $\sim \beta$ and a quadrupole component $\sim \beta^2$. The same equation shows that higher multipoles are generated in higher orders in v/c. Consider first the highest-amplitude term v/c in Eq. (5.22). For this term the mean-square perturbations amplitude $\langle(\Delta T(\theta)/T^2)\rangle$ is of the order $\beta^2/2$. For the sake of evaluation we assume $\langle(\Delta T/T^2)\rangle_{\text{dip}} \sim \beta^2/2 \sim 10^{-6}$ and evaluate the level of quadrupole anisotropy generated by local movements of matter. Eq. (5.22) implies that

$$\frac{\Delta T(\theta)}{T_0}\bigg|_q \simeq \frac{2}{3}\beta^2 P_2(\cos \theta), \tag{5.23}$$

where $P_2(\cos \theta)$ is the Legendre polynomial of index $l = 2$. The order of magnitude of the level of the quadrupole component due to the local speed of the observer is close to 10^{-6} ($\sim \beta^2$). Let us now analyse the observational data on the dipole anisotropy of the primordial radiation. The best precision in measuring the amplitude and orientation of $(\Delta T/T)|_{\text{dip}}$ was achieved in the COBE project, with not only the quantity $\Delta T|_{\text{dip}} = 3.35 \pm 0.024\,\text{mK}$, but also the temperature $T_0 = 2.725 \pm 0.020$ K measured successfully (Kogut *et al.*, 1996a). In dimensionless units, therefore, $(\Delta T/T)|_{\text{dip}} \simeq 1.23 \times 10^{-3}$, and its orientation in galactic coordinates (l, b) corresponds to $l = 264.26° \pm 0.33°$; $b = 48.22° \pm 0.13°$.

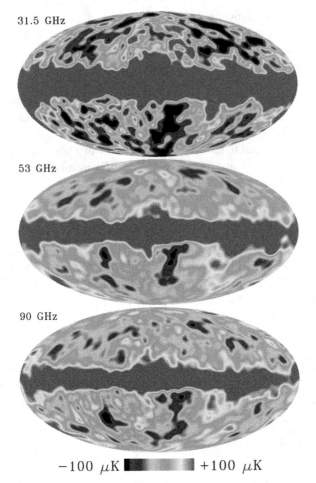

31.5 GHz

53 GHz

90 GHz

$-100\ \mu\mathrm{K}$ ▰▰▰▰ $+100\ \mu\mathrm{K}$

Figure 5.1 Large-scale ($\theta > 7°$) CMB anisotropy (the COBE data are from Bennett *et al.*, 1996). The aggregate radio sky maps were obtained by combining three frequencies: 31, 53 and 90 GHz. The grey band across the middle represents the smoothed emission of the Galaxy.

In equatorial coordinates (α, δ) this results in the values $\alpha = 11^{\mathrm{h}}12^{\mathrm{m}}.2 \pm 0.8^{\mathrm{m}}$ and $\delta = -7.06° \pm 0.16°$ (the 2000° epoch). Figure 5.1 maps the distribution of anisotropy measured by the COBE satellite (Bennett *et al.*, 1996). In galactic coordinates, the distribution of dipole anisotropy is characterized by not one but three quantities, namely ΔT_x, ΔT_y and ΔT_z, in correspondence with the work of Kogut *et al.* (1996a):

$$\Delta T(l, b) = \Delta T_x \cos(l) \cos(b) + \Delta T_y \sin(l) \cos(b) + \Delta T_z \sin(b). \qquad (5.24)$$

The corresponding amplitudes ΔT_x, ΔT_y and ΔT_z measured in the COBE project are listed in Table 5.1 for the frequency ranges 31, 53 and 90 GHz; these data ignore the correction for the galactic emission. Tables 5.2 and 5.3 represent contributions from the galactic component (the synchrotron and free–free emission and the emission by dust) and the amplitude of the observer's velocity relative to the CMB radiation.

Table 5.1. *The observed parameters of dipole anisotropy* $\Delta T(l, b) = \Delta T_x \cos(l) \cos(b) + \Delta T_y \sin(l) \sin(b) + \Delta T_z \sin(b)$

All parameters were renormalized to the antenna temperature. Correction due to the contribution of the Galaxy is not included.

Channel	Type	ΔT_x (mK)	ΔT_x (mK)	ΔT_z (mK)	Amplitude mK	l^{II} (deg)	b^{II} (deg)
31A	mean	−200	−2216	2406	3277	264.82	47.25
	noise	21	31	23	27	0.56	0.49
	gain	5	55	60	57	0.00	0.00
	systematics	16	22	14	18	0.43	0.34
	total error	27	67	66	66	0.71	0.60
31B	mean	−190	−2180	2396	3245	265.00	47.60
	noise	24	35	26	31	0.65	0.56
	gain	4	50	55	52	0.00	0.00
	systematics	21	29	27	28	0.56	0.50
	total error	32	68	67	67	0.86	0.75
53A	mean	−198	−2082	2314	3120	264.56	47.89
	noise	7	10	8	9	0.21	0.18
	gain	1	14	16	15	0.00	0.00
	systematics	9	17	10	13	0.25	0.27
	total error	11	24	21	22	0.33	0.32
53B	mean	−199	−2067	2353	3139	264.48	48.56
	noise	8	12	9	10	0.23	0.20
	gain	1	14	16	15	0.00	0.00
	systematics	7	11	10	10	0.22	0.20
	total error	11	22	21	21	0.31	0.29
90A	mean	−180	−1820	2058	2753	264.33	48.37
	noise	13	19	15	17	0.42	0.37
	gain	3	36	41	39	0.00	0.00
	systematics	8	17	11	14	0.27	0.32
	total error	16	44	45	45	0.50	0.49
90B	mean	−174	−1830	2029	2738	264.56	47.82
	noise	9	13	10	12	0.29	0.26
	gain	2	23	26	25	0.00	0.00
	systematics	6	13	11	12	0.20	0.26
	total error	11	30	30	30	0.35	0.37

Converting to the system of coordinates with an angle θ between the vector $\vec{v}$ and the direction of arrival of quanta, we can transform the COBE data to the form of Eq. (5.22). Figure 5.2 plots the antenna temperature obtained by processing the experimental data (Kogut *et al.*, 1996a) as a function of angle θ. We see from Fig. 5.2 that the behaviour of $T_A(\theta)$ agrees perfectly well with theoretical predictions.

Figure 5.3 illustrates the accuracy of measuring the amplitude and orientation of the dipole component, with a correction for the galactic emission in all frequency ranges of the COBE. For comparison, the data obtained by other observational groups in a number of frequencies are also shown. Note that the COBE results correspond with high accuracy to ΔT_{dip},

Table 5.2. *Contribution of the radiation from the galaxy to the dipole component*

Type of radiation	ΔT_x (μK)	ΔT_y (μK)	ΔT_z (μK)	Amplitude (μK)	l^{II} (deg)	b^{II} (deg)
Synchrotron	3.8 ± 1.2	1.2 ± 0.4	-1.5 ± 0.5	4.3 ± 1.1	18 ± 8	-21 ± 8
Free–free	-1.3 ± 8.7	-8.1 ± 21.0	-11.6 ± 20.8	14.2 ± 20.8	261 ± 64	-55 ± 84
Dust	0.3 ± 0.1	0.3 ± 0.1	-0.2 ± 0.1	0.5 ± 0.1	45 ± 13	-25 ± 9
Combined	2.8 ± 8.8	-6.6 ± 21.0	-13.3 ± 20.8	15.1 ± 20.5	293 ± 92	-62 ± 75

Table 5.3. *Relative motion velocities*

Type	Velocity (km/s)	l^{II} (deg)	b^{II} (deg)	Reference
Sun–CMB	369.5 ± 3.0	264.4 ± 0.3	48.4 ± 0.5	Kogut *et al.* (1993)
Sun–LSR	20.0 ± 1.4	57 ± 4	23 ± 4	Kerr and Lynden-Bell (1986)
LSR–GC	222.0 ± 5.0	91.1 ± 0.4	0	Fich, Blitz and Stark (1989)
GC–CMD	552.2 ± 5.5	266.5 ± 0.3	29.1 ± 0.4	
Sun–LG	308 ± 23	105 ± 5	-7 ± 4	Yahil, Tammann and Sandage (1977)
LG–CMB	627 ± 22	276 ± 3	30 ± 3	

Note: LSR: local system of coordinates; GC: galactic centre; LG: local group.

not depending on frequency, as should be the case for the Planck blackbody radiation. Figure 5.3(b) indicates the direction of motion of the observer relative to the CMB on the diagram (l, b). As we see from this diagram, the COBE data have the smallest errors and fall within the error intervals of the earlier experiments. However, dipole localizations obtained using the differential radiometer (DMR) and FIRAS – a COBE instrument – are different at the 68% confidence level. At the 95% confidence level this difference lies within the measurement error range.

We need to emphasize again that the CMB anisotropy is related to the observer's motion in the local group relative to the CMB. This motion is caused by local inhomogeneities of the gravitational potential. The dipole anisotropy for non-perturbed CMB temperature $T_0 = 2.73$ K corresponds to the parameter $\beta = (1.23 \pm 0.01) \times 10^{-3}$, which is equivalent to the velocity modulus $v \simeq 370 \pm 3$ km s^{-1}. Using this value of velocity combined with the measured dipole orientation makes possible the evaluation of the velocity and the direction of motion of the local group: $v_{LG} \simeq 627 \pm 22$ $(l^{II}, b^{II}) = (276° \pm 3°, 30° \pm 3°)$.

Note, nevertheless, that owing to large errors this prediction is rather difficult to compare with the dipole anisotropy data obtained by other techniques, for instance the anisotropy of the x-ray background measured by the HEAO-1 satellite: $v_\alpha \simeq 475 \pm 165$ and $(l^{II}, b^{II}) \simeq (280°, 30°)$. Furthermore, it is very difficult to separate the contribution of each component in view of the interference between the cosmological dipole component and local inhomogeneities of the gravitational potential. This is the reason why the principal information on the behaviour and distribution of inhomogeneities in the Universe on a scale above 10^2–3×10^2 Mpc is carried by the harmonics $l \geq 2$; we analyse these harmonics in the following.

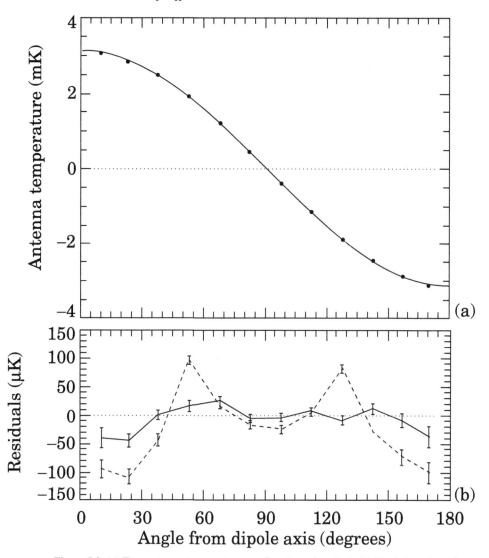

Figure 5.2 (a) True antenna temperature as a function of angle θ. (b) Deviations from the dipole distribution from the data of Kogut *et al.* (1993, 1996a). Adapted from Kogut *et al.* (1993).

5.2.2 *Quadrupole anisotropy of the CMB and higher harmonics*

In the quadrupole anisotropy mode of the CMB the contribution from our Galaxy is found to be comparable with the amplitude of the primordial (cosmological) component (Bennet *et al.*, 1996). If we use the COBE results, the corresponding root-mean-square value of the quadrupole component is found to be $Q_{\rm rms} = 10.0^{+7}_{-4}$ μK (68% confidence level).

At the 95% significance level the quadrupole amplitude lies within the interval 4 $\mu K \leq Q_{\rm rms} \leq 28$ μm (Kogut *et al.*, 1996a) and can be easily matched by any realistic model of the adiabatic perturbations spectrum. We wish to emphasize that the idea of comparing the

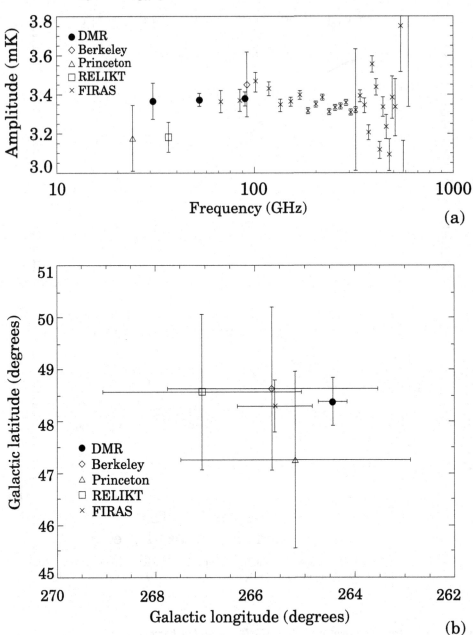

Figure 5.3 Amplitude orientation of the dipole. Adapted from Kogut *et al.* (1993).

amplitude of each specific multipole with theoretical predictions is fraught with an evident internal contradiction. In principle, the source of CMB anisotropy on angular scales above 1–3° could be any type of perturbations (adiabatic or isopotential), including gravitational waves. The free parameters that implement the 'matching' of the perturbation spectrum $P(k)$ and the corresponding characteristics $\Delta T(\theta, \varphi)$ are, in a natural way, the amplitude and shape of the spectrum $P(k)$. One could fix the type of perturbations or their superposition, and aim at bringing together as close as possible the theoretically predicted behaviour of the correlation function $\xi_{th}(\theta)$ and its experimentally determined value $\xi_{ex}(\theta)$, taken, for instance, from the COBE data. If we assume that the spectrum shape for the potential perturbations and gravitational waves is a free parameter, then this problem of minimizing the deviations $\xi_{th}(\theta)$ and $\xi_{ex}(\theta)$ is always solvable. However, this solution would be practically useless because the experimental values $C_{l(ex)}$ for each l may incorporate unequal error levels, including systematic effects that would be immediately transformed into errors of spectrum reconstruction. Furthermore, we remind the reader that the large-scale anisotropy $\Delta T(\theta, \varphi)$ corresponds to the scales of spatial perturbations $r_{LS} > 100$–$300\,\text{Mpc}$ (depending on the values of the parameters h, Ω_{tot}, Ω_b, Ω_{dm}, Ω_Λ, etc.). For these scales we have no reliable observational techniques that would allow us immediately to test the spectra $P(k)$ using the data on the large-scale distribution of matter. Consequently, the cosmological value of this 'technique' of minimizing $\xi_{th}(\theta)$ and $\xi_{ex}(\theta)$ would be very nearly zero because its results could not be verified to an acceptable degree. We believe that a more constructive, and hence preferable, approach is one in which the bridge between typical galactic scales ($\sim 1\,\text{Mpc}$) and r_{LS} is imposed by the spectral model $P(k)$, whose parameters are verifiable by using the data on the spatial distribution both in the range $r \ll r_{LS}$ and with the data on the CMB anisotropy ($r \geq r_{LS}$). One of the most successful models of this type is the power law $P(k) = Ak^n$ already discussed for the adiabatic mode of perturbations density, preset over the entire domain of the modulus of $\vec{k}$.

The CDM model with $\Omega_{tot} = 1$, $\Omega_\Lambda = 0$ and the exponents of adiabatic perturbations $n = 1$ is the simplest model for an analytic investigation of the large-scale anisotropy of the CMB.

With this model, it is possible to monitor analytically the behaviour of the function $\xi_{th}(\theta)$, and then easily generalize the results to more complex cosmological models that include a non-zero cosmological constant or non-adiabatic perturbation modes. Following Kofman and Starobinsky (1985), we rewrite the correlation function $\xi_{th}(\theta)$ for the 'standard' CDM model in the form

$$\xi_{th}(\theta) = \overline{A} \sum_{l=2}^{\infty} \frac{2l + 1}{l(l + 1)} P_l(\cos \theta), \tag{5.25}$$

where A is a normalization constant proportional to the amplitude of the Harrison–Zeldovich spectrum for adiabatic inhomogeneities ($n = 1$). The monopole and dipole components were subtracted from Eq. (5.25). The expression for the correlation function (5.25) can be transformed to a combination of trigonometric functions (Kofman and Starobinsky, 1985):

$$\xi_{th}(\theta) = \overline{A} \left\{ -\frac{3}{2} \cos \theta - 1 - 2 \ln \left(\sin \frac{\theta}{2} \right) \right\}. \tag{5.26}$$

Formally this expression has a singularity at $\theta \to 0$, of clearly understood nature. Summation in Eq. (5.25), no matter how large l is, falls beyond the limits of applicability of the theory based on the Sachs–Wolfe effect. Formally we need to use l_{rec} corresponding to the condition $l_{rec}\theta_{rec} \simeq 1$ (where θ_{rec} is the angular measure of the recombination horizon recalculated to the current age of the Universe) as the upper bound on the sum in Eq. (5.25).

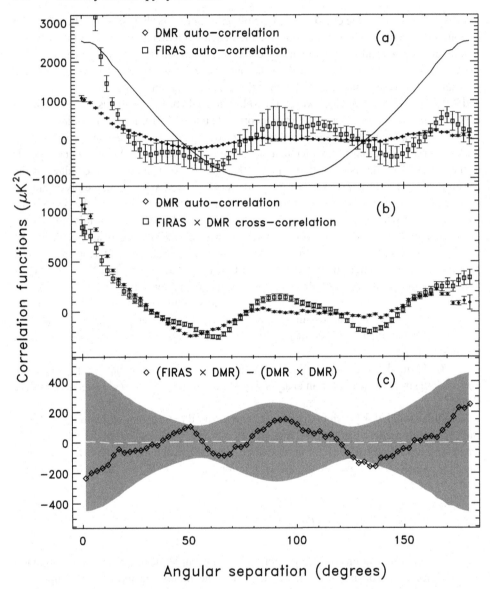

Figure 5.4 Auto- and cross-correlation of the signal on COBE maps. (a) Autocorrelation functions. The solid curve is the contribution of the quadrupole component. (b) Cross-correlation of the DMR × FIRAS data. (c) The difference (FIRAS × DMR) − (DMR × DMR). Adapted from Fixsen *et al.* (1997).

Another factor that must also be remembered when comparing theoretical predictions with observations is the angular resolution of the antenna of the receiving equipment, which can be worse than θ_{rec}. One illustration of this situation is the COBE experiment, in which $\theta_A \sim 10° \gg \theta_{\text{rec}}$ (see Fig. 5.4). With this correction of theoretical predictions taken into account, the correlation function of fluctuations of $\Delta T(\theta)$ on angular scales imposed by the Sachs–Wolfe effect has an extremely simple structure, Eq. (5.26), that reflects two most important factors: selection of a flat-space model of the Universe and the scale-invariant spectrum of metric perturbations $P_g(k)$ related to the spectrum of density perturbations by a simple formula $P_g(k) = k^{-4}P(k) \sim k^{-3}$. Note that the variance of the perturbations metric in the interval k–$k + \Delta k$,

$$\langle h^2 \rangle \sim \int_k^{k+\Delta k} dk \cdot k^2 P_g(k) \sim \int_k^{k+\Delta k} \frac{dk}{k}, \tag{5.27}$$

is practically independent of the choice of the interval width.

Making use now of the results of the analysis of the spectrum $C(l)$ and the correlation function $\xi_{th}(\theta)$ in the framework of the standard CDM model, it is not difficult to interpret what causes changes in these characteristics in more complex cosmological models. When we were discussing the observational status of the hidden mass problem in Chapter 1, we mentioned several times the importance of taking the cosmological constant into account for the dynamics of expansion of the Universe. The emergence of the vacuum density Ω_Λ in the anisotropy distribution of the CMB is characterized by two main effects.

First, the law dictating the growth of perturbations at the post- hydrogen-recombination stage changes, especially as $z \to 0$ (Peebles, 1983). Secondly, owing to the effect of vacuum density changes, the Universe expands faster than in models with $\Omega_\Lambda = 0$. This results in the renormalization of angular scales that correspond to the spatial scales of inhomogeneities ΔT (Kofman and Starobinsky, 1985; Zabotin and Naselsky, 1983). Taken together, these factors result in the following modification of the spectrum of large- scale modes of $\Delta T(\theta)$ (Kofman and Starobinsky, 1985) for adiabatic metric perturbations with the Harrison–Zeldovich spectrum:

$$C(l) = \frac{\overline{A}}{l(l+1)} K_l^2(\Omega_\Lambda), \tag{5.28}$$

where

$$K_l^2(\Omega_\Lambda) = 1 + D_l \left[\left(\frac{\Omega_\Lambda}{1 - \Omega_\Lambda} \right)^{1/6} - d \right], \quad l \geq 5, \tag{5.29}$$

$d \simeq 1.04$, $D_2 = 1.58$, $D_3 = 1.31$, $D_4 = 1.12$. If $l \geq 5$, the approximation for the spectrum transformation coefficients K_l changes as follows:

$$K_l^2(\Omega_\Lambda) = 1 + \frac{B(\Omega_\Lambda)}{l + 1/2}, \tag{5.30}$$

so the corresponding coefficients B for the preferable values of the parameter $\Omega_\Lambda = 0.7$–0.9 are: 1.53, 2.707 and 5.325. The behaviour of the function a_l related to the spectrum $C(l)$ by the condition $a_l = C^{1/2}(l)$ for various ΛCDM models was investigated in detail by Gorski, Silk and Vittorio (1992). The condition common to these models is $\Omega_{tot} = 1 = \Omega_0 + \Omega_\Lambda$. The general feature of ΛCDM cosmological models is an increase in the level of large-scale CMB anisotropy as Ω_Λ increases. This result has a fairly simple explanation. The higher the vacuum energy density, the earlier the moment of equality of cold matter densities

$\rho_{dm} = \Omega_{dm}\rho_{cr}(1+z)^3$ and $\rho_\Lambda = \Omega_\Lambda\rho_{cr}$:

$$1 + z_{eq}^* = \left(\frac{\Omega_\Lambda}{\Omega_{dm}}\right)^{1/3} = \left(\frac{\Omega_\Lambda}{1-\Omega_\Lambda}\right)^{1/3}. \tag{5.31}$$

Note that the increase in adiabatic perturbations terminates at the stage of vacuum density domination (Peebles, 1983). Therefore, for non-linear structures to form in matter distribution in a ΛCDM model, it is necessary for the initial amplitudes to be the higher, the higher z^* and with it the higher Ω_Λ are. Owing to the weak dependence of z^* on Ω_Λ this effect becomes automatically small in amplitude but is nevertheless substantial if we consider temperature fluctuations of the CMB.

Note that adiabatic perturbations are not the only source of large-scale CMB anisotropy. Primordial isopotential perturbations or gravitational waves would result in an anisotropy of the CMB intensity distribution on the celestial sphere as a consequence of perturbations of the gravitational potential that they produce and the ensuing gravitational shift in the frequency of quanta. The role of gravitational waves in the formation of large-scale angular variations of ΔT will be discussed in detail in Section 5.2.3.

5.2.3   *Gravitational waves as sources of large-scale anisotropy of the CMB*

One of the most important predictions of today's theories of the early stages of evolution of the Universe based on inflation models is the prediction of the background of gravitational waves that are generated in the process of restructuring of the vacuum of physical fields (Starobinsky, 1979). The qualitative differences between the characteristics of anisotropy generated by adiabatic perturbations in the epochs prior to and after hydrogen recombination and by gravitational waves can be understood by analysing specifics of their evolution. First of all, gravitational waves differ from adiabatic perturbations of the metric independent from the mass distribution as a result of the tensor nature of the former: low-amplitude gravitational waves do not cause redistribution of density and velocity of matter and undergo independent evolution. This feature was pointed out in the pioneering paper Lifshits (1946). Owing to the expansion of the Universe, the gravitational wave amplitude is a function of time (Grischuk, 1974):

$$h_\alpha^\beta(\xi, \vec{x}) = \sum_{\vec{k}} \frac{\mu_{\vec{k}}(\xi)}{a(\xi)} G_\alpha^\beta(\vec{k}, \vec{x}) + k \cdot c. \tag{5.32}$$

For simplicity, we consider a flat-space model of the Universe in which the evolution of the amplitude $\mu_{\vec{k}} \equiv \mu_{\vec{k}}(\xi)$ is described by a familiar equation (Grischuk, 1974):

$$\mu_{\vec{k}}'' + \left(k^2 - \frac{a''}{a}\right)\mu_{\vec{k}}(\xi) = 0. \tag{5.33}$$

Asymptotics of the function $\mu_{\vec{k}}(\xi)$ in the limit of long $(k\xi \ll 1)$ and short $(k\xi \gg 1)$ gravitational waves follow from Eq. (5.33). In the former case $(k\xi \ll 1)$ the function $\mu_k(\xi)$ is implied by a well known solution (Grischuk, 1974):

$$\mu_{\vec{k}} \simeq C_1 \cdot a + C_2 \cdot a \int \frac{d\xi}{a^2}. \tag{5.34}$$

In the short-wavelength limit the equation for amplitudes $\mu_{\vec{k}}(\xi)$ corresponds to the equation for the harmonic oscillator and has a simple analytic solution,

$$\mu_{\vec{k}} \simeq \tilde{C}_1 \sin k\xi + \tilde{C}_2 \cos k\xi. \tag{5.35}$$

Note that if the equation of state of the matter in the early Universe corresponds to the ultrarelativistic limit $P = \varepsilon/3$, $a''/a \equiv 0$ implies that the high-frequency approximation, Eq. (5.35), corresponds to the exact solution, Eq. (5.33) (Grischuk, 1974).

We now make use of Eqs (5.34) and (5.35) and trace the changes in the metric perturbation amplitude in each mode k, depending on the ratio of the gravitational wavelength and the scale of the cosmological horizon. We assume that $\xi \gg \xi_{\mathrm{rec}}$ and that the dynamics of expansion of the Universe is dictated by the hidden mass with $\Omega_{\mathrm{tot}} = 1$ and $\Omega_{\Lambda} = 0$. The corresponding generalization to models with $\Omega_{\Lambda} \neq 0$ can be constructed by analogy with the adiabatic modes discussed in the preceding section.

Consequently, if $k\xi \ll 1$ then in view of Eq. (5.34) the growing mode is given by the expression $\mu_{\vec{k}} \propto C_1(\vec{k}) \cdot \xi^2 \simeq g_{\vec{k}} \cdot k^2 \xi^2$, while the decreasing mode can be ignored. The function $\mu_{\vec{k}}$ reaches a maximum $\mu_{\vec{k}} \simeq g_{\vec{k}}$ at $k\xi \sim 1$ and switches from the evolution regime in Eq. (5.34) to the oscillation regime in Eq. (5.35). These oscillations are of constant amplitude that does not exceed $g_{\vec{k}}$:

$$\tilde{C}_{1(\vec{k})} = g_{\vec{k}} k^2 \xi_*^2 \left(\sin k\xi_* + 2 \frac{\cos k\xi_*}{k\xi_*} \right), \tag{5.36}$$

$$\tilde{C}_{2(\vec{k})} = g_{\vec{k}} k^2 \xi_*^2 \left(\cos k\xi_* - 2 \frac{\sin k\xi_*}{k\xi_*} \right),$$

where $k\xi_* \simeq 2\pi$ corresponds to the condition of equality of the gravitational wavelength and the horizon size. The metric perturbations created by the gravitational wave are of order $h_k \sim \mu_k/a \sim g_k/\xi^2$ and fall off as the conformal time ξ increases. Therefore, the main contribution to gravitational potential perturbations, and hence to CMB anisotropy, comes from gravitational waves with $k\xi_* \simeq 2\pi$ when n_k reaches a maximum. As is the case with adiabatic modes, the CMB angular anisotropy distribution is completely determined by choosing the spectrum of gravitational waves $\Phi_g(k)$. In the simplest models of inflation that predict a scale-invariant spectrum of metric perturbations $\Phi_g \sim k^{-3}$, the spectrum of gravitational waves is also scale-invariant: $\Phi_g(k) \propto k^{-3}$ (Starobinsky, 1985a,b). This means that for angular scales $\theta \gg \theta_{\mathrm{rec}}$, where θ_{rec} is the angular scale of the recombination horizon, the multipole structure $C(l)^{\mathrm{gw}}$ should not be very different from the structure $C(l)$ for adiabatic modes. A detailed numerical analysis of the transfer equation for quanta in the presence of gravitational waves leads to a dependence of $C(l)^{\mathrm{gw}}$ on l as shown in Fig. 5.5. As we see from this figure, the peculiarity of gravitational waves is pronounced when $l = 2$ (the quadrupole component) and when $l > 10^2$, where the contribution of gravitational waves (gw) to CMB anisotropy is suppressed.

Note that any inflationary model that uses the condition of slow variation of the inflaton potential and the relation between the contributions of the gravitational waves and scalar (adiabatic) perturbations to the CMB anisotropy is conveniently characterized in terms of the so-called T/S relation (Polarski and Starobinsky, 1994):

$$\frac{T}{S} = \frac{C(l)^{\mathrm{gw}}}{C(l)^{\mathrm{A}}} = \vec{k}_l |n_{\mathrm{T}}|; \qquad n_{\mathrm{T}} < 0, \tag{5.37}$$

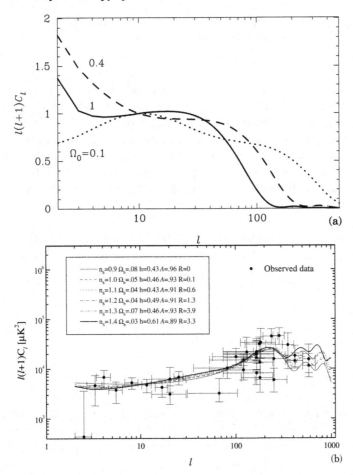

Figure 5.5 (a) Anisotropy spectrum generated by gravitational waves in the two CDM open models with $h = 0.7$, $\Omega_b h^2 = 0.00125$ and $\Omega_0 = 0.1$ (dotted line), and $\Omega_0 = 0.4$ (dashed line). The solid line corresponds to the standard CDM model without cosmological constant. Adapted from Challinor (2000). (b) CMB anisotropy power spectrum for ΛCDM cosmological models with corresponding background of the gravitational waves from inflation; A is the amplitude of the CMB anisotropy power spectrum normalized on the amplitude of the COBE correlation function at $l = 10$, $R = T/S$. Adapted from Melchiori *et al.* (1999).

where $n_T = \mathrm{d}(\log k^3 \Phi_g^2(k))/\mathrm{d}(\log k)$ is the deviation of the power spectrum of gravitational waves from the scale-invariant spectrum

$$\vec{k}_l = 6.2 \quad \text{for} \quad l \gg 1 \quad \text{and} \quad |n_T| \ll 1 \tag{5.38}$$

and $\vec{k}_2 = 6.93$, $\vec{k}_3 = 5.45$, $\vec{k}_5 = 6.10$, $\vec{k}_{10} \simeq 5.3$ for $|n_T| \ll 1$ (Polarski and Starobinsky, 1994).

As we see from Eqs (5.37) and (5.38), it is most likely that the contribution of gravitational waves does not exceed that of adiabatic perturbations. We should not dismiss the fact, however, that for special inflationary models it may be comparable to, and even greater than, the contribution of adiabatic modes (for details, see Lukash and Mikheeva (1998)).

5.3 The Silk and Doppler effects and the Sakharov oscillations of the CMB spectrum

When discussing the mechanisms of generation of the primordial CMB anisotropy in Section 5.2, we emphasized that it is related to three most important physical processes: the gravitational shift in the frequency of quanta (the Sachs–Wolfe effect), the adiabatic confinement of radiation before the hydrogen recombination epoch in the zones of elevated and reduced density (the Silk effect) and the scattering of quanta by free electrons moving in adiabatic perturbations both before and during recombination (the Doppler effect).

It has become fashionable to split each of the above mechanisms into 'subclasses' depending on which part of the fluctuations spectrum $C(l)$ is to be described and what physical processes in combination provide the main contribution to $C(l)$ for a given value of l. This may be correct in principle when we deal with a detailed theory of formation of CMB anisotropy (we can also add of CMB polarization formation) based on a serious mathematical foundation. From the aesthetics standpoint, however, and paying its due to the history of CMB research, we consider it natural to retain the older classification. On one hand, it has already nearly 'celebrated' its 35th anniversary, and on the other hand it is a visual reflection of the nature of the fundamental sources of generation of the CMB anisotropy that we find on the right-hand side of Eq. (2.10). As before (see Sections 5.1 and 5.2), we use the Newton gauge for metric perturbations, as proposed in Ma and Bertshinger (1995) and Mukhanov, Feldman and Branderberger (1992); the spatial interval of this gauge is given by Eq. (4.2).

A distinctive feature of the high-frequency approximation to be analysed in this section is that it takes into account the effect of induced transparency (a change in optical depth, τ, in the course of hydrogen recombination). Comparing this approximation with long-wavelength asymptotics of the equation of transfer of quanta in a weakly non-uniform medium, we can state that the high-frequency approximation describes the angular distribution of the CMB anisotropy on scales that are smaller than or of the order of magnitude of the angular scale of the cosmological horizon r_{rec} (taken at the moment of hydrogen recombination and recalculated to the current moment, taking into account the expansion of the Universe).

Using the same method of expanding temperature perturbations $\theta = \Delta T / T$ as used in analysing the Sachs–Wolfe effect (see Section 5.2), we give the corresponding generalization of the set of equations for the multipole components θ_l, taking into account the variation of the optical depth of plasma in (conformal) time. For convenience we first consider, as before, the flat-space cosmological model ($E = 0$) and then generalize these results to more complicated models with $E \neq 0$. The corresponding modification of transfer equations for each l mode, θ_l, is as follows (Hu and White, 1996):

$$\theta_0' = -\frac{k}{3}\theta_1 - \Phi',$$

$$\theta_1' = k\left[\theta_0 + \Psi - \frac{2}{5}\theta_2\right] - \tau'(\theta_1 - v_{\text{b}}),$$

$$\theta_2' = k\left[\frac{2}{3}\theta_1 - \frac{3}{7}\theta_3\right] - \tau'\left(\frac{3}{4}\theta_2\right),$$
(5.39)

$$\theta_2' = k\left[\frac{l}{2l-1}\theta_{l-1} - \frac{l+1}{2l+3}\theta_{l+3}\right] - \tau'\theta_l), \ l > 2.$$

The reader will remember that, as before, the relation between the kth temperature fluctuation mode in the direction of arrival of quanta $\vec{\gamma}$ and $\theta(\xi, \vec{k}, \vec{\gamma})$ is given by the expression

$$\theta(\xi, \vec{k}, \vec{\gamma}) = \sum_l (-\mathrm{i})^l \theta_l P_l(\vec{k} \cdot \vec{\gamma}), \tag{5.40}$$

where v_b denotes the kth harmonic in the expansion of the velocity field perturbations in matter (see Eqs (5.39)) and $\tau' = n_\mathrm{e} \sigma_\mathrm{T} a$ is the derivative of the optical depth of plasma with respect to time, $\xi = \int \mathrm{d}t/a$, where a is the scale factor, n_e is the concentration of free electrons and σ_T is the Thomson cross-section. The set of equations (5.39) needs to be complemented with an equation for V_b (see Hu and White (1996)). Furthermore, Eqs (5.39) were derived making an important assumption, namely that polarization effects in the scattering of quanta by electrons do not play any significant role in the formation of the CMB anisotropy. The validity of this assumption is confirmed *a posteriori* (see Chapter 6 for details).

Starting with hydrodynamics equations for baryonic fractions of matter, we write the corresponding relation between density and velocity perturbations in the following form (Hu and White, 1996):

$$\delta'_\mathrm{b} = -k v_\mathrm{b} - 3\Phi'; \qquad v'_\mathrm{b} = -\frac{a'}{a} v_\mathrm{b} + k\Psi + \frac{\tau'}{R}(\theta_1 - V_\mathrm{b}), \tag{5.41}$$

where $R = (3\rho_\mathrm{b})/(4\rho_\gamma)$. As before, $\theta_0 \equiv \Delta T/T = \frac{1}{4}(\delta \rho_\gamma/\rho_\gamma)$, where ρ_γ is the density of CMB quanta. Furthermore, the relation between metric perturbations and density and velocity perturbations in multicomponent matter is given by the following relations (Bardeen, 1980):

$$k^2 \Phi = 4\pi G a^2 \sum_i \left[\rho_i \delta_i + 3\frac{a'}{a}(\rho_i + P_i)\frac{V_i}{k} \right];$$

$$k^2(\Psi + \Phi) = -8\pi G a^2 \sum P_i \Pi_i, \tag{5.42}$$

where the index i provides correspondence to each species of matter in the Universe (including the hidden mass, massive and massless neutrinos, the baryonic fraction and the CMB), and Π_i is the anisotropic part of the pressure.

The set of equations (5.39)–(5.42) fully describes the process of generation of primary anisotropy of the CMB in the expanding Universe in the epoch before and after hydrogen recombination. Obviously, it does not take into account the possible distortions of the characteristics $\theta(\vec{k}, \vec{\gamma}, \xi)$ at the later stages of evolution when the first galaxies and quasars, capable of changing the ionization balance in the plasma and causing additional scattering of quanta by electrons, had formed. This aspect of the problem will be discussed in detail in Chapter 6. We need to emphasize in particular that an analysis of the solutions of Eqs (5.39)–(5.42) is a rather complicated mathematical problem. We have already mentioned that if the primordial adiabatic perturbations of the metric, velocity and density of the plasma emerged from quantum fluctuations of the vacuum during the inflation stage, it is only natural to expect that the random quantities Θ_l, V_b, δ_i, Φ and Ψ have the normal (Gaussian) distribution. In this case, all statistical properties of random Gaussian fields are fully determined by the appropriate power spectra.

We have already introduced a definition of the spectrum $C(l)$ for the CMB anisotropy. It is therefore clear that if we are interested in the behaviour of the function $C(l)$ at $l \simeq 5000$, then to find $C(l)$ from Eqs (5.39)–(5.42) we need to take into account roughly 5000 equations that belong to the class of 'rigid' systems which involve 'rapid' variables, such as the optical plasma depth or the quantities $\Theta_l(\xi)$ at $l \gg 10^2$, along with slowly varying variables (for example the scale factor $a(t)$ or CMB temperature). In the general case, a correct analysis of Eqs (5.39)–(5.42) for various cosmological models is possible only in the framework of a numerical simulation which constitutes a sufficiently complex mathematical problem. At the same time, it is possible to generate analytically a number of asymptotes that agree perfectly well with the results of numerical experiments, and thereby reveal the main factors that dictate the formation of the primordial CMB anisotropy and its relation to the perturbations of the metric, velocity and density of matter. This relation is unique in that the corresponding scales of perturbations that generate ΔT are found to be quite close to the spatial scales of today's structures. Therefore, we can look at the very beginning of the process of gravitational growth of perturbations that spawned structure in the cosmos by analysing the angular distribution of ΔT in the sky. And since the modern stage of structure evolution in the Universe is strongly non-linear, there is no need to justify in detail to what extent it is important for us to know the initial conditions of this process in order to understand the laws dictating how it progresses. These 'initial' conditions can (and must!) be obtained by analysing the CMB anisotropy.

A substantial simplification of the analytical investigation of the solutions of Eqs (5.39)–(5.42) is connected with analysing the so-called high-frequency approximation in which the parameter k/τ' is assumed to be large: $k/\tau' \gg 1$. This approximation was first used to analyse the process of generating ΔT in a baryonic model of the Universe (Peebles and Yu, 1970; Zeldovich and Sunyaev, 1970). When the plasma is highly opaque, $\tau \gg 1$ (but $\tau/\tau' \sim \xi$), the high-frequency approximation $k/\tau' \gg 1$ is equivalent to the condition $k\xi \gg \tau$. In this approximation, following Peebles and Yu (1970), we can retain only the multipole (θ_0) and dipole (θ_1) components, with which Eqs (5.39) transform to the following form:

$$\theta_0' = -\frac{k}{3}\theta_1 - \Phi'; \qquad \theta_1' = -\frac{R}{1+R}\frac{a'}{a}\theta_1 + \frac{k}{1+R}\theta_0 + k\Psi. \tag{5.43}$$

The set of equations (5.43) has an important property. We see that the characteristics of the CMB anisotropy (θ_0) and (θ_1) depend on the behaviour of the gravitational potentials Φ and Ψ. In the standard baryonic model that was carefully investigated at the beginning of the 1970s, the contribution of Φ and Ψ to the set of equations can be neglected for the following reason. Let us take a look at the set of equations (5.42). It immediately implies that both potentials, Φ and Ψ, are related to CMB temperature perturbations via the expressions

$$\Phi, \Psi \sim \frac{4\pi \rho_\gamma G \theta_0 a^2}{k^2} \simeq \theta_0 (k^2 \xi^2)^{-1} \ll \theta_0.$$

The reason why the generated metric perturbations are small is very simple. High-frequency density perturbations of the CMB and the baryonic matter create perturbations whose amplitude decreases with wavelength. Formally, this signifies that one can neglect the terms Φ' and Ψ Eqs (5.43) for the baryonic Universe (with the hidden mass not taken into account); after this, seeking the solution becomes trivial. Substituting θ_1 from the first equation of (5.43) into the second, we obtain

$$\frac{\mathrm{d}}{\mathrm{d}\xi}[(1+R)\theta_0'] + \frac{k^2}{3}\theta_0 = 0. \tag{5.44}$$

Obviously, solutions of this equation are oscillatory in nature and describe sound waves in the baryon–photon gas.

In models that take into account the cosmological role of non-baryonic matter that manifests itself as hidden mass, the situation with metric perturbations becomes somewhat more complicated compared with the one described above. The point is that along with perturbations of the density of the CMB and baryons, perturbations of the hidden mass density also take part in the formation of perturbations of the gravitational potential. Formally, compared with the preceding estimate of the perturbation level of Φ and Ψ, a new parameter emerges in the theory, namely δ_x, which corresponds to density perturbations in the hidden mass, and a parameter $\rho_x/\rho_b \gg 1$, which is a measure of the extent to which the hidden mass density exceeds the density of the baryonic fraction of matter. This means that the formation of both Φ and Ψ is driven by perturbations in dark matter and, hence, that they act as external sources of anisotropy in the set of equations (5.43). With this factor taken into account, Eq. (5.44) generalizes to

$$\frac{d}{d\xi}[(1+R)\theta_0'] + \frac{k^2}{3}\theta_0 = -\frac{k^2}{3}(1+R)\Psi - \frac{d}{d\xi}[(1+R)\Phi']. \tag{5.45}$$

We need to point out that further analysis of Eq. (5.45) depends on the accuracy of approximation that we aim to achieve. As the zero approximation illustrating the qualitative aspect of the problem, we can ignore (as we described above) the effect of perturbations in the baryon–photon gas on the gravitational potential, assuming it to be responding only to hidden mass perturbations.[1] In the general case, which also covers the effect of CMB temperature perturbations and perturbations of matter density on the potentials Φ and Ψ, Eqs (5.45) and (5.41)–(5.42) imply the well known equation for the combination of the zero-order and first momenta (Hu and White, 1996), namely

$$\Delta = \theta_0' + \frac{a'}{ak}\theta_1, \tag{5.46}$$

of the following type:

$$\left[1 + \frac{6}{y^2}(1+R)\right]\left[\Delta' - \frac{y'\Delta}{y(1+R)}\right] + \frac{1}{3}\left[1 - 3\frac{y''}{y} + 6\left(\frac{y'}{y}\right)^2\right]\theta_1 = \frac{y''}{y}\Psi_S - \Phi_S', \tag{5.47a}$$

$$\theta_1' + \frac{y'}{y}\theta_1 - \left[1 - \frac{6}{y^2}(1+R)^2\right]\frac{\Delta}{(1+R)} = \Psi_S. \tag{5.47b}$$

Here an apostrophe indicates a derivative with respect to the variable $x = k\xi$ with k fixed, $y = (\Omega_\gamma H_0^2)^{1/2}ak$, Ω_γ is today's density of CMB quanta in units of critical density, and Ψ_S and Φ_S correspond to the potentials Ψ and Φ that depend only on density and perturbations in hidden matter.

Let us consider the behaviour of θ_1 and Δ in the limit $z \gg 10^3$ when we can ignore the role of the baryonic component in comparison with the density of the electromagnetic radiation. Formally, this corresponds to the parameter R tending to zero and, since $a(\xi) \propto \xi$, the parameter y becoming proportional to x. In this approximation we obtain from Eqs (5.47) that

$$\Delta' - \frac{1}{x}\Delta + \frac{\theta'}{3} = \frac{x^2}{x^2+6}\left[\frac{\Psi_S}{x} - \Psi_S'\right]; \quad \theta' + \frac{\theta_1}{x} - \left[1 - \frac{6}{x^2}\right]\Delta = \Psi_S. \tag{5.48}$$

Equations (5.48) imply that (Hu and White, 1996)

---

[1] Simultaneously, we assume that the contribution due to neutrinos is also negligible.

$$\Delta'' + \frac{1}{3}\left[1 - \frac{6}{x^2}\right]\Delta = S(x), \tag{5.49}$$

where

$$S(x) = -\left[\frac{1}{3} - \frac{12}{(x^2 + 6)^2}\right]\Psi_S - \frac{x^2(x^2 + 18)}{(x^2 + 6)^2}\Phi'_S + \frac{x}{x^2 + 6}\Psi' - \frac{x^2}{x^2 + 6}\Phi''_S. \tag{5.50}$$

Note that in the limit of small x the gravitational influence on the behaviour of radiation and baryons in the gas by the hidden mass becomes small, so that formally we can assume $S(x) \simeq 0$ in Eq. (5.49). In this approximation the equation for Δ becomes the Bessel equation and its solutions can be obtained in terms of trigonometric functions (Doroshkevich, 1985; Kodama and Sasaki, 1984; Naselsky and Novikov, 1993; Starobinsky, 1988):

$$\Delta = A\Delta^+ + B\Delta^-,$$

$$\Delta^+ = -\cos\left(\frac{x}{\sqrt{3}}\right) + \frac{\sqrt{3}}{x}\sin\left(\frac{x}{\sqrt{3}}\right), \tag{5.51}$$

$$\Delta^- = -\sin\left(\frac{x}{\sqrt{3}}\right) - \frac{\sqrt{3}}{x}\cos\left(\frac{x}{\sqrt{3}}\right),$$

where A and B are constants. Using Eq. (5.51) as a fundamental solution, Hu and White suggested using the Green's functions method for finding a solution in the case of an external source $S(x)$ (Hu and White, 1996). The main idea boils down to the assumption that the constants A and B in Eq. (5.51) are now regarded as functions of the parameter x, as follows:

$$\Delta(x) = A(x)\Delta^+(x) + B(x)\Delta^-(x),$$

where

$$A(x) = A(x_{in}) - \sqrt{3}\int_{x_{in}}^{x}\Delta^-(x')S(x')\,dx',$$

$$B(x) = B(x_{in}) - \sqrt{3}\int_{x_{in}}^{x}\Delta^+(x')S(x')\,dx', \tag{5.52}$$

where x_{in} corresponds to the moment of time for which the amplitude of the initial fluctuations $A(x_{in})$ and $B(x_{in})$ is prescribed.

This approach makes it possible to simplify greatly the mathematical investigation of the problem, even though numerical analytical techniques are still required. Note that once we switch over to numerical analysis of solutions, the preferable approach in our opinion is to discuss the behaviour modes of temperature perturbations of the CMB in the framework of the rigorously formulated problem, which naturally cover the asymptotics (5.49)–(5.52) as a particular case.

At the same time, the interpretation of a simple analytical solution for eigenmodes of perturbations Δ in the form of Eq. (5.51) is perfectly transparent. It describes the evolution of acoustic waves in the cosmological plasma long before the onset of hydrogen recombination. As the derivation of Eq. (5.51) immediately assumed the smallness of the parameter $R = \frac{3}{4}(\rho_b/\rho_\gamma)$, it is clear that the speed of sound in the baryon–photon gas is $1/\sqrt{3}$ of the speed of light in vacuum and that the corresponding perturbations evolve in the acoustic wave mode. As the age of the Universe increases and its temperature decreases, the parameter $R \sim a$

becomes gradually more and more important. This means that the speed of sound in the medium changes with time – the factor that must be taken into account in analysing the high-frequency approximation of temperature perturbations in the CMB. This approach can be illustrated using as an example the set of equations (5.47) and limiting the analysis to the 'crudest' high-frequency approximation $x \gg 1$. In this approximation Eq. (5.47a) yields

$$\Delta' + \frac{1}{3}\theta_1 = 0 \tag{5.53}$$

and Eq. (5.47b) implies

$$\theta_1' - \frac{\Delta}{1+R} = 0. \tag{5.54}$$

Substituting Θ_1 from Eq. (5.53) into Eq. (5.54), we finally obtain

$$\Delta'' + \frac{1}{3(1+R)}\Delta = 0. \tag{5.55}$$

We see that Eq. (5.55) describes acoustic waves in the baryon–photon gas; the wave velocity is $C_S = 1/\sqrt{3(1+R)}$ of the speed of light in vacuum. The baryonic matter enters equations through the parameter $R = \frac{3}{4}(\rho_b/\rho_\gamma)$, which increases as the process approaches the moment of hydrogen recombination, $z \simeq 10^3$. We can evaluate this parameter assuming $\Omega_b h^2 \simeq 0.02$, $\Omega_x \sim 0.25$, $z_{rec} \simeq 10^3$, as follows:

$$R \simeq \frac{3}{4}\frac{\Omega_b}{\Omega_x}\frac{z_{eq}}{z_{rec}} \simeq 0.37, \tag{5.56}$$

where $z_{eq} \simeq 1.2 \times 10^4 \Omega_x h^2$ is the redshift in the epoch when the densities of radiation and hidden mass were equal.

We see from Eq. (5.56) that the effect of the baryonic component on the speed of sound at the moment when the plasma becomes transparent is fairly strong ($\sim$40%).

Furthermore, this variation of the speed of sound results in variations of wavelength for acoustic perturbations in the plasma; hence, at the onset of plasma transparency, the low- and high-multipole parts of the spectrum are sensitive to acoustic modes with unequal wavelengths. However, one of the most important processes for the formation of primordial anisotropy of the CMB is obviously the process of transparency onset in the course of recombination of the cosmological hydrogen. In this period, which timewise constitutes a relatively small fraction ($\sim$3%–10%) of the age of the Universe at the moment $z = 10^3$, a sharp drop in the concentration of free electrons destroys the connection between radiation and baryons; after this, CMB quanta propagate freely from the surface of 'last scattering' ($z = 10^3$) to an observer whose position corresponds to $z = 0$. In fact, it is at the moment of hydrogen recombination that the phase transition takes place in the radiation–baryonic matter system and the elasticity of the medium drops sharply (by almost 4.5 orders of magnitude). We have already discussed this situation in Chapter 4 in connection with the Sakharov oscillations in the distribution of baryonic matter immediately after the plasma becomes transparent. We remarked then that the Sakharov modulations of perturbation amplitude in the baryonic component will be smoothed in models where the dark matter density sharply dominates that of the baryonic fraction. However, they survive intact in the primordial electromagnetic radiation, which takes the role of a unique probe of the properties of the cosmic plasma and of the pregalactic inhomogeneity of the matter!

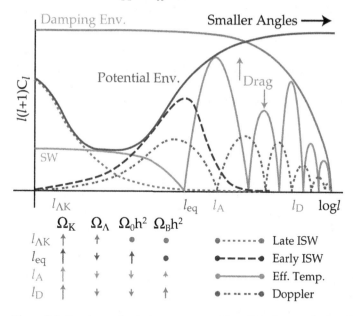

Figure 5.6 Spectrum of anisotropy generated by adiabatic perturbations in the ΛCDM model. Different curves indicate different mechanisms of formation of the CMB anisotropy and point to the ranges l for which these mechanisms prove to be the most efficient. Arrows indicate the direction of change in the appropriate scales as the values of the cosmological parameters increase. Filled circles mark independence of scales from the corresponding parameters. Adapted from Hu, Sugiyama and Silk (1996).

Figure 5.6 plots the qualitative dependence of the spectrum of primordial anisotropy of the CMB as a function of multipole number, l, for different values of the parameters $\Omega_k \equiv (1 - \Omega_\Lambda - \Omega_0)$, Ω_Λ, $\Omega_0 h^2$, $\Omega_b h^2$, where Ω_k denotes curvature, Ω_Λ is the density of the vacuum, Ω_0 is the density of the hidden mass tied up in the structural elements of matter (galaxies and galaxies clusters) and Ω_b is the density of baryons.

The parameters $l_{\Lambda k}$, l_{eq}, l_A and l_D denote the ranges for the appropriate multipoles where the effects of curvature and vacuum are significant ($l_{\Lambda k}$), the Sachs–Wolfe effect (l_{eq}), the acoustic Sakharov oscillations (l_A), and dissipative damping and non-instantaneous onset of transparency of the plasma for the electromagnetic radiation (l_D). Incidentally, there is a relatively narrow stretch in the range $l \leq l_{eq}$ where the Doppler scattering of quanta by electrons contributes significantly to the formation of $C(l)$ as compared with the effect of adiabatic clustering of radiation (the Silk effect; see Silk (1968)). If $l \geq l_A$ and $l \sim l_A$, the Silk effect is predominant and both effects (the Doppler effect and the Silk effect) are observed jointly in a number of zones within the range $l > l_A$. As could be expected, the $C(l)$ spectrum generated by the Doppler mechanism is phase-shifted relative to the Silk effect. This is the familiar phase shift between density and velocity perturbations in acoustic waves.

As we see from Fig. 5.6, the anisotropy spectrum amplitude drops considerably beginning with the multipoles $l \geq l_D$. This effect is caused by dissipative processes influencing the dynamics of evolution of perturbations in the baryon–photon gas. We will concentrate on the details of this phenomenon in Section 5.3.1.

5.3.1   *Dissipation of perturbations in the hydrogen recombination epoch*

To analyse the CMB anisotropy spectrum in the limit $l > l_D$, we turn to the set of equations (5.42) that takes into account explicitly the anisotropy of the energy–momentum tensor Π_i. Obviously, the electromagnetic radiation acts as one of the sources of this anisotropy (Chibisov, 1972a,b; Hu and Sugiyama, 1995; Weinberg, 1972). Thus following Hu and White (1996) we choose the following expression for estimating the anisotropic part of the energy–momenum tensor:

$$\Pi_\gamma \simeq \frac{8}{5}\left(\frac{k}{\tau'}\right) f_2^{-1}\theta_1. \tag{5.57}$$

Here $f_2 \simeq 1$ for non-polarized radiation (Chibisov, 1972a,b; Weinberg, 1972). Using Eq. (5.57), which relates the hydrodynamic peculiar velocity of baryons and the θ_1 components of perturbations, we can rewrite the expression for the temperature of the CMB in the following form (see Eq. (5.42) and Hu and White (1996)):

$$V_b - \theta_1 = -(\tau')^{-1}R\left[I\omega\theta_1 - k\Psi\right] - (\tau')^{-2}R^2\omega^2\theta_1. \tag{5.58}$$

Our derivation of Eq. (5.58) used the high-frequency approximation for acoustic waves both for the Θ_1 and the Δ components (see Eq. (5.46)) in the form $\exp(i\int \omega\,d\xi)$. Then Eqs (5.42), (5.57) and (5.58) yield a dispersion equation in the following form (Hu and White, 1996):

$$\omega = \pm kc + \frac{ik^2(\tau')^{-1}}{6}\left[\frac{R^2}{(1+R)^2} + \frac{4}{5}f_2^{-1}(1+R)^{-1}\right]. \tag{5.59}$$

As we see from Eqs (5.52), the imaginary part of the acoustic wave frequency describes the damping of oscillations in the course of scattering of quanta by electrons, with the damping decrement equal to (Hu and White, 1996)

$$k_D^{-2} \simeq \frac{1}{6}\int \frac{d\xi}{\tau'}\frac{R^2 + \frac{4}{5}f_2^{-1}(1+R)}{(1+R)^2}. \tag{5.60}$$

The decrement k_D from Eq. (5.60) corresponds to the damping factor of the amplitude of acoustic oscillations of the type $\exp(-k^2/k_D^2)$ that describes the diffusion-related attenuation of acoustic waves. We need to underline the fact that the dissipative scale k_D depends on the ionization history of the cosmic plasma. Equation (5.60) shows that in the hydrogen recombination epoch the degree of ionization[2] diminishes rapidly and the optical depth relative to the Thomson scattering falls off rather steeply from $\tau \sim 10\text{--}10^2$ at the start of recombination ($z \sim 3 \times 10^3$) to $\tau \ll 1$ at $z \ll 800\text{--}900$.

Formally, a decrease in the optical depth τ is accompanied with a reduction in the rate of change of τ', and hence the characteristic diffusion scale $\lambda_D \sim k_D^{-1}$ grows sharply. Does this mean that all peculiarities in the distribution of the CMB anisotropy on the surface of last scattering will be smoothed to the scale of the acoustic horizon? The answer is obvious: no, it does not, since the diffusion approximation itself is constrained by quite specific bounds on applicability.

We already mentioned at the beginning of this subsection that the approximation of 'strong bonding' between photons and electrons is valid only for waves whose wave vector modulus satisfies the relation $\tau'/k \gg 1$. Since τ falls off sharply in the course of hydrogen recombination, it is clear that for each k we can find a moment of time ξ_* at which the diffusion approximation becomes invalid ($\tau'/k \simeq 1$).

---

[2] Note that the ionization history meant here is the standard one, ignoring the effects of reionization at small z.

It is only natural that with this numerical computation approach the modes switch automatically; there is no need to go into details of this process. However, in the case of an analytical investigation of the situation, it is very useful to propose a simple algorithm to evaluate the effect that offers the means for checking the predictions of numerical approaches and to analyse in greater detail the dependence of the dissipation scale on the parameters of the cosmological model. We therefore introduce a parameter $\Delta\eta_{\rm rec}/\eta_{\rm rec} = \xi_{\rm rec}$ that characterizes the length of time required for the plasma to become transparent to the CMB at the time when $\tau(\eta_{\rm rec}) \simeq 1$. The equation

$$\tau'/k = 1 \tag{5.61}$$

determines, within an order of magnitude, the moment $\xi_{\rm rec}$ at which the diffusion approximation becomes invalid, that is

$$k\eta_k \simeq \tau(\eta_k)\left(\frac{\Delta\eta_k}{\eta_k}\right)^{-1}, \tag{5.62}$$

where $\Delta\eta_k/\eta_k = \tau(\eta_k)/\eta_k$. Equation (5.62) shows that the moment η_k is inversely proportional to k. In the vicinity of the transparency onset, $\Delta\eta_k/\eta_k \sim \eta_{\rm rec}$ and $\tau(\eta_k) \sim \tau(\eta_{\rm rec}) \sim 1$. This means that $\eta_k \sim k^{-1}\xi_{\rm rec}^{-1}$. We substitute this estimate of η_k into the upper limit of the integral (5.60) to obtain

$$k_{\rm D}^{-2} \sim \int_0^{\eta_k} \frac{{\rm d}\eta}{\tau'} \sim \frac{\Delta\eta_{\rm rec}}{k} = \xi_{\rm rec}\frac{\eta_r}{k}. \tag{5.63}$$

As we see from this estimate, the characteristic scale of dissipation of acoustic modes now depends on k and the corresponding decrement $D = k^2/k_{\rm D}^2$ is now a linear function of k (Doroshkevich *et al.*, 1978):

$$D(k) \simeq k\xi_{\rm rec}\eta_{\rm rec}. \tag{5.64}$$

This means that for acoustic modes with $k\xi_{\rm rec}\eta_{\rm rec} \leq 1$, the influence of dissipative processes can be ignored while the modes with $k\xi_{\rm rec}\eta_{\rm rec} \sim 1$ will first decay exponentially in k, but will follow the laws of diffusion as k keeps growing. Quantitatively the decrement is an explicit function of the length of the period of hydrogen recombination, $\Delta\eta_{\rm rec}$. Therefore, any 'delays' in recombination that are accompanied by an increase in $\Delta\eta_{\rm rec}$ should weaken the high-multipole part of the spectrum, $C(l)$. At the same time, any 'delays' in recombination will result in shifting $\eta_{\rm rec}$ to smaller z and will also shift the spectrum $C(l)$ towards smaller l. Consequently, an analysis of the CMB anisotropy spectrum helps to extract unique information on the parameters of the cosmological model and on the ionization history of the Universe that are based on the characteristic features of $C(l)$. It may also provide answers to the questions on the meaning of these features and with what accuracy the modern theory of generation of primordial anisotropy can predict their numerical characteristics. Section 5.4 is devoted to answering these questions.

5.4 $C(l)$ as a function of the parameters of the cosmological model

One of the most important predictions of today's theory of generation of the CMB anisotropy is based on the understanding that perturbations of density, velocity and metric that evolve in the sound wave mode existed at the time of hydrogen recombination. The CMB perturbation amplitude after recombination is modulated by acoustic modes before the recombination phase, in exact accordance with the effect predicted by Sakharov (1965). One

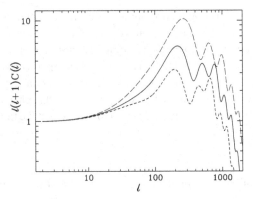

Figure 5.7 Anisotropy spectrum $C(l)$ as a function of the parameter $\Omega_0 h^2$ assuming $\Omega_b h^2 = 0.015$. The curves correspond to the following values of the parameter $\Omega_0 h^2$: solid curve, 0.25 with $h = 0.5$; long-dash curve, 0.09 with $h = 0.3$; short-dash curve, 0.64 with $h = 0.8$. Adapted from a figure by W. Hu, whose website may be found at http://background.uchicago/edu/whu/metaanim.html.

of the most important characteristics of the CMB anisotropy is the ΔT spectrum in the space of multipoles that was studied analytically in Hu and Sugiyama (1995), Jørgensen *et al.* (1995), Peebles (1980) and Starobinsky (1983), and numerically in Bond and Efstathiou (1984), Silk and Wilson (1981), Vittorio and Silk (1984), Wilson and Silk (1981) and Zaldarriaga, Seljak and Bertshinger (1998), amongest other papers. It is logical that when analysing the dynamics of perturbations, one has to choose concrete parameters of the cosmological model that serve as a scene on which plays out the evolution of small perturbations that lead both to the formation of galaxies and to the anisotropy of the CMB. We will attempt here to classify the main parameters of the cosmological model, evaluating the extent of their influence on various perturbations; we follow the approach of Hu and Sugiyama (1995) (see also Mukhanov (2003) and Weinberg (2001a,b)).

5.4.1   *Dark matter density in galaxies and galaxy clusters*

The role played by the dark matter density is characterized by the parameter $\Omega_{dm} h^2$; it makes itself felt mainly in the following three aspects. First, the density of dark matter dictated directly the dynamics of expansion of the Universe during hydrogen recombination. Secondly, the potential perturbations in dark matter interact with the electron–photon–baryon plasma only through gravitation. Hence, they are not subject to damping in the process of plasma becoming transparent for the primordial radiation and thus play the role of 'generator' of the anisotropy. At the same time, they are responsible for the behaviour of the degree of ionization of the plasma through their influencing the rate of expansion. Thirdly, by 'surviving' till today they affect the current scale of particles' horizons and determine the relation between spatial dimensions of fluctuations at the moment of recombination and their current angular size.

Figure 5.7 plots the spectrum $C(l)$ as a function of the multipole number l for a number of values of the parameter $\Omega_0 \equiv \Omega_{dm}$. The figure demonstrates that as the value of $\Omega_0 h^2$ decreases, the amplitude of the first Sakharov peak ($l \sim 200$) increases and at the same time its position undergoes a shift. The Harrison–Zeldovich spectrum of primordial adiabatic perturbations was used to plot Fig. 5.9. The density of the baryonic fraction of matter was not varied and is assumed equal to $\Omega_b h^2 = 0.015$.

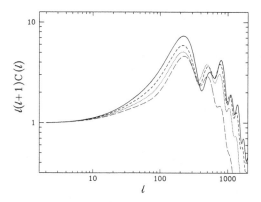

Figure 5.8 $C(l)$ as a function of the parameter $\Omega_b h^2$ assuming that $\Omega_0 = 1$ and $h = 0.5$. The curves correspond to the following values of the parameter $\Omega_b h^2$: solid curve, 0.025; dotted curve, 0.0075; long-dash curve, 0.0025; short-dash curve, 0.015. Adapted from a figure by W. Hu, whose website may be found at http://background.uchicago/edu/whu/metaanim.html.

5.4.2 Density of the baryonic fraction of matter

The density of the baryonic fraction of matter is, along with that of dark matter, one of the most important parameters of the thermal and ionization history of the cosmic plasma. It also determines, although to a lesser degree (in view of the smallness of the ratio $\Omega_b/\Omega_0 \sim 10^{-1}$) the dynamics of the expansion of the Universe, but it affects much more strongly the kinetics of recombination and especially the amplitudes and positions of the Sakharov peaks. To illustrate the dependence of the anisotropy spectrum $C(l)$ on the parameter $\Omega_b h^2$, we plot in Fig. 5.8 the distribution of $C(l)$ over multipoles in a flat-space cosmological CDM model with adiabatic perturbations (the Harrison–Zeldovich spectrum). We see in this figure that as the density of the baryonic fraction of matter decreases, the spectrum $C(l)$ undergoes considerable restructuring. The position of the first peak ($l \sim 200$) is practically independent of the parameter $\Omega_b h^2$, even though its amplitude changes significantly. Other peaks with $l \gg 200$ both shift their positions and change their amplitudes. Note also that as the parameter $\Omega_b h^2$ decreases, the damping decrement of acoustic waves grows sharply, which is accompanied by a faster fall-off of the spectrum in the range (≥ 300–500). Incidentally, the differences between $C(l)$ amplitudes in the high-multipole part of the spectrum for different values of the parameter $\Omega_b h^2$ reach one to two orders of magnitude, and sometimes higher. This part of the spectrum is thus an excellent indicator of the value of the parameter $\Omega_b h^2$.

5.4.3 The cosmological constant

As we remarked at the very beginning of this book, attention to possible observational manifestations of the cosmological constant is stimulated not only by the current observational data on the CMB anisotropy spectrum, but also in connection with cosmological SNIa supernovas. The effect of the vacuum energy density that reveals itself in the cosmological constant in Einstein's equations results in a corrected calculated age of the Universe and, hence, a change in the angular dimension of the recombination horizon.

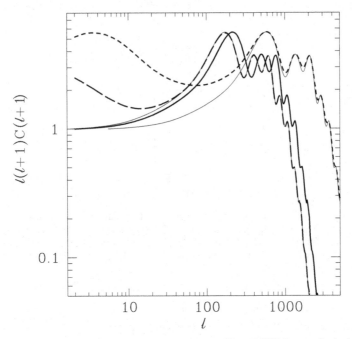

Figure 5.9 The anisotropy spectrum generated in a ΛCDM cosmological model $C(l)$ as a function of $\Omega_\Lambda h^2$ assuming that $\Omega_0 h^2 = 0.25$ and $\Omega_b h^2 = 0.0125$. The curves correspond to the following values of the parameter Ω_Λ: solid curve, 0 for $\Omega_0 = 1$; long-dash curve, 0.9 for $\Omega_0 = 0.1$; short-dash curve, 0 for $\Omega_0 = 0.1$. Thin solid lines correspond to $\Omega_K = 1 - \Omega_\Lambda - \Omega_0$ CDM models. Adapted from a figure by W. Hu, whose website may be found at http://background.uchicago/edu/whu/metaanim.html.

In its turn, this leads to angular redistribution of the anisotropy power spectrum and, hence, to redistribution in the space of multipoles l. This factor was pointed out in Hu and White (1996), Kofman and Starobinsky (1985) and Zabotin and Naselsky (1982a,b).

Figure 5.9 plots $C(l)$ as a function of l for a number of values of the parameter Ω_Λ. For the sake of comparison, this figure also shows the distribution $C(l)$ in an open model with $\Omega_0 = 0.1$ and a flat-space model with $\Omega_\Lambda = 0$ and $\Omega_0 = 1$. It is remarkable that the presence of the cosmological constant firstly modifies the behaviour of the power spectrum for $l \leq 10$. We have discussed this effect in Section 5.1.

Secondly, it essentially modifies the position of the first Sakharov peak owing to the effect of Ω_Λ, for instance in comparison with the standard CDM model or the open CDM model.

Thirdly, the general distribution of peaks over l is considerably shifted – as a result of renormalization of angular scales and of the corresponding l. This shift is one of the most significant factors that transform the spectrum $C(l)$ in ΛCDM cosmological models.

5.4.4   *The exponent of the spectrum of adiabatic perturbations*

When analysing $C(l)$ as a function of cosmological parameters in Sections 5.4.1–5.4.3, we assumed that the spectrum of primordial adiabatic perturbations agrees with the scale-invariant Harrison–Zeldovich spectrum.

Regarding matter density perturbations, this spectrum is characterized by the power law $P_\delta(k) \propto k^n$, with $n = 1$. Note that this form of the spectrum is predicted, to within logarithmic factors, by the simplest inflation models (Linde, 1990), although the spectral

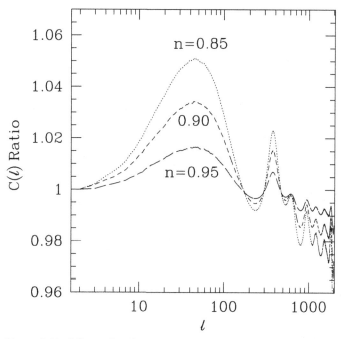

Figure 5.10 $C(l)$ as a function of the parameter n for the CDM model. Adapted from a figure by W. Hu, whose website may be found at http://Background.uchicago/edu/whu/metaanim.html.

exponent in more complex models may be a function of k, at least in a certain range of wave vectors (see, for example, Ivanov *et al.* (1994) and Starobinsky (1992a,b)). Assuming that $n(k)$ hardly change in the range of Sakharov peaks on the spectrum $C(l)$, we make use of the data of Hu *et al.* (1997) in order to answer the question to what extent the anisotropy spectrum $C(l)$ is sensitive to the exponent n. Figure 5.10 plots the ratio $C(l)(n)/C(l)(n = 1)$ in the standard CDM cosmological model, which gives a qualitatively and quantitatively correct representation of the trend of the process for any model. A change of slope in the spectrum from $n = 1$ to $n = 0.85$ results in increasing the level of fluctuations on the large-scale segment ($l < 10^2$) of the spectrum. Note from Fig. 5.10 that as n decreases, the power spectrum in the zone of the first Sakharov peak increases, after which the power of $C(l)$ decreases systematically in the region of $l > 200$. The nature of this effect is obvious. The fall-off in the spectrum's exponent results in power concentration in the region of small l (large spatial scales), which clearly weakens the high-multipole range of the spectrum. Obviously, the behaviour of $C(l)(n)$ at $n > 1$ should be the absolute opposite. Most of the power will concentrate at high l, while the low-multipole segment of the spectrum will be weakened.

5.4.5 *Isopotential (isocurvature) initial perturbations*

The initial perturbations in matter, generated in the course of inflation, could in principle be not of adiabatic but of isopotential type. Furthermore, it cannot be ruled out that, along with 'pure' cases, mixed modes could also be formed in which, for instance, an adiabatic mode could be mixed with an isopotential one. Let us turn to analysing the difference in the spectra $C(l)$ for the adiabatic and isopotential primordial inhomogeneity in the Universe using the axion hidden mass as an example (Hu and White, 1996).

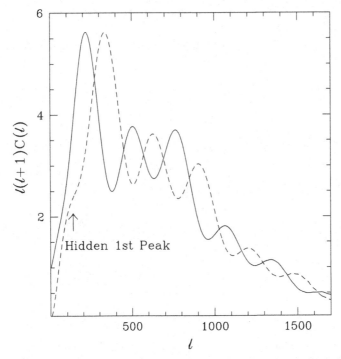

Figure 5.11 A comparison of $C(l)$ for adiabatic and isopotential (isocurvature) perturbations. The curves correspond to the following models: solid curve, inflationary CDM model; dashed curve, isopotential (isocurvature) perturbations. Adapted from White and Hu (1996).

If the initial irregularity of matter on spatial scales covering galaxies and their clusters up to $r \sim 10^3$ Mpc was of isopotential type, then the spectrum $C_{iso}(l)$ for this mode would be radically different from the spectrum of an adiabatic irregularity (see Fig. 5.11). First, the structure of the signal changes drastically. The first Sakharov peak, which is clearly pronounced in $C_a(l)$, is almost totally smoothed out in the isopotential mode. Also, the first maximum of $C_{iso}(l)$ is found not at $l = 200$ but at $l \simeq 300$–350. Furthermore, all other peaks are rigorously in antiphase with respect to the adiabatic mode.

5.4.6   *Role played by massless neutrinos*

The main effect of massless neutrinos reduces to shifting the moment of equilibration of the hidden mass density and that of the CMB plus neutrinos towards smaller redshifts. Since the neutrino density is proportional to the number of neutrino flavours N_ν, it is clear that an increase in N_ν must result in increased amplitude of $C(l)$. The higher the value of N_ν, the lower the redshifts in the epoch of equal densities of the non-relativistic and ultrarelativistic matter and the higher are the amplitudes of the primordial fluctuations required for the formation of structures by the time $z = 0$. This tendency is clearly pronounced in Fig. 5.12, where two model spectra with $N_\nu = 2$ and $N_\nu = 4$ are plotted, especially in the region of the first Sakharov peak. At the same time, massless neutrinos are a source of anisotropy of the energy–momentum tensor by changing the structure of peaks in the high-multipole part of the spectrum $C(l)$. Figure 5.12 demonstrates that if we are interested in determining the shape of the spectrum with accuracy $\sim$5–10% or better, the role of this component in the formation of $C(l)$ becomes very significant.

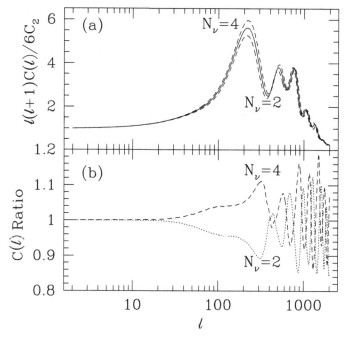

Figure 5.12 (a) Spectra $C(l)$ for adiabatic perturbations for different flavours of massless neutrinos. Solid curve corresponds to $N_\nu = 3$. (b) Ratio of the spectra $C(N_\nu/C(N_\nu = 3))$ as a function of l for $N_\nu = 4$ (dashed curve) and $N_\nu = 2$ (dotted curve). Adapted from Hu *et al.* (1995a,b).

5.4.7 Massive neutrinos

We mentioned at the very beginning of this chapter the role played by the idea of non-zero mass of the neutrino in shaping the current status of the theory of generation of the CMB anisotropy. Unfortunately, the first (alas erroneous) estimates of the neutrino mass are by now mere small print in physics history. There is an impression, nevertheless, that the concept of massive neutrinos will gain its rightful place in modern cosmology, if not now then in the near future. Our optimism is based on the observational data of the flux of atmospheric neutrinos recorded at the Super-Kamiokande (Fukuda *et al.*, 1998) and MACRO (Ambrosio *et al.*, 1998, 2001) detectors. According to these data, it seems to be very probable that neutrino oscillations $\nu_\mu \leftrightarrow \nu_\tau$, first discussed by Bruno Pontekorvo, are a reality. Two characteristics of ν_μ and ν_τ are of principal importance for describing the effects of $\nu_\mu \leftrightarrow \nu_\tau$ oscillations. First of all, we point to the mean-square shift in the masses of these particles. The data of the above experiments for Δm^2 yield very definite observational constraints: $5 \times 10^{-4} \mathrm{eV}^2 \leq \Delta m^2 \leq 6 \times 10^{-3} \mathrm{eV}^2$. The second important characteristic is the mixing angle, which is found to be close to 40–$45°$ or, more precisely, $\sin^2 2\theta \geq 0.82$. The experimental data of LSND (Athanassopoulos *et al.*, 1998), $\Delta m^2 \leq 0.2 \mathrm{eV}^2$, and solar neutrino data (Bahcall, Kravtsov and Smirnov, 1998) confirm the conclusions. of the Super-Kamiokande and MACRO collaborations. If we accept the hypothesis that all three flavours of neutrino (the electron, muon and τ neutrinos) are massive, then their effect on the dynamics of expansion of the Universe in the hydrogen recombination epoch is found to be the same as in the case of $m_\nu = 0$, provided $\sum_i m_{\nu_i} \ll 1 \mathrm{eV}$.

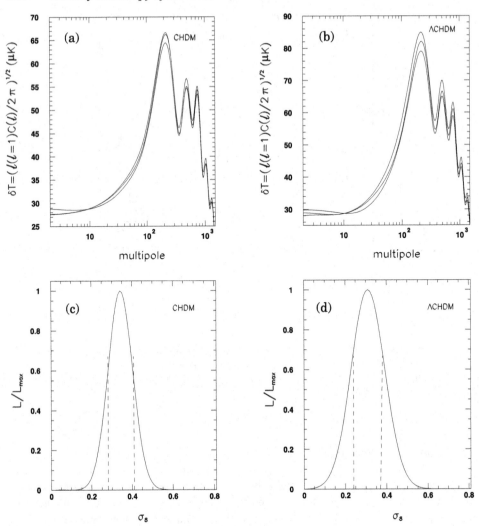

Figure 5.13 The spectra $C(l)$ for a model with neutrino oscillations. (a) CDM + HDM models; (b) ΛCHDM models. In (a) from the top to the bottom (at the first peak) $\sigma_8 = 0.2, 0.3, 0.5$; (b) shows the same as (a), but for the ΛCHDM models. (c), (d) Likelihood dependence on σ_8. Adapted from Popa *et al.* (2000).

Unfortunately, neutrino oscillations experiments place limits only on the squared difference between the masses of different particles, and reveal practically nothing about the masses themselves. We can conclude with a degree of caution that $m_{\nu_i}^2 \geq \Delta m^2 \geq 5 \times 10^{-4}$ eV^2 at least for one flavour of neutrinos and hence that $m_{\nu_i} \geq 2.2 \times 10^{-2}$ eV. Clearly, these particles did affect the acoustic horizon in the hydrogen recombination epoch by influencing the rate of expansion of the Universe and thereby changed the position and amplitude of Sakharov peaks in the $C(l)$ spectrum. Figure 5.13 shows the results of calculations of the $C(l)$ spectrum in a model with massive neutrinos taken from Popa, Burigana and Mandolesi (2000). This figure shows that the effect of massive neutrinos proves to be sufficiently strong after taking this factor into account when processing experimental data in the search for the CMB anisotropy.

6

Primordial polarization of the cosmic microwave background

6.1   Introduction

The primordial density perturbations result not only in the variation of temperature of the primordial radiation background, but also in its polarization (Basko and Polnarev, 1979; Crittenden, Coulson and Turok, 1995; Hu, 2003; Hu and White, 1997a,b; Kaiser, 1983; Kosowsky, 1999; Naselsky and Polnarev, 1987; Ng and Ng, 1995, 1996; Polnarev, 1985; Rees, 1968; Zaldarriaga, 2004, Zaldarriaga and Harari, 1995). Measurements of the CMB polarization make it possible to extract additional information, which helps in avoiding ambiguities in reconstructing cosmological parameters from the observational data (see Readhead *et al.* (2004)).

For instance, polarization is very sensitive to the presence of tensor perturbations (gravitational waves) and, as shown in Zaldarriaga and Harari (1995), deviations from zero of the so-called pseudo-scalar or 'magnetic' component of polarization would be an irrefutable indication of the presence of gravitational waves.

Let us consider the properties of the polarization field generated on the surface of the last scattering of quanta by electrons. In the general case, we can introduce, for elliptically polarized waves, an expansion of the electric field vector into two orthogonal components (see Fig. 6.1) (Hu and White, 1997a,b). We denote them by E_ξ and E_τ, assuming that the reference frame system has been chosen. We denote the corresponding components of the vector $\vec{E}$ after a quantum is scattered by an electron by E_ξ^S and E_τ^S. The components before the scattering, E_ξ and E_τ, depend on time as follows:

$$
\begin{aligned}
E_\xi &= E_\xi^0 \sin(\omega t - e_1), \\
E_\tau &= E_\tau^0 \sin(\omega t - e_2),
\end{aligned}
\tag{6.1}
$$

where E_ξ^0 is the wave amplitude, ω is the wave frequency, and e_1 and e_2 are the initial phases. Starting with Eqs (6.1), we introduce the characteristics of the radiation field known as the Stokes parameters as follows:

$$
\begin{aligned}
I &= \left(E_\xi^0\right)^2 + \left(E_\tau^0\right)^2 \equiv I_\xi + I_\tau, \\
Q &= E_{0\xi}^2 - E_{0\tau}^2 \equiv I_\xi - I_\tau, \\
U &= 2E_{0\xi} E_{0\tau} \cos(e_1 - e_2), \\
V &= 2E_{0\xi} E_{0\tau} \sin(e_1 - e_2),
\end{aligned}
\tag{6.2}
$$

where I_ξ and I_τ are the intensities of the ξ and τ components. The definitions in Eqs (6.2) imply an obvious inequality, $I^2 \geq Q^2 + U^2 + V^2$. The question arises as to how the Stokes

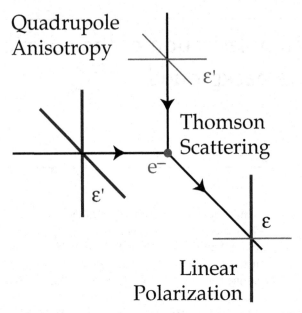

Figure 6.1 Scattering of a quantum by an electron, where θ is the scattering angle. Adapted from Hu and White (1997a,b).

parameters are transformed when the coordinate system is rotated by an angle α. By virtue of definitions (6.2) the I and V components do not change in response to clockwise rotation of the coordinate system by this angle while the Q and U components are transformed according to the law that is defined as the effect produced by the rotation operator:

$$
\tilde{L}(\alpha) = \begin{pmatrix} \cos^2 \alpha & \sin^2 \alpha & \frac{1}{2}\sin 2\alpha & 0 \\ \sin^2 \alpha & \cos^2 \alpha & -\frac{1}{2}\sin 2\alpha & 0 \\ -\sin 2\alpha & \sin 2\alpha & \cos 2\alpha & 0 \\ 0 & 0 & 0 & 1 \end{pmatrix}
\tag{6.3}
$$

on the vector $\vec{I} = (I_\xi, I_\tau, U, V)$.

Let us consider how the Stokes parameters change after a scattering of a quantum by an electron. In a single-collision event (see Fig. 6.1) the relation between the vector components $\vec{I}^{\mathrm{S}}$ and $\vec{I}$ is given by

$$
\vec{I}^{\mathrm{S}} = \sigma_{\mathrm{T}} \hat{R} \times \vec{I},
\tag{6.4}
$$

where

$$
\hat{R} = \frac{3}{2} \begin{pmatrix} \cos^2 \theta & 0 & 0 & 0 \\ 0 & 1 & 0 & 0 \\ 0 & 0 & \cos \theta & 0 \\ 0 & 0 & 0 & \cos \theta \end{pmatrix}.
$$

To analyse the changes in the Stokes parameters in the course of scattering of a quantum by an electron, we choose the laboratory coordinate system as shown in Fig. 6.2. In this reference frame, the Stokes vector after scattering is expressed via the pre-scattering components of the

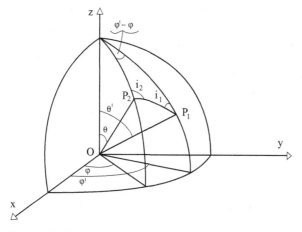

Figure 6.2 Polarization in laboratory reference frame. For notation of angles, see text. Adapted from Melchiori and Vittorio (1996).

vector $\vec{I}(\theta', \varphi')$ (Balbi *et al.*, 2002; Hu and White, 1997a,b; Melchiori and Vittorio, 1996):

$$\vec{I}^S(\theta, \varphi) = \frac{1}{4\pi} \int_{4\pi} \left[\hat{P}(\theta, \varphi, \theta', \varphi') \times \vec{I}(\theta', \varphi') \right] d\Omega', \tag{6.5}$$

where

$$\hat{P} = \hat{Q} \times \left[\hat{P}^0(\mu, \mu') + \sqrt{(1-\mu^2)}\sqrt{(1-\mu'^2)} \hat{P}^1(\mu, \varphi, \mu', \varphi') + \hat{P}^2(\mu, \varphi, \mu', \varphi') \right], \tag{6.6a}$$

$$\hat{Q} = \frac{3}{2} \begin{pmatrix} 1 & 0 & 0 & 0 \\ 0 & 1 & 0 & 0 \\ 0 & 0 & 2 & 0 \\ 0 & 0 & 0 & 2 \end{pmatrix}, \tag{6.6b}$$

$$\hat{P}^0 = \frac{3}{4} \begin{pmatrix} 2(1-\mu^2)(1-\mu'^2) + \mu^2\mu'^2 & \mu^2 & 0 & 0 \\ \mu'^2 & 1 & 0 & 0 \\ 0 & 0 & 0 & 0 \\ 0 & 0 & 0 & \mu\mu' \end{pmatrix}, \tag{6.6c}$$

$$\hat{P}^1 = \frac{3}{4} \begin{pmatrix} 4\mu\mu'\cos(\varphi-\varphi') & 0 & 2\mu\sin(\varphi-\varphi') & 0 \\ 0 & 0 & 0 & 0 \\ -2\mu'\sin(\varphi-\varphi') & 0 & \cos(\varphi-\varphi') & 0 \\ 0 & 0 & 0 & \cos(\varphi-\varphi') \end{pmatrix}, \tag{6.6d}$$

$$\hat{P}^2 = \frac{3}{4} \begin{pmatrix} \mu^2\mu'^2\cos 2(\varphi'-\varphi) & \mu^2\cos 2(\varphi'-\varphi) & \mu^2\mu'\sin 2(\varphi'-\varphi) & 0 \\ -\mu'^2\cos 2(\varphi'-\varphi) & \cos 2(\varphi'-\varphi) & -\mu'\sin 2(\varphi'-\varphi) & 0 \\ -\mu'^2\mu\sin 2(\varphi-\varphi') & \mu\sin 2(\varphi-\varphi') & \mu\mu'\cos 2(\varphi'-\varphi) & 0 \\ 0 & 0 & 0 & 0 \end{pmatrix}, \tag{6.6e}$$

where $\mu = \cos\theta$ and $\mu' = \cos\theta'$.

We need to point out an important property of polarization, namely that it is generated beginning only with quadrupole modes for $I(\theta', \varphi)$ in Eq. (6.5). This is due to the property of the polarization operator $\hat{P}$ for quantum scattering by electrons. Furthermore, it is important to emphasize that the 'last' scattering does not in itself result in the generation of polarization unless there were peculiar motions of plasma and perturbations of density and gravitational potential in the epoch of hydrogen recombination: the primordially non-polarized CMB radiation remains non-polarized after electromagnetic radiation separates from electrons. However, the situation changes drastically in the presence of small fluctuations of density, velocity and metric during the hydrogen recombination epoch. We have already mentioned that weak anisotropy of the CMB is the source of formation of the angular distribution of polarization whose properties depend directly on the characteristics of the 'source'.

Following Peebles and Yu (1970) (see also Peebles (1980)) we introduce fluctuations of Stokes parameters,

$$
\begin{pmatrix} I \\ Q \\ U \\ V \end{pmatrix} = \frac{\rho_\gamma(t)}{4\pi} \begin{pmatrix} 1+i \\ q \\ u \\ v \end{pmatrix},
\tag{6.7}
$$

and make use of the kinetic Boltzmann equations to find the angular dependence of these parameters. Solutions of these equations in general form were studied both analytically and numerically in Peebles and Yu (1970) and later in Crittenden *et al.* (1995); those readers interested in details are referred to these original publications. The most important characteristic of the CMB polarization is, as in the case of anisotropy, the power spectrum of fluctuations arising in the space of multiples l. In contrast to the anisotropy power spectrum, the determination of the spectrum in the case of polarization requires the knowledge of certain details. As we saw at the beginning of this section (see Eqs (6.2)), the Stokes parameters Q and U are transformed into each other by rotating the reference frame by an angle α in the polarization plane.

Following Crittenden *et al.* (1995), we find the correlation function of the Q component as[1]

$$
\left\langle \frac{Q(\vec{\gamma}_1)Q(\vec{\gamma}_2)}{T_0^2} \right\rangle = A(\theta) + B(\theta, \varphi),
\tag{6.8}
$$

where

$$
A(\theta) = P_l(\cos\theta),
\tag{6.9}
$$

$$
B(\theta, \varphi) = \frac{1}{4\pi} \sum_l (2l+1) B_l \cos(4\varphi) \cdot P_l(\cos\theta),
\tag{6.10}
$$

$C(l)^Q$ is the multipole spectrum of the Q component and B_l is the so-called spectrum of the UQ correlations (Crittenden *et al.*, 1995). Note that the spectra for the U component and gravitational waves are found in a similar way. A new element of the theory arises when we take into account the cross-correlation between anisotropy and polarization of the primordial radiation. According to Crittenden *et al.* (1995) and Zaldarriaga and Seljak

---

[1] We give the expression for $A(\theta)$ and $B(\theta, \varphi)$ in the approximation of small angles θ and φ.

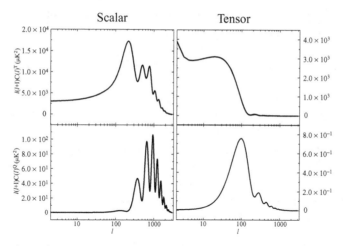

Figure 6.3 The spectral distribution of anisotropy (above) and polarization (below) for scalar and tensor modes. Adapted from Melchiori and Vittorio (1996).

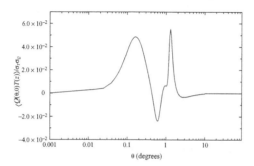

Figure 6.4 Cross-correlation function for anisotropy and polarization as a function of angle θ. Adapted from Melchiori and Vittorio (1996).

(1997), the corresponding cross-correlation function is given by the expression

$$\left\langle \frac{\delta T(\vec{\gamma}_1)Q(\vec{\gamma}_2)}{T_0^2} \right\rangle = \frac{1}{4\pi} \sum_l (2l+1)C(l)^{QT} \cos(2\varphi)P_l^2(\cos\theta), \qquad (6.11)$$

where $P_l^2(\cos\theta)$ are associated Legendre polynomials.

Figure 6.3 gives the results of numerical simulation of anisotropy and polarization spectra generated in the cosmological SCDM model by adiabatic perturbations of gravitational waves. The parameters of the model were chosen as follows: $\Omega_b = 0.05$; $\Omega_\Lambda = 0$; $\Omega_0 = 0.95$; $n_s = 1$ (the Harrison–Zeldovich spectrum), $h = 0.5$ and the hydrogen recombination is standard. Note that purely tensor modes (gravitational waves) illustrate the tendency to modify the structure of anisotropy and polarization spectra. As we see from comparing the polarization spectra for adiabatic modes and gravitational waves, the latter generate extremely small polarization. Figure 6.4 illustrates the behaviour of the cross-correlation function of anisotropy and polarization in the adiabatic mode of perturbations as functions of the angle θ. We see that appreciable correlations are found in the range 0.1–$1°$ while there are practically no correlations outside of this range.

6.2   Electric and magnetic components of the polarization field

As we stressed in Section 6.1, the Stokes Q and U parameters are transformed into each other, by virtue of the nature of the polarization field, by a rotation of the coordinate system through an angle α in the plane perpendicular to the vector $\vec{n}$ pointing in the direction of the arrival of quanta.

We introduce a simplification by assuming that the angular scale we investigate for polarization is so small that the corresponding area of the sky can be regarded as flat. Under this approximation the polarization field in the sky can be treated as a two-dimensional field in the (x, y) plane. The photon polarization is described by a tensor of a_{ij} of rank 2 in a plane perpendicular to the trajectory of photons. The trace of this tensor is, by definition, zero, which corresponds to zero polarization and can be included in the total radiation intensity. It is convenient to express this term through the Pauli matrices σ_α, $\alpha = 1, 2, 3$, that form a complete set of the 2×2 traceless matrix space:

$$a = \xi_\alpha \sigma_\alpha. \tag{6.12}$$

The parameter ξ_2 equals the amplitude of circular polarization, which does not arise under Thomson scattering. We assume therefore that $\xi_2 = 0$. In this case the matrix a is symmetrical and formed by two functions:

$$a = \begin{pmatrix} Q & U \\ U & -Q \end{pmatrix}. \tag{6.13}$$

The functions Q and U depend on the coordinate system. They are components of the tensor a_{ij} and obey the corresponding rule of tensor transformation

$$a'_{ij} = T_i^k T_j^l a_{kl}, \tag{6.14}$$

where transformation coefficients are given by the formula $x_i = T_i^k x_k$. Thus a rotation of the coordinate system gives

$$T = \begin{pmatrix} c & s \\ -s & c \end{pmatrix}, \tag{6.15}$$

where $c = \cos \varphi$, $s = \sin \varphi$, and φ is the angle of rotation. The parameters Q and U transform as follows:

$$Q' = Q \cos 2\varphi + U \sin 2\varphi; \qquad U' = -Q \sin 2\varphi + U \cos 2\varphi. \tag{6.16}$$

In many cases it is more convenient to work with invariant quantities or at least with vectors whose directions are clearly recognizable on polarization maps. The following invariants (or scalars, which are the same) exist that can be constructed of tensors of rank 2. The first invariant is obviously the trace of the matrix: $\mathrm{Tr}\, a = \sum a_{ii}$; it vanishes in the case we consider here. The second invariant is the determinant of the matrix a,

$$\det a = Q^2 + U^2. \tag{6.17}$$

The maximum value of polarization is given by the expression $\sqrt{Q^2 + U^2}$. The direction of maximum polarization is determined by one of the eigenvectors of the matrix a_{ij} (see Dolgov *et al.* (1999)).

These are well known algebraic invariants that exist in the space of arbitrary dimensionality. Using the vector differentiation operator we can construct two more invariants. We can choose them as follows:

$$S = \partial_i \partial_j a_{ij}; \qquad P = \varepsilon_{kj} \partial_k \partial_i a_{ij}, \tag{6.18}$$

where $j = 1, 2$ and $\partial_j = \partial/\partial x^j$. These invariants can be expressed in terms of Q and U as follows:

$$S = \left(\partial_1^2 - \partial_2^2\right) Q + 2\partial_1 \partial_2 U; \qquad P = \left(\partial_1^2 - \partial_2^2\right) U - 2\partial_1 \partial_2 Q. \tag{6.19}$$

The first scalar invariant exists in the space of any dimensionality while the second pseudo-scalar exists only in two-dimensional space, owing to the presence of the antisymmetric pseudotensor ε_{rj} (a similar antisymmetric tensor in D-dimensional space has D indices). The values taken by S and P coincide up to a scalar factor with the corresponding values E and B introduced in Seljak (1996a,b) and Zaldarriaga and Seljak (1997). This E is, to use the terminology of those authors, an analogue of the electric field component of the electromagnetic wave, and B is, correspondingly, an analogue of the magnetic component. From our standpoint, it would be more natural to denote them by S and P in order to emphasize their scalar and pseudo-scalar nature, rather than the electric and magnetic parts of polarization, since these quantities have no relation to vectors. Therefore we use the terminology from Stebbins (1996) (see also Kamionkowski, Kosowsky and Stebbins (1997a,b)).

An important property of the pseudo-scalar P is that it vanishes if polarization is caused only by scalar density perturbations. In this case the Stokes matrix can be expressed in terms of derivatives of one scalar function:

$$a_{ij} = (2\partial i \partial j - \delta_{ij} \partial k \partial k) \, \Psi. \tag{6.20}$$

It can be readily confirmed that in this case P is indeed zero. We do not share the opinion and terminology chosen in Kamionkowski and Kosowsky (1998), where it is stated that the corresponding field has zero vorticity. It was pointed out in Dolgov *et al.* (1999) that this is not so, and that in the general case the eigenvectors of the Stokes matrix are not of zero vorticity. The correctness of this general attitude may be tested in simple situations. This means, among other things, that the flux of maximum polarization direction lines may have non-zero vorticity.

If tensor perturbations (gravitational waves) are present, then the polarization matrix has a general form determined by two independent functions. It is well known that an arbitrary three-dimensional vector can be written in terms of scalar and vector potentials:

$$\vec{V} = \text{grad } \Phi + \text{curl } \vec{A}. \tag{6.21}$$

In the case of two dimensions, an arbitrary vector can be defined via derivatives of a scalar and a pseudo-scalar:

$$V_j = \partial_j \Phi_1 + \varepsilon_{jk} \partial_k \Phi_2. \tag{6.22}$$

In direct analogy to this, an arbitrary traceless symmetric 2×2 matrix can be expressed through scalar and pseudo-scalar potentials:

$$a_{ij} = \left(2\partial_i \partial_j - \delta_{ij} \partial^2\right) \Psi + \left(\varepsilon_{ik} \partial_k \partial_j + \varepsilon_{jk} \partial_k \partial_i\right) \Phi. \tag{6.23}$$

Now the scalar P described by Eq. (6.18) does not vanish, and this property makes it possible to study possible tensor perturbations by measuring CMB polarization. If $\Psi = 0$ then the

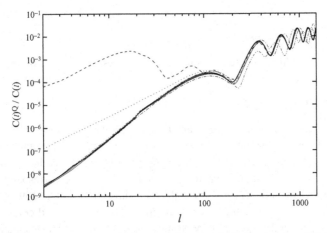

Figure 6.5 The spectra of the E component of polarization in a CDM model (thick solid curve), ΛCDM (thin solid curve) and ΛCDM with $\Omega_b = 0.02$ (dashed curve). For the parameters of the models, see text. Adapted from Melchiori and Vittorio (1996).

scalar S vanishes. Unfortunately, this does not signify that tensor perturbations dominate: they contribute both to Φ and to Ψ.

Let us ask ourselves how high is the sensitivity of the E component of polarization field to the parameters of the cosmological model and type of primordial perturbations that generate both the anisotropy and polarization of the primordial radiation. We have already seen in Chapter 5, devoted to analyzing the CMB anisotropy, that for $C(l)$ the dependence on the factors listed above is quite well pronounced. This conclusion is also applicable fully to the polarization field, for which Fig. 6.5 plots the spectra for the E component of polarization generated by adiabatic perturbations in the SCDM model ($\Omega_b = 0.05; \Omega_{tot} = 0.95; h = 0.5$), in a ΛCDM model ($\Omega_b = 0.05; \Omega_{dm} = 0.3; \Omega_\alpha = 0.65; h = 0.5$) and in the CDM model with low baryonic fraction density ($\Omega_{dm} = 0.2$), but with $\Omega_\alpha = 0.98$ so that $\Omega_{tot} = 1$. As we see from this figure, the polarization spectrum is found to be very sensitive to the choice of parameters of the cosmological model.

6.3   Local and non-local descriptions of polarization

Establishing which of the components of the polarization field would be easier to measure – the local or the global (averaged over a certain area of the sky) – would be an interesting observational problem.

We believe that a solution would strongly depend on the properties of noise. For instance, if the noise of the polarization field in measurements of the CMB originates from randomly distributed pointlike sources with average separation on the sky greater than the antenna resolution, the less labour-consuming (and hence less difficult) would be the measurements of local quantities; that is, direct measurements of S and P using Eq. (6.18). However, it may be the case that it is easier to suppress sources of noise than to measure quantities averaged over that part of the sky. To go into details, we introduce expressions for averaged values of S and P (or E and B) (Seljak and Zaldarriaga, 1998). First we find the Fourier transforms of the fields Q and U, as follows:

$$\tilde{Q}(\vec{l}) = \int y\, e^{-i\vec{l}\,\vec{y}}\, Q(\vec{y})\, dy^2 \tag{6.24}$$

and similarly for U. The Fourier transforms for the scalar and pseudo-scalar fields can be written as follows (φ_l is the polar angle in the Fourier space of coordinates $\vec{l}$):

$$
\widetilde{S_N}(\vec{l}) = N(l^2) \int e^{-i\vec{l}\,\vec{y}} \left[Q(\vec{y}) \cos 2\varphi_l + U(\vec{y}) \sin 2\varphi_l \right] d^2 y,
$$

$$
\widetilde{P_N}(\vec{l}) = N(l^2) \int e^{-i\vec{l}\,\vec{y}} \left[U(\vec{y}) \cos 2\varphi_l - Q(\vec{y}) \sin 2\varphi_l \right] d^2 y.
$$

(6.25)

The scalar function $N(l^2)$ is arbitrary; it conserves the scalar and pseudo-scalar properties of S and P. So that Eqs (6.25) and (6.18) are equivalent, we choose the function $N(l)$ in the form $N(l^2) = l^2$. The definition used in Seljak and Zaldarriaga (1998) corresponds to choosing $N(l^2) = 1$. This means that non-locality is introduced into the coordinate space by the inverse Laplace operator $1/\partial^2$, that is, by the Green's function for the Laplacian. To obtain the functions S_N and P_N we apply the inverse Fourier transform:

$$
S_N(\vec{x}) = \int \frac{d^2 l}{(2\pi)^2} N(l^2) \int e^{i\vec{l}(\vec{x}-\vec{y})} \left[Q(\vec{y}) \cos 2\varphi_l + U(\vec{y}) \sin 2\varphi_l \right] d^2 y,
$$

$$
P_N(\vec{x}) = \int \frac{d^2 l}{(2\pi)^2} N(l^2) \int e^{i\vec{l}(\vec{x}-\vec{y})} \left[-Q(\vec{y}) \sin 2\varphi_l + U(\vec{y}) \cos 2\varphi_l \right] d^2 y,
$$

(6.26)

where φ_l is the angle between the vector $\vec{l}$ and a certain fixed direction which is conveniently assumed to be the direction of the vector $\vec{x}$, that is, $\varphi_l = \varphi_{xl}$.

Now we can carry out complete integration over all directions of the vector $\vec{l}$. To simplify notation, we introduce the vector

$$
\vec{\rho} = \vec{x} - \vec{y}
$$

(6.27)

and three angles, $\varphi_{l\rho}$, φ_{px} and φ_{xl} between the directions $\vec{\rho}$, $\vec{x}$ and $\vec{y}$. Obviously,

$$
\varphi_{l\rho} + \varphi_{px} + \varphi_{xl} = 0.
$$

(6.28)

The integral over angular variables reduces to

$$
\int_0^{2\pi} e^{il\rho \cos \varphi_{l\rho}} \left(A \cos 2\varphi_{l\rho} + B \sin 2\varphi_{l\rho} \right) d\varphi_{l\rho},
$$

(6.29)

where the functions A and B are independent of $\varphi_{l\rho}$. The second term vanishes and the first yields

$$
\int_0^{2\pi} e^{il\rho \cos \varphi_{l\rho}} \cos 2\varphi_{l\rho} \, d\varphi_{l\rho} = -2\pi J_2 (l\rho),
$$

(6.30)

where $J_2(z)$ is the Bessel function (see Gradstein and Ryzhik (1994)).

Integration over l depends on the type of the function $N(l^2)$, and the result of this integration is a function of the modulus of the vector $\vec{p}$:

$$
\int_0^\infty l N(l^2) J_2 (l\rho) \, dl = F_N (\rho).
$$

(6.31)

In a particular case of $N(l^2) = 1$ chosen in Seljak and Zaldarriaga (1998) the integral can be taken in the following manner. Formally it diverges and a regularization procedure is required. This can be achieved by adding a small imaginary part to l in order to ensure convergence (in other words, we need to shift the integration contour to the upper half of the l plane). Using the relation

$$z J_2(z) = J_1(z) - z J_1'(z) \tag{6.32}$$

and integrating by parts, we obtain

$$F_1(\rho) = \frac{1}{\rho^2} \int_0^\infty z J_2(z)\, dz = \frac{1}{\rho^2} \left[2 \int_0^\infty J_1(z) - z J_1 \big|_0^\infty\, dz \right] = \frac{2}{\rho^2}, \tag{6.33}$$

and finally, gathering together all contributions, we obtain

$$S_N(\vec{x}) = \frac{1}{2} \int_0^\infty \rho F_N(\rho)\, d\rho \int_0^{2\pi} \left[Q\, (\vec{x} - \vec{\rho}) \cos 2\varphi + U\, (\vec{x} - \vec{\rho}) \sin 2\varphi \right] d\varphi,$$

$$P_N(\vec{x}) = \frac{1}{2} \int_0^\infty \rho F_N(\rho)\, d\rho \int_0^{2\pi} \left[-Q\, (\vec{x} - \vec{\rho}) \sin 2\varphi + U\, (\vec{x} - \vec{\rho}) \cos 2\varphi \right] d\varphi.$$

$$\tag{6.34}$$

In the particular case of $F_N(\rho) = F_1(\rho) = 2/\rho^2$ chosen in Seljak and Zaldarriaga (1998) we arrive at almost the same result as in this reference but with one difference: we see no reason to consider the window-function $F_1(\rho) = 2/\rho^2$ equal to zero at $\rho = 0$. In any case this is a measure-zero difference and does not affect the value of the integral in Eq. (7.23). Therefore we can ignore the difference. We are of the opinion that the conclusion made in Seljak and Zaldarriaga (1998) is more significant; it states that in order to avoid a difficult (or even impossible) integration of data over the entire sky, we can use a modified smoothing function

$$F(\rho) = -g(\rho) + \frac{2}{\rho^2} \int_0^\rho \rho' g(\rho')\, d\rho', \tag{6.35}$$

where the function $g(\rho)$ obeys the condition

$$\int \rho g(\rho)\, d\rho = 0, \tag{6.36}$$

and where the last integral is taken over the entire sky.

We believe that any smoothing function can be used legitimately and that no additional conditions must be imposed that would constrain its structure. In order to demonstrate this, we calculate the functions $S_N(\vec{x})$ and $P_N(\vec{x})$ for a particular case of scalar perturbations when a Stokes matrix is given by Eq. (6.9). The calculation of derivatives in polar coordinates is simple, and after straightforward algebraic manipulations we obtain

$$S_N(\vec{x}) = \frac{1}{2\pi} \int_0^\infty \rho W(\rho)\, d\rho \int_0^{2\pi} \left(\Psi_{\rho,\rho}\, (\vec{x} - \vec{\rho}) - \frac{\Psi_\rho(\vec{x} - \vec{\rho})}{\rho} \right) d\varphi,$$

$$P_N(\vec{x}) = \frac{1}{2\pi} \int_0^\infty \rho W(\rho)\, d\rho \int_0^{2\pi} \left(\frac{2\Psi_{\rho,\varphi}\, (\vec{x} - \vec{\rho})}{\rho} - \frac{2\Psi_\varphi(\vec{x} - \vec{\rho})}{\rho^2} \right) d\varphi,$$

$$\tag{6.37}$$

where the lower indices p or φ denote differentiation over the corresponding variable and $W(\rho)$ is an arbitrary smoothing function. The second expression in Eqs (6.26) shows that P does indeed vanish for any smoothing function. Therefore, in order to prove the absence of tensor perturbations, one needs either to make sure that the local value of $P(\vec{x})$ given by Eq. (6.8) does vanish, or that the non-local quantity yielded by Eqs (6.26), with an arbitrary function $W(\rho)$, does vanish. It depends on the properties of noise which of the methods proves more efficient.

6.4   Geometric representation of the polarization field

To study the properties of the CMB polarization field, and in particular the distribution of this polarization over the celestial sphere, it is necessary to develop a technique (or techniques) of its visualization (mapping). Mapping causes no difficulties for anisotropy distribution because the values of $\Delta T(\theta, \varphi)$ on the sphere are characterized by a scalar function and are radially visualized using a linear or colour scale that puts a value of ΔT in correspondence with a certain colour. This method obviously fails for the CBM polarization as it is definitely not a linear function. Moreover, it is clear that several methods can be suggested that would represent the fundamental properties of a polarized signal. For instance, to map the polarization field, Bond and Efstathiou (1987) used the 'vector' $\vec{P}(\theta, \varphi)$ whose length $|\vec{P}|$ equals $|\vec{P}| = P = \sqrt{Q^2 + U^2}$ and whose direction is given by the condition $\tan 2\varphi = U/Q$. Zaldarriaga and Seljak (1997) used the maps $E(\theta, \varphi)$ and $B(\theta, \varphi)$ of the components of the polarization fields.

Dolgov *et al.* (1999) suggested using a 'vector' field $\vec{n}$, corresponding to eigenvectors of the Stokes matrix for visualizing the peculiarities of the polarization field. Note that any transformation of the quantities Q and U can (and must) reflect specifics of their distribution on the celestial sphere. Since the magnitude of the polarization vector $P = \sqrt{Q^2 + U^2}$ is invariant under rotations of the polarization plane, its properties will inevitably affect any transformation of the type $\tilde{L}(\theta, \varphi)$. Particular interest is connected with the field structure in the neighbourhood of the so-called singularities of the polarization field at which $P = 0$ by virtue of the vanishing of the Q and U components at the same point of space.

Of course, the measurement of the CMB polarization in the neighbourhood of points where polarization vanishes is an extremely difficult observational problem. However, there is no need to approach $Q^2 + U^2 = 0$ very closely. The type of singularity can be identified from a visualization of the flux line pattern in the region where polarization has not yet completely vanished.

An analysis of singularities of such fields is given in Dolgov *et al.* (1999) and Naselsky and Novikov (1998). Dolgov *et al.* (1999) discovered that the types of singularities are not described by the familiar classification of singularities of vector fields in the standard theory of dynamic systems. Because of the non-analytical behaviour of vectors in the neighbourhood of zero points $Q^2 + U^2 = 0$, the separatrixes terminate at a singularity, while in the ordinary case they continue smoothly through these points. This unusual behaviour is clearly traceable on polarization maps computed in Seljak and Zaldarriaga (1997, 1998). Following Naselsky and Novikov (1998) and Dolgov *et al.* (1999), we consider the problem of singularities of the polarization field in more detail. We will use the eigenvectors of the Stokes matrix found in Dolgov *et al.* (1999) as characteristics of the angular distribution of polarization on the celestial sphere.

The eigenvectors of the matrix (6.13) are expressed in terms of the Stokes parameters Q and U as follows:

$$\vec{n}^+ \sim \{U, \lambda - Q\},$$
$$\vec{n}^- \sim \{-U, \lambda + Q\}, \tag{6.38}$$

where $\lambda = \sqrt{Q^2 + U^2}$ is the eigenvalue and $\vec{n}^\pm$ correspond to the positive and negative eigenvalues $\pm\lambda$. The vector $\vec{n}^+$ is parallel to the direction of polarization maximum, while the vector $\vec{n}^-$ points along the polarization minimum. This is also evident in the base of eigenvectors that diagonalizes the polarization matrix, $a = \text{diag}(\lambda, -\lambda)$. The total intensity of light polarized along $\vec{n}^\pm$ is given by $I_\pm = I_0 \pm \lambda$. Consequently, the intensity along the direction $\vec{n}^+$ is found to be higher than that along the direction of $\vec{n}^-$.

To avoid ambiguity, we choose to consider the field of directions of the vector $\vec{n}^+$ and the singular points of this field. We are looking at the problem of singular points of the vector field in the case when the direction of the components $[x(t), y(t)]$ of this two-dimensional vector field obeys the following equation:

$$\frac{dy}{dx} = \frac{F_1(x, y)}{F_2(x, y)}. \tag{6.39}$$

Singularities may arise when both functions $F_{1,2}(x, y)$ vanish. In this case the condition of unicity of the solution of the differential equation is not met, and more than one integral curve can be traced through the same point. The standard theory is valid when the functions $F_{1,2}(x, y)$ are analytical in the neighbourhood of the zeros, and the first order of the Taylor expansion for them has the form

$$F_j = a_j(x - x_0) + b_j(y - y_0). \tag{6.40}$$

In this situation there exist three types of singularities: nodes, saddles and foci (see Bronshtein and Semendyaev (1955)). The separatrixes of these solutions are three intersecting lines that degenerate to straight lines in the linear approximation. However, in the case of the vector polarization field, the main equations change to

$$\frac{dy}{dx} = \frac{n_y^+}{n_x^+} = \frac{\lambda - Q}{U}. \tag{6.41}$$

As before, the singular points may arise if both the numerator and the denominator vanish. This is equivalent to the condition $Q = U = 0$. The main difference from the standard case is that the numerator is not analytical in the vicinity of zero. This fact leads to a very different behaviour of the integral curves close to such points. In this case the standard theory does not work. In what follows we investigate the structure of solutions in the neighbourhood of these points directly. Assume now that the functions Q and U are analytical close to $Q = U = 0$ and can be described by the equations

$$Q \simeq q_1 x + q_2 y, \qquad U \simeq u_1 x + u_2 y. \tag{6.42}$$

For brevity we assume that Q and U vanish for $x = y = 0$. For convenience we introduce new coordinates:

$$\xi = q_1 x + q_2 y; \qquad \eta = u_1 x + u_2 y. \tag{6.43}$$

Since a transformation of these coordinates consists of rotation and a change of scale, the type of singularities remains unchanged. We introduce polar coordinates on a plane as follows:

$$\xi = r \cos \varphi; \qquad \eta = r \sin \varphi. \tag{6.44}$$

In these coordinates, Eq. (6.41) becomes

$$\frac{d\ln r}{d\phi} = \frac{N}{D} \equiv \frac{q_2 t^3 + (q_1 - 2u_2)t^2 - (q_2 + 2u_1)t - q_1}{u_2 t^3 + (u_1 + 2q_2)t^2 + (2q_1 - u_2)t - u_1},$$

(6.45)

where $t = \tan(\varphi/2)$.

In the general case, the denominator D has three roots t_j, $j = 1, 2, 3$. With no loss of generality we can assume that $u_2 = 1$, so that these roots meet the following conditions:

$$t_1 t_2 t_3 = u_1, \quad t_1 t_2 + t_2 t_3 + t_3 t_1 = 2q_1 - 1, \quad t_1 + t_2 + t_3 = -(u_1 + 2q_2).$$

(6.46)

The integration of Eq. (6.45) is simplified if we expand the right-hand side into elementary factors,

$$\frac{d\ln r}{d\varphi} = q_1 + \sum_j^3 \frac{B_j}{t - t_j},$$

(6.47)

where we immediately see that none of the indices $D_j = N(t_j)/(t_j - t_k)(t_j t_l)$ in j, k, l is equal to another. It can be easily confirmed that

$$B_1 = -\frac{(1 + t_2 t_3)\left(1 + t_1^2\right)^2}{2(t_1 - t_2)(t_1 - t_3)}.$$

(6.48)

The remaining parameters B_2 and B_3 are found by using the cyclic permutation of indices in Eq. (6.48). Since $d\ln r/d\varphi = (d\ln r/dt)(1 + t^2)/2$, the equation can finally be rewritten in the form

$$\frac{d\ln r}{dt} = \frac{2}{1 + t^2}\left(q_1 + \sum_j^3 \frac{B_j}{t - t_j}\right),$$

(6.49)

and this makes integration a simple operation. The corresponding solution has the form

$$r = r_0(1 + t^2)\prod_j^3 (t - t_j)^{2\nu_j},$$

(6.50)

where r_0 is an arbitrary constant and ν_j (the power-law exponents) are given by

$$\nu_j = \frac{B_j}{1 + t_j^2},$$

(6.51)

with constants B_j obtained from Eq. (6.48). It is not difficult to confirm that the ν_j satisfy the conditions

$$\sum_l^3 \nu_j = -1,$$

(6.52)

$$\sum_j^3 \nu_j t_j = -\frac{1}{2}\left(\sum_j^3 t_j + \prod_j^3 t_j\right) = q_1,$$

(6.53)

$$\prod_j^3 \nu_j = \frac{\left(1 + t_1^2\right)\left(1 + t_2^2\right)(1 + t_3)}{8(t_1 - t_2)^2(t_2 - t_3)^2(t_3 - t_1)^2}(1 + t_1 t_2)(1 + t_2 t_3)(1 + t_3 t_1).$$

(6.54)

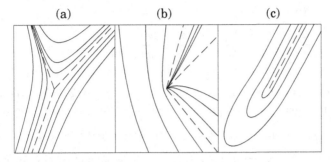

Figure 6.6 Line flux for three different types of singularity: (a) saddles, (b) beaks and (c) comets. Dashed curves correspond to singular solutions (separatrixes). Adapted from Dolgov *et al.* (1999).

The last three factors in Eq. (6.54) are proportional to the determinant $D = q_1 u_2 - q_2 u_1$, and

$$(1 + t_1 t_2)(1 + t_2 t_3)(1 + t_3 t_1) = 2(q_1 u_2 - u_1 q_2)/u_2^2 \equiv 2d/u_2^2. \tag{6.55}$$

If all roots t_j are real, then the sign of the product $\prod_j^3 v_j$ coincides with that of the determinant D. However, if one of the roots t_1 is real while the other two are complex conjugate, then the determinant and the product have opposite signs.

Now we can start to classify the points of singularity. Following Dolgov *et al.* (1999) we first consider the case of three real roots t_j. The behaviour of the solutions is dictated by the signs of the power-law exponents v_j. As follows from Eq. (6.52), at least one of these exponents must be negative. In order to find which other sign combinations are possible, we assume (without loss of generality) that

$$t_1 > t_2 > t_3. \tag{6.56}$$

In this case the following relations hold:

$$\begin{aligned}
\text{sign}[v_1] &= \text{sign}\left[-(1 + t_2 t_3)\right], \\
\text{sign}[v_2] &= \text{sign}\left[(1 + t_1 t_3)\right], \\
\text{sign}[v_3] &= \text{sign}\left[-(1 + t_1 t_2)\right].
\end{aligned} \tag{6.57}$$

Therefore, if $t_3 > 0$ then the following signs arise for v_j: $(-, +, -)$. If $t_3 < 0$ and $t_2 > 0$, then $v_3 < 0$ and one or both of v_1 and v_2 are negative. They cannot both be positive because if $(1 + t_1 t_3) > 0$ then at the same time $(1 + t_2 t_3) > 0$ and $v_1 < 0$. The situation is similar in the case of $t_1 > 0$ and $t_2 < 0$, in which case the sign sequence $(-, +, +)$ is impossible. If all $t_j < 0$, the sign sequence becomes $(-, -, +)$. Therefore only two combinations of signs are allowed for v_j: these are $(-, -, -)$ and $(-, -, +)$. The former combination is realized in the case $d < 0$ in correspondence with Eqs (6.54) and (6.55). If, however, the determinant is positive then the sign combination for v_j is $(-, -, +)$.

If $d < 0$, the solution does not pass through zero in the neighbourhood of the point of singularity, and its behaviour resembles the behaviour of a conventional saddle point; the only difference is that in this case we have three linear asymptotes, not four as in the conventional case (see Fig. 6.6(a)). Following Dolgov *et al.* (1999), we also refer to these points as 'saddle' points.

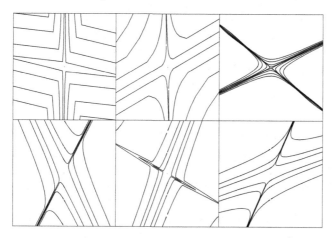

Figure 6.7 Transformation of the flux lines of the 'vector' $\vec{V}$ near the beak (upper panel) and the comet (lower panel) singularities of $\vec{n}^+$ plotted in Figs 6.6(b) and (c), respectively. The maps from left to right correspond to rotation by 45 and 90°, respectively, with respect to the first one. Adapted from Dolgov *et al.* (1999).

The fact that in our case, in contrast to ordinary singularities, the separatrixes do not pass through points of singularity corresponds to a non-analytical behaviour of Eq. (6.41) by virtue of the square root. If $d > 0$, the sign combination becomes $(-, -, +)$ and the solution vanishes along one of the directions while it tends to infinity along the other two. The form of the solution greatly differs from the standard one. The field lines cannot be continuous from $\varphi = \varphi_1$ to $\varphi = \varphi_1 + \pi$, which is possible in the ordinary case. We refer to this type of singularity as a 'beak' (see Fig. 6.6(b)). If only one of the roots t_j is real and the other two are complex conjugate, the solution has the form

$$\frac{r}{r_0} = (t^2 + 1) \mid t - t_2 \mid^{4\text{Re }\nu_2} \exp(4\beta \text{ Im } \nu_2)(t - t_1)^{2\nu_1}, \tag{6.58}$$

where $\beta = \tan^{-1}[\Im t_2/(t - \Re t_2)]$. The real root is negative, $\nu_1 < 0$, as we see from Eq. (6.58), and therefore r does not vanish in the neighbourhood of this singularity. The flux of lines of the polarization field is shown for this case in Fig. 6.6(c). This type of singularity can be referred to as a 'comet'. This situation occurs for $d > 0$.

Note that the first classification of singular points of the CMB polarization map was performed by Naselsky and Novikov (1998), in which the equation $dy/dx = Q/U$ was used to describe the behaviour of flux lines of the 'vector' $\vec{V} = [U, Q]$. However, under a coordinate transformation (for example, rotation) U and Q are not transformed as components of a vector. For a detailed comparison of the two methods, see Dolgov *et al.* (1999). Here we would like to note that, in accordance with the standard classification, there can be three possible types of singular points of the 'vector' $\vec{V}$ in a fixed reference frame: knots, foci and saddles (see Figs 6.6 and 6.7).

The probability of realization of the various types of singularity was calculated in the following manner. Obviously, the probability of a 'saddle' point forming is 50%, since the saddle arises only for $d < 0$. The probability that 'comets' and 'beaks' arise was calculated numerically provided $d > 0$ and there is one real root of the equation $D = 0$ (for comets) or three real roots for beaks. Here D is the denominator in Eq. (6.45) (see Dolgov *et al.* (1999)).

The probabilities of the formation of saddles, beaks and comets with randomly chosen q_1, q_2, u_1 and u_2 equal $W_s = 0.5$, $W_b \simeq 0.116$ and $W_c \simeq 0.384$, respectively.

We can also evaluate the density of singularities on the map (see Bond and Efstathiou (1987) and Naselsky and Novikov (1998)). All points of singularity arise when $Q = 0$ and $U = 0$. The density of these points is proportional to

$$\mathrm{d}\,Q\,\mathrm{d}U = |d|\,\mathrm{d}x\,\mathrm{d}y \qquad (6.59)$$

and therefore the density is defined in terms of the mean values of the determinant $d = q_1 u_2 + q_2 u_1$. It can be shown that saddles are responsible for 50% of all singularities $\langle n_s \rangle = 0.5n$, where n is the concentration of all singular points (Dolgov *et al.*, 1999). The calculation of the concentration of beaks and comets is more complicated and is necessarily numerical. According to our estimates, the surface densities of beaks and comets are given by $\langle n_b \rangle \simeq 0.052n$ and $\langle n_s \rangle \simeq 0.448n$, respectively. Deviations from this and the above-found values for $W_{s,b,c}$ may indicate that perturbations deviate from the Gaussian type.

Such deviations can be caused by various factors. For instance, this could be the presence on CMB polarization maps of various types of noise due to the synchrotron radiation of galaxies, the background of discrete sources, etc. An important disturbing factor for the primordial signal is also the noise of receiver electronics, which is inevitable in all observational data. All these sources distorted the general properties of the polarization field in the neighbourhood of the singularity points where the Q and U components vanish. At first glance, the situation corresponding to the study of the properties of the signal in the zones surrounding such peculiar points looks extremely unpromising. Indeed, the Q and U components are formally a realization of a random Gaussian process (with a natural assumption that the perturbations we seek are also Gaussian), and therefore the behaviour of the fields Q and U in the vicinity of the points $Q = 0$, $U = 0$ is also random. Furthermore, the local properties of the field must be measured at high angular resolution, which is quite a difficult task in most experiments. At the same time, we need to note that a point that plays an important part in analysing the properties of the polarization field is that the characteristic correlation scale of both the Q and U components, θ_c, is close to 10–15 minutes of arc for the most realistic cosmological models. This means that the zones of peculiarity around the points $Q = 0$, $U = 0$ will have practically the same characteristic size, to within an order of magnitude. If this is true, any noise, for instance the measurement electronics noise with correlation scale below θ_c, will result in distortion (destruction) of field behaviour in the neighbourhood of the singularity points; that is it will reduce their concentration and change the signal structure within that neighbourhood. Consequently, owing to the random nature of this noise, this manifestation can be identified. We will take up this aspect in detail in the following chapter.

7

Statistical properties of random fields of anisotropy and polarization in the CMB

7.1   Introduction

In this chapter we concentrate on testing the statistical properties of anisotropy and polarization fields generated by primary perturbations of density, velocity and gravitational potential in the hydrogen recombination epoch. The main working hypothesis on which both the modern theory of structure formation in the Universe and the theory of generation of primordial anisotropy and polarization are based is the assumption of the random nature of the distribution of values of amplitude and phase of the primordial perturbations.

We need to emphasize that the idea of the random nature of primordial perturbations was introduced into cosmology long before it was ultimately explained in terms of inflationary models. In a certain sense, this idea reflects the fact that the distribution of matter in the Universe on spatial scales exceeding 10^2 Mpc is on average uniform and isotropic. If the formation of structures in different parts of the Universe separated by distances greater than 10^2–10^3 Mpc occurred independently of one another, it is then only natural to assume that the primordial distributions of the density, velocity and gravitational potential perturbations were once equally independent. Moreover, we know that non-linear in amplitude perturbations of density resulting from condensation of matter on smaller scales also evolved from small fluctuations. If there is no primordially favoured separation line between the scales of galaxies and clusters on the one hand, and the scale of uniformity in the Universe on the other, then it is natural to assume that primordial perturbations had a random distribution of amplitudes and phases and amplitudes on the entire scale of spatial scales.

The randomness of the spatial distribution of perturbations – provided they are small – is automatically transferred to the Fourier amplitudes. In the simplest case, two more assumptions remain: the form of the distribution function of Fourier amplitudes, which is assumed to be Gaussian, and the equidistribution of metric perturbation amplitudes over the values of the wave vector. After this we arrive at the scale-invariant Harrison–Zeldovich spectrum, which is well known in cosmology and widely used to process observational data on the CMB anisotropy. It is the Harrison–Zeldovich spectrum that corresponds to the spectrum of primordial metric perturbations that are, according to most theories of inflationary expansion of the early Universe, a consequence of quantum fluctuations of the vacuum of physical fields (see, for example, Linde (1990)). Nevertheless, we should mention that the progress in inflation models and, among other things, the inclusion of several fields that model more complicated properties of the vacuum led to the prediction of, along with the possibility of generation of primary Gaussian metric perturbations, more complicated non-Gaussian statistical properties of fluctuations. By virtue of the linear relation between the perturbations of metric, velocity and plasma density during the hydrogen recombination epoch, the statistical properties of the CMB anisotropy and polarization distributions on the sky is in complete agreement with the statistics of primordial perturbations. Therefore, by analysing these specifics of the CMB anisotropy and polarization, we analyse also the properties of inflationary models that describe the birth of the Universe.

Note that along with this fundamental problem there exist a number of important 'applications' problems that require a knowledge of the characteristics of distribution of anisotropy and polarization on the celestial sphere. We again stress the dependence of multiple anisotropy and polarization spectra on the most important cosmological parameters that are discussed in detail in this chapter. However, Gaussian random processes are the only ones for which the knowledge of all spectral parameters completely determines all statistical properties. For non-Gaussian statistics the situation is more complicated (see, for example, Coles and Barrow (1987)). We remind the reader that observations of the CMB anisotropy and polarization are inevitably made in the presence of a number of non-Gaussian sources of radio background.

For instance, the radiation of our Galaxy is clearly visible on CMB anisotropy maps obtained by the COBE mission. Non-Gaussian signals that manifest themselves as additional sources of noise against the background of the useful signal include radio emission of active galactic nuclei, pulsars, dust and hot gas clouds, and synchrotron radiation. Therefore, it is most important to subtract the contributions of these components from the primary signal; knowledge of the statistical nature of the primordial anisotropy and the polarization of the CMB is crucial in solving this problem. Moreover, an insufficient or incorrect account of the factors listed above results in a false structure of anisotropy and polarization power spectra and hence in erroneous determination of parameters of the cosmological model. Finally, the non-Gaussian character of the observed signal of anisotropy and polarization may 'arise' during the experiment as a result of specific properties of the antenna of the radio telescope, of the strategy of observations, insufficient accuracy of establishing the directivity of the antenna, delays in bolometer response, etc.

All these 'instrument' factors require detailed analysis focused on a single goal: the minimization of the level of systematic errors introduced by the experiment into the structure and amplitude of the signal. The study of these specifics is inseparable from studying the characteristics of the primordial signal itself – they are the features that are distorted in the course of the experiment.

The list of problems arising in connection with studying the statistical nature of anisotropy and polarization signals can be continued almost *ad infinitum*, but even the examples given above are sufficient to illustrate the importance and urgency of this type of work. This chapter concentrates on analysing the most important results obtained by modern cosmology since the 1980s of theoretical and experimental studies of statistical properties of the CMB anisotropy and polarization. Several new statistical methods for the study of the CMB anisotropy will be described in chapter 8 in connection with the analysis of the WMAP observational data.

7.2   Spectral parameters of the Gaussian anisotropy field

In this section we introduce the main characteristics of the random Gaussian field of anisotropy, leaving the statistical properties of the polarization field to the subsequent sections of this chapter. Our analysis is based on the generalization of Rice's idea (Rice, 1944, 1945) of describing the properties of one-dimensional noise to three-dimensional noise (developed in Bardeen *et al.* (1986)) and to two-dimensional Gaussian fields (developed in Bond and Efstathiou (1987)).

To facilitate the reading of this section, we begin with a summary of the main definitions and properties of Gaussian statistics of the CMB anisotropy that we used in part earlier. Following Bond and Efstathiou (1987), we consider the temperature distribution of the CMB radiation of the celestial sphere. Let us assume that this field is a random, two-dimensional Gaussian field on a sphere. This field is completely described by the power spectrum $C(l)$. Using this description, we can choose the familiar expression for the CMB temperature,

$$T(\vec{q}) = \langle T(\vec{q}) \rangle + \sum_{l=1}^{\infty} \sum_{m=-l}^{l} a_l^m C(l)^{\frac{1}{2}} Y_l^m(\vec{q}), \qquad (7.1)$$

where q is a unit vector tangential to the direction of propagation of photons, a_l^m are independent random Gaussian numbers, $\langle T(\vec{q}) \rangle$ is the mean temperature of the CMB radiation such that $\langle T(\vec{q}) \rangle = (1/4\pi) \int T(\vec{q}) \, d\Omega$, and Y_l^m are spherical harmonics. We introduce the following expression for the CMB anisotropy: $\Delta T(\vec{q}) = (T(\vec{q}) - \langle T(\vec{q}) \rangle)/\langle T(\vec{q}) \rangle$. The two-point correlation function, $C(\theta)$, can be found by averaging $T(\vec{q}) \times T(\vec{q}')$ over the entire sky, provided the angle between the directions $\vec{q}$ and $\vec{q}'$ remains constant:

$$C_{\text{obs}}(\theta) = \langle \Delta T(\vec{q}) \cdot \Delta T(\vec{q}') \rangle, \quad \vec{q} \cdot \vec{q}' = \cos \theta. \qquad (7.2)$$

Taking into account Eq. (7.2) and also the equality $\langle a_l^m a_{l'}^{m'} \rangle = \delta_{ll'} \delta_{mm'}$, we obtain

$$C_{\text{obs}}(\theta) = \frac{1}{4\pi} \sum_{l=2}^{\infty} \sum_{m=-l}^{l} \left(a_l^m \right)^2 C(l) P_l(\cos \theta). \qquad (7.3)$$

The mean value of the observational correlation function is given by

$$C(\theta) = \overline{C_{\text{obs}}(\theta)} = \frac{1}{4\pi} \sum_{l=2}^{\infty} (2l + 1) C(l) P_l(\cos \theta) . \qquad (7.4)$$

Along with an analysis of properties of the ΔT signal on the celestial sphere, a crucial factor in studying the CMB anisotropy is the so-called 'flat-sky' approximation, in which a small area of the sphere is considered to be flat. This approximation makes it possible to apply the Fourier analysis technique on the plane, which greatly simplifies the mathematical aspect of the problem. However, the flat-sky approximation introduces certain errors into the statistical characteristics of the signal which must be taken into account, even if there is no external noise.

Following Abbott and Wise (1984), we will consider the effect of the final size of an area of the sky on the properties of the correlation function under observation. We take into account that this correlation function deviates from its mean value over the ensemble approximately by the value of variance, due to the 'cosmic variance':[1]

$$D_0(\theta) = \overline{C_{\text{obs}}^2}(\theta) - \overline{C_{\text{obs}}}^2 = \left(\frac{1}{4\pi} \right)^2 \sum_l (2l + 1) C_l^2 P_l^2(\cos \theta). \qquad (7.5)$$

---

[1] This term reflects the fact that when analysing the statistical properties of the ΔT signal, we only have a single realization on the celestial sphere but cannot carry out averaging over the ensemble of realizations.

The zero subscript 0 on D indicates that this value was obtained by averaging over the entire sky. The solution $D_0(\theta)$ is sufficiently small if $\theta \sim 1°$, but if we consider only a small part of the sky this quantity increases to

$$D_\Omega(\theta) \sim \sqrt{\frac{4\pi}{\Xi}} D_0(\theta), \tag{7.6}$$

where Ξ is the area of this small region in units of 4π. We see from Eq. (7.6) that as $[\Xi] \to 0$ the uncertainty in the behaviour of $C(l)$ increases as $\Xi^{1/2}$, which automatically leads to the error $\delta C(l)/C(l) \simeq (\Xi l)^{-1/2}$ (Abbott and Wise, 1984; Knox, 1995). Therefore, the flat-sky approximation describes the general characteristics of the spectrum $C(l)$ with an error $\delta C(l)/C(l) \ll 1$ only for multipoles with numbers $l\Xi \gg 1$.

If we investigate only a small patch of the sky, then its geometry is very nearly flat and we can introduce Cartesian coordinates (x, y) and rewrite $\Delta T(x, y)$ as the sum of the Fourier series (Bond and Efstathiou, 1987)

$$\Delta T(x, y) = \sum_{ij} a_{ij} C^{\frac{1}{2}}(k) \cos\left(2\pi \frac{ix + jy}{L} + \varphi_{ij}\right), \tag{7.7}$$

where $C(k)$ is the power spectrum, $k = \frac{2\pi}{L}\sqrt{i^2 + j^2}$, a_{ij} are independent random Gaussian quantities, φ_{ij} are random phases equidistributed in the interval $(0, 2\pi)$, and $L \simeq \Xi^{1/2}$ is the angular size of the investigated region.

The correlation function $C_{\text{obs}}(r) = \langle \Delta T(x, y) \cdot \Delta T(x', y') \rangle$ can be obtained by averaging over the square $L \times L$ similarly to Eq. (7.2):

$$C_{\text{obs}}(r) = \frac{1}{2} \sum_{ij} a_{ij}^2 C(k) J_0(kr), \tag{7.8}$$

where $r = \sqrt{(x - x')^2 + (y - y')^2}$. Formally, we average over the ensemble of realizations to obtain

$$C(r) = \overline{C_{\text{obs}}(r)} = \frac{1}{2} \sum_{ij} C(k) J_0(kr), \tag{7.9}$$

where k was defined earlier as a function of i and j. Equation (7.9) is in good argeement with Eq. (7.4) since if $\theta \ll \pi$, then $P_l(\cos\theta) \simeq J_0(l\theta)$ and $l\theta \simeq kr$.

Consequently, perturbations ΔT can be described by Eq. (7.7), where $C(k) \simeq C(l)$, $k \sim l/\xi_n$ (ξ_n is the current horizon for $z = 0$). Obviously the observational correlation function deviates from the mean value over the ensemble by the value of variance:

$$D(r) = \overline{C_{\text{obs}}^2(r)} - (\overline{C_{\text{obs}}(r)})^2 = \frac{1}{2} \sum_{ij} C^2(k) J_0^2(kr). \tag{7.10}$$

The correlation function for $r \ll L$ and its variance can be written in the following form:

$$C(r) = \pi \int k C(k) J_0(kr)\, dk,$$
$$D(r) = \pi \int k C^2(k) J_0^2(kr)\, dk, \tag{7.11}$$
$$C_{\text{obs}}(r) \sim C(r) \pm \sqrt{D(r)}.$$

We now use Eqs (7.11) and introduce spectral parameters, as in Bond and Efstathiou (1987):

$$\sigma_0^2 = \pi \int k C(k) \, dk,$$
$$\sigma_1^2 = \pi \int k^3 C(k) \, dk,$$
$$\sigma_2^2 = \pi \int k^5 C(k) \, dk,$$
$$R_* = \sigma_1/\sigma_2, \qquad r_c = \sigma_0/\sigma_1, \qquad \gamma = \sigma_1^2/(\sigma_0\sigma_2). \tag{7.12}$$

It is clear from Eqs (7.12) that the spectral parameters are completely defined by the values of the correlation function and of its second and fourth derivatives with respect to r evaluated at a point $r = 0$:

$$\sigma_i^2 = (-1)^i (i!) 2^{2i} \frac{d^i C(\omega)}{d\omega^i} \bigg|_{\omega=0}. \tag{7.13}$$

Therefore, the spectral parameters σ_i^2 are moments of the spectrum $C(l)$ and completely describe, as shown in Bardeen *et al.* (1986) and Bond and Efstathiou (1987), the local topology of the CMB anisotropy maps. Let us take a closer look at this aspect of the problem.

7.3 Local topology of the random Gaussian anisotropy field: peak statistics

We remind the reader that a Gaussian random field is a field whose joint Gaussian probability of distribution of random variables x_i is given by the expression

$$P(x_1, \ldots, x_n) \, dx_1 \ldots dx_n = \frac{e^{-Q}}{((2\pi)^n \det M)^{1/2}} \, dx_1 \ldots dx_n,$$
$$2Q = \sum_{ij} \Delta x_i (M^{-1})_{ij} \Delta x_j. \tag{7.14}$$

To determine the covariant matrix M_{ij} in Eqs (7.14) we require only the mean values of random variables $\langle x_i \rangle$ and their variances, as follows:

$$M_{ij} = \langle \Delta x_i \Delta x_j \rangle, \qquad \Delta x_i = x_i - \langle x_i \rangle. \tag{7.15}$$

Let us assume now that a random Gaussian process has already been realized on a sphere as it occurs in data on the CMB anisotropy measured by COBE. On the whole, an anisotropy realization appears as a sequence of light and dark zones corresponding to maxima (light zones) and minima (dark zones) of the signal $\Delta T/T$. How is the number of these zones related to the properties of the spectrum of the Gaussian distribution $\Delta T/T$? What are the largest (positive) and smallest (negative) values of $\Delta T/T$ on the map? What is the structure of the signal in the neighbourhood of the points of maxima and minima of $\Delta T/T$? These and other questions constitute the gist of the problem of studying the local topology of the anisotropy field, and the modern theory of random fields provides quite definite answers to these questions. Why is it so important to study the local topology of the signal? Obviously, testing its topological features can confirm or reject the hypothesis of the normal (Gaussian) distribution of anisotropy. Furthermore, it could be possible in some cases to classify the sources of non-Gaussian distortions of the signal and eliminate some of them (effects of systematics, manifestations of galactic and extragalactic noise, etc.), thus coming very close to solving the problem of determining the statistical nature of primordial non-uniformity of the metric, density and velocity of matter calculated from the data on the CMB anisotropy.

A new question arises: why not make use of, say, the analysis of the three-point correlation function, or bispectrum, or higher-order moments which are standard tests of detecting that a signal is non-Gaussian?[2] The answer is surprisingly simple. It is not only possible but also necessary to make use of these standard tests when studying the statistical properties of the realization of random fields. In practice, however, the negative and positive results of applying these tests can hardly be considered final, since a hypothesis of the Gaussian nature of the signal can be confirmed only by using an infinite number of n-point correlation functions. A distribution may look Gaussian up to very high moments but then reveal its non-Gaussian side. Examples of this type of anomaly are well known (see, for example, Kendall and Stuart (1977)). For this reason, any additional statistics and tests that prove to be sensitive to various properties of a Gaussian process mutually complement one another and make it possible to come very close to solving the problem.

As a first important step in studying the local structure of anisotropy maps of the CMB, we consider the statistics of $\Delta T / T$ peaks; this was first suggested as one of the critical tests for the Gaussian nature of the primordial signal in Sazin (1985) and Zabotin and Naselsky (1985). A detailed theory of peaks in the $\Delta(T\theta, \varphi)$ distribution (for a Gaussian signal) was perfected in Bond and Efstathiou (1987) and generalized to specific non-Gaussian fields in Coles and Barrow (1987). Following Bardeen *et al.* (1986) and Bond and Efstathiou (1987) we treat the peak distribution $\Delta T / T$ on a sphere as a random pointlike process characterized by the probability density

$$n_{\mathrm{pk}} = \sum_p \delta(\vec{q} - \vec{q}_p), \tag{7.16}$$

where $\vec{q}$ are the coordinates of an arbitrary point on a sphere, $\vec{q}_p$ are the positions of extreme points of the distribution $\Delta T(\vec{q})$: $\vec{\nabla}(\Delta T(\vec{q})) = 0$; $\vec{\nabla}$ is the gradient operator on a sphere.

Let us consider the behaviour of the anisotropy fields in the neighbourhood of an extreme point. Following Bardeen *et al.* (1986) we introduce the notation $\xi(\vec{q}) \equiv \nabla(\Delta T(\vec{q}))$ and expand $(\Delta T(\vec{q})$ into a Taylor series:

$$\Delta T(\vec{q}) = \Delta T(\vec{q}_p) + \frac{1}{2} \sum_{ij} \xi_{ij}(\vec{q} - \vec{q}_p)_i(\vec{q} - \vec{q}_p)_j. \tag{7.17}$$

Now we carry out similar operations for the field gradient,

$$\eta(\vec{q}) \simeq \sum_{ij} \xi_{ij}(\vec{q} - \vec{q}_p)_j, \tag{7.18}$$

where ξ_{ij} is the matrix of second derivatives of $\Delta T(\vec{q})$ with respect to the variable $\vec{q}$. Assuming the matrix ξ_{ij} to be non-singular at a point $\vec{q} = \vec{q}_p$ we find the components of the vector $(\vec{q} - \vec{q}_p)$ from Eq. (7.18) to be given by

$$(\vec{q} - \vec{q}_p)_i \simeq \xi^{-1}(\vec{q}_p)\eta_i(\vec{q}), \tag{7.19}$$

where $\xi^{-1}(\vec{q}_p)$ is a matrix inverted with respect to the matrix ξ_{ij}. Combining Eqs (7.16) and (7.19) and taking into account the properties of the δ function, we obtain

$$\langle n_{\mathrm{pk}}(\vec{q})\rangle = \langle|\det\xi^{-1}(\vec{q})|\delta^{(2)}[\vec{\eta}(\vec{q})]\rangle = \int \mathrm{d}(\mathrm{d}T)\,\mathrm{d}^6\xi\,|\det\xi^{-1}(\vec{q})|P(\Delta T; \vec{\xi} = 0; \xi),$$

$$\tag{7.20}$$

---

[2] A detailed analysis of these techniques is given in Heavens and Sheth (1999) and Peebles (1983).

where $P(\Delta T, \vec{\xi} = 0, \xi)$ is a joint distribution function of anisotropy ΔT and of its first and second derivatives evaluated at a point $\vec{\xi} = \{\xi_i\} = 0$. The uniformity and isotropy of the field ΔT signifies on average that the mean number of peaks is independent of coordinates on a sphere. Following Bond and Efstathiou (1987) we introduce a dimensionless variable $\nu \equiv \Delta T/\sigma_0$, where σ_0 is defined in Eqs (7.12)–(7.13), and equals the square root of the variance ΔT.

We define the functions $N_{\max}(\nu)\,d\nu$ and $N_{\min}(\nu)\,d\nu$ as the density of ΔT maxima and minima, respectively, on the anisotropy map, whose amplitude lies within the interval $\nu \div (\nu + d\nu)$. We refer the reader interested in details of calculations to the original publications, Bardeen *et al.* (1986) and Bond and Efstathiou (1987), and give the final expressions for $N_{\max}(\nu)\,d\nu$ and $N_{\min}(\nu)\,d\nu$ (Novikov and Jørgensen, 1996a,b) as follows:

$$N_{\max}(\nu)\,d\nu = \frac{1}{2\pi\theta_*^2} \exp\left(-\frac{\nu^2}{2}\right) \frac{d\nu}{(2\pi)^{1/2}} G(\gamma, \gamma\nu), \tag{7.21a}$$

$$G(\gamma, x_*) \equiv (x_*^2 - \gamma^2)\left[1 - \frac{1}{2}\mathrm{erfc}\left\{\frac{x_*}{[2(1-\gamma^2)]^{1/2}}\right\}\right] + x_*(1-\gamma^2)\frac{\exp\{-x_*^2/[2(1-\gamma^2)]\}}{[2\pi(1-\gamma^2)]^{1/2}}$$

$$+ \frac{\exp[-x_*^2/(3-2\gamma^2)]}{(3-2\gamma^2)^{1/2}}\left[1 - \frac{1}{2}\mathrm{erfc}\left\{\frac{x_*}{[2(1-\gamma^2)(3-2\gamma^2)]^{1/2}}\right\}\right], \tag{7.21b}$$

where $\mathrm{erfc}(x) = (1 - (2/\sqrt{\pi}))\int_o^x dt\, e^{-t^2}$, $\theta_* = \sqrt{2}\sigma_1/\sigma_2$, and σ_1, σ_2 and γ are defined in Eqs (7.12) and (7.13). So,

$$(N_{\max} + N_{\min})(\nu)\,d\nu = \frac{\gamma^2}{2\pi\theta_*^2} \frac{\exp(-\nu^2/2)}{(2\pi)^{1/2}}\left\{\nu^2 - 1 + \frac{\exp[-x_*^2/(3-2\gamma^2)]}{\gamma^2(3-2\gamma^2)^{1/2}}\right\}. \tag{7.22}$$

Integration of Eqs (7.21) and (7.22) from ν_t to ∞, where $\nu = \nu_t$ is an arbitrarily chosen level, gives the mean number of maxima and minima whose amplitude is above the threshold ν_t (Bardeen *et al.*, 1986; Bond and Efstathiou, 1987):

$$n_{\max}(\nu_t) + n_{\min}(\nu_t) = \frac{\gamma^2}{(2\pi)^{3/2}\theta_*^2} \nu_t e^{-\nu_t^2/2} + \frac{1}{4\pi\sqrt{3}\theta_*^2}\mathrm{erfc}\left\{\frac{\nu_t}{[2(1-2\gamma^2/3)]^{1/2}}\right\}. \tag{7.23}$$

If $\nu_t \to -\infty$ then the total number of extrema on the anisotropy map is given by (Bond and Efstathiou, 1987) as

$$n_{\mathrm{pk}}(-\infty) = \frac{1}{4\pi\sqrt{3}}\theta_*^{-2}(\mathrm{ster}^{-1}). \tag{7.24}$$

Equations (7.21)–(7.24) open up the study of the Gaussin nature of the signal, based on calculating the concentration of maxima and minima on anisotropy maps in different cosmological models.

At the same time, Eqs (7.21)–(7.24) need a slight modification that should take into account the actual parameters of the experiment. In the 'simple' case, when the properties of the signal are dictated by the primordial anisotropy of the CMB radiation smoothed by the receiving antenna and instrument noise, the spectrum parameters change. We assume that the transmission function of the antenna has a Gaussian profile with a characteristic angular scale θ_A, and that the instrument noise is a 'white' noise whose spectrum $C_{l(\mathrm{noise})}$ is independent of

Table 7.1.

FWHM (arcmin)	Ω	θ_c (arcmin)	γ	θ_* (arcmin)	θ_c (arcmin)	γ	θ_* (arcmin)	θ_c (arcmin)	γ	θ_* (arcmin)
5	0.1	8.6	0.53	4.5	5.7	0.44	2.5	3.6	0.61	2.2
5	0.3	11.1	0.45	5.0	5.9	0.40	2.3	3.5	0.61	2.2
5	1.0	15.0	0.40	6.1	6.1	0.37	2.3	3.6	0.60	2.1
10	0.1	15.0	0.49	7.4	12.8	0.45	5.7	8.4	0.54	4.5
10	0.3	17.0	0.52	8.9	13.7	0.42	5.8	8.5	0.53	4.5
10	1.0	22.0	0.43	9.4	16.0	0.34	5.5	8.9	0.50	4.4
20	0.1	32.3	0.44	14.2	29.7	0.42	12.3	21.1	0.45	9.6
20	0.3	30.2	0.48	14.6	28.0	0.45	12.7	20.5	0.48	9.7
20	1.0	35.9	0.49	17.5	32.9	0.42	13.9	23.0	0.43	9.9

l. Following Barreiro *et al.* (1997), we define the total correlation function for a mixture of primordial anisotropy and noise as follows:

$$C(\alpha, \sigma) = \frac{1}{4\pi} \sum_l (2l + 1) \left[C(l) e^{-l(l+1)\theta_A^2} + C_{noise} \right] P_l(\cos\theta). \tag{7.25}$$

Then, in view of the definition in Eq. (7.13), all spectral parameters of the random Gaussian CMB anisotropy field are functions of θ_A and C_{noise}, including also the external instrument noise. Again following Barreiro *et al.* (1997), we introduce the noise amplitude $A_{noise}(10')$ that characterizes the level of noise smoothed on a scale $10'$ using a Gaussian filter. Without discussing the specifics of the experiments, we choose three values,[3] $A_{noise}(10') = (0, 1, 3) \times 10^{-5}$. The case of $A_{noise}(10') = 0$ is that of an 'ideal' experiment that does not distort the properties of the primordial signal on scales of $10'$ and higher. Two other values simulate the effect of noise on a map at levels close to the amplitudes of the primordial signal. We now fix three models of the Universe with $\Omega_{tot} = 1$, $\Omega_{tot} = 0.3$ and $\Omega_{tot} = 0.1$ with the baryonic density $\Omega_b = 0.05$ and the Hubble constant $h = 0.5$. We consider the hidden mass to be 'cold' and assume that initial perturbations are adiabatic and have the Harrison–Zeldovich spectrum.

What is the behaviour of the concentration of anisotropy field maxima $N_{max}(\nu_t)$ for these three cosmological models? Table 7.1 (Barreiro *et al.*, 1997) summarizes the spectral parameters for various values of θ_A and $A_{noise}(10')$. The relation between θ_A and the width of the directivity diagram of the antenna at half-amplitude (FWHM) is given by $\theta_A = 0.425$ (FWHM).

We need to emphasize that as the resolving power of the antenna decreases (the FWHM increases), the number of peaks on the sphere increases by almost three orders of magnitude (Table 7.2). At the same time, the statistical spread in the number of peaks, $\Delta n \sim \sqrt{N}$, for FWHM $= 5'$ is found to be unimportant for an 'ideal' experiment ($A_{noise}(10') = 0$) but becomes decisive if $A_{noise}(10') = 10^{-5}$ and $A_{noise}(10') = 3 \times 10^{-5}$. Note that the parameters of the PLANCK space mission will be very close to the parameters of an 'ideal' experiment. In other models the signal properties are determined by the noise, which in fact results in an approximate equality of the number of peaks on the sphere.

---

[3] Note that the parameter $A_{noise}(10')$ characterizes the level of fluctuations $\Delta T/T$ and is therefore dimensionless.

Table 7.2.

FWHM (arcmin)	ν	$\Omega = 0.1$	$\Omega = 0.3$	$\Omega = 1$	$\Omega = 0.1$	$\Omega = 0.3$	$\Omega = 1$	$\Omega = 0.1$	$\Omega = 0.3$	$\Omega = 1$
5	3	4541	2912	1657	11 019	10 965	10 459	25 147	25 883	25 962
5	3.5	1011	636	357	2401	2362	2228	5674	5842	5794
5	4	174	108	60	407	397	271	986	1016	1007
10	3	1518	1106	753	2192	1953	1615	4727	4660	4295
10	3.5	335	258	164	479	423	341	1055	1038	949
10	4	57	44	28	81	72	56	182	179	163
20	3	335	379	267	419	452	339	795	832	690
20	3.5	74	84	59	91	99	73	174	183	150
20	4	13	14	10	15	17	12	30	31	25

Table 7.2 summarizes the results in these models as a function of ν_t, FWHM and the noise level (Barreiro *et al.*, 1997).

7.4 Signal structure in the neighbourhood of minima and maxima of the CMB anisotropy

The theory makes it possible not only to predict the average number of extrema of the random Gaussian anisotropy field, but also to calculate the most probable structure of the field ΔT in the neighbourhood of a maximum or a minimum (Bardeen *et al.*, 1986; Bond and Efstathiou, 1987). Following Bond and Efstathiou (1987), we choose a polar system of coordinates $(\overline{\omega} = 2 \sin \theta/2; \varphi)$ with origin at the point of maximum of the field $\Delta T(q_i)$ and resort to the flat-sky approximation to describe the structure of ΔT in the neighbourhood of this point. We assume that the peak height equals ν. Then the field distribution in a neighbourhood of the maximum is elliptical, as follows:

$$\Delta_t(\overline{\omega}) = \sigma_0 \left\{ \nu - \frac{1}{2} \gamma x \left(\frac{\overline{\omega}}{\theta_c} \right)^2 [1 + 2e_1 \cos(2\varphi)] \right\}, \tag{7.26}$$

where $x = \nabla^2(\Delta T)/\sigma_2$ is the radial curvature, e_1 is the symmetry, and the coordinate system is oriented along the major and minor axes of the ellipsoid. The radial curvature x and the asymmetry e in Eq. (7.26) are random parameters that change from one realization to another. Using the results from Bond and Efstathiou (1987), we can compare the characteristic sizes of zones covered by these peaks at half-maximum. In the former case ($\nu = 1$) the corresponding zone radius is found to be close to θ_*, while for the $\nu = 3$ peak it exceeds θ_* by a factor of 3 to 4. At the same time, the mean area of the zone at the level ν_t in the peak neighbourhoods of height $\nu > \nu_t$ is approximately equal to

$$S(\nu_t) \simeq 2\pi \left(\frac{\theta_*}{\gamma \nu_t} \right)^2 \left(1 - \frac{1}{\nu_t^2} \right); \qquad \nu_t \gg 1 \tag{7.27}$$

and decreases as the level ν_t is increased. As a result, the high peaks of the Gaussian field ΔT have sharp tops and sufficiently extended pedestals. Their shape approaches elliptical, at least for high ($\nu > 2$) peaks, but the pedestals of each peak are of random shape.

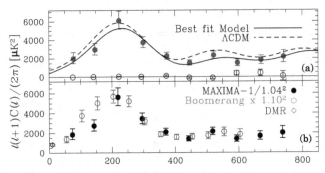

Figure 7.1 (a) The CMB anisotropy spectrum obtained in the MAXIMA-1 experiment (Hanany *et al.*, 2000a,b) (top), as compared to (b) the BOOMERANG data (de Bernardis *et al.*, 2000). Adapted from Hanany *et al.* (2000a,b).

Taking into account the above-listed properties of peak statistics in the random Gaussian field ΔT, we can turn to possible practical uses of them. The next section mostly deals with analysing peaks on ΔT maps obtained in the BOOMERANG (De Bernardis *et al.*, 2000) and MAXIMA-1 (Hanany *et al.*, 2000a,b) experiments.

7.5   Peak statistics on anisotropy maps

We mentioned in Section 7.1 that the observational data obtained by the BOOMERANG and MAXIMA-1 missions (De Bernardis *et al.*, 2000; Hanany *et al.*, 2000a,b) marked the advent of a new era in studying the spectra of the CMB radiation. The measured angular spectrum (see Fig. 7.1) clearly shows a peak on an angular scale corresponding to the spectral harmonic with the multipole number $l \simeq 200$. This is a Sakharov peak. However, the structure of the CMB anisotropy spectrum for $l > 400$ is not yet clear, and a new series of experiments is required, such as MAP and PLANCK missions.

There are several factors of major importance for future experiments. For instance, the PLANCK experiments will cover a much greater area of the sky than BOOMERANG and MAXIMA-1. Moreover, PLANCK works on two HFI channels with $v \simeq 545$ GHz and $v \simeq 857$ GHz, which provide the resolution FWHR= $5'$.

In this section we analyse peak statistics (statistics of maxima and minima) on the CMB anisotropy map. We compare these statistics using maps obtained by BOOMERANG and MAXIMA-1 with those of future observations by PLANCK and we predict certain properties and shapes of peaks.

Let us consider a model, very nearly realistic, obtained by virtue of successful observations by MAXIMA-1. According to Hanany *et al.* (2000), the map obtained on the basis of these observations shows a high-amplitude peak on the $\Delta T/T$ distribution (Fig. 7.2). The coordinates of this peak are: declination $\simeq 58.6°$, right ascension $15^{\mathrm{h}}35^{\mathrm{m}}$. The amplitude of this peak on a map filtered with the Wiener filter is $\Delta T \sim 2.3\text{--}2.5\sigma$, and the value of anisotropy decreases monotonously to the level of 1σ at $15.2^{\mathrm{h}} < \alpha \leq 15.4^{\mathrm{h}}$ and $58.5° < \delta < 60°$.

In the following, we examine the salient features in the structure of the peaks in terms of the measurements of the CMB achievable with the PLANCK experiment; these measurements will have better resolution than those currently available. A similar prediction was made in Bunn, Hoffman and Silk (1996) for the experiment conducted on Tenerife (in the canary Islands) in which the COBE DMR data were used, although the technique was somewhat

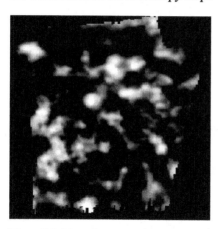

Figure 7.2 Map of an area of the sky obtained by the MAXIMA-1 mission (the Wiener filter was applied). The map accumulates three 150 GHz channels and one 240 GHz channel. Adapted from Hanany *et al.* (2000a,b).

different. For instance, we show that more accurate measurements do not resolve the internal structure (new high peaks $> 1.5\sigma$) inside the area mentioned above.

In an actual experiment, an antenna resolution is finite and the spectral parameters depend on the antenna diagram θ_A and hence on the number of Sakharov peaks that can be resolved by this antenna. This means that the structure of high peaks of $\Delta T / T$ of future PLANCK maps may differ from the corresponding structure of peaks on the maps generated using BOOMERANG and MAXIMA-1 data.

On all cosmological models, the CMB anisotropy spectrum $C(l)$ can be described as a sum of Gaussian peaks with centres at the points of maxima l_n ($l \geq 30$) (Kotok *et al.*, 2001),

$$\frac{l(l+1)C(l)}{2N\pi} = \left\{ \sum_n A_n \exp\left[-\frac{(l-l_n)^2}{2d_n^2} \right] + 1 \right\} e^{-l^2 s^2}, \qquad (7.28)$$

where n is the number of peaks, d_n is the peak width, l_n is its location and N is the normalizing factor for low multipoles (for example normalization based on the COBE data). The last term of Eq. (7.28) takes into account the Silk attenuation on the angular scale s. Note that we have not introduced into Eq. (7.28) any low- or high-multipole filters (transmission factors and the antenna). This means that Eq. (7.28) describes the initial spectrum of perturbations on the sky without any smoothing. Indeed, the last two factors are very important and their effect on $\Delta T / T$ maps is decisive. A spectrum of the type (7.28) provides information on the effect of each Sakharov peak on the topology of $\Delta T / T$ maps. Therefore, by using the Eq. (7.28) approximation, we can study the effect of the first, second and subsequent Sakharov peaks on the spectral parameters of future maps obtained using the WMAP, PLANCK and other missions.

The next question concerns the transmission function, $W(l)$, of a given experiment. We simulate the basic properties of the function $W(l)$ in the approximation of small angular size of the map:

$$G(l) = \frac{W(l)}{l} = \exp\left[-l(l+1)\theta_A^2 \right] \begin{cases} l^m & \text{for} \quad l \ll 30, \\ l^{-1} & \text{for} \quad l \gg 30, \end{cases} \qquad (7.29)$$

($m = 2$ for a two-beam and $m = 3$ for a three-beam circuit of low-multipole filtration). The exponential in Eq. (7.29) describes an antenna with $\theta_A \simeq 7.45 \times 10^{-3}$ ($\theta_{\text{FWHM}}/1°$). To describe the asymptotes of the cofactor in Eq. (7.29) we introduce a function which corresponds to both limits, as follows:

$$G(l) \simeq \frac{(lR)^{m+1}}{l[1 + (lR)^{m+1}]},\tag{7.30}$$

where R is a characteristic angular scale (see Eq. (7.29)) for low-multipole filtration. Therefore, the spectral parameters for this model are given by the relation

$$\sigma_i^2 = \int_0^\infty dl\, l^{2i} g(l) \left[1 + \sum_n A_n \exp\left(-\frac{(l - l_n)^2}{2d_n^2}\right)\right] e^{-l^2(s^2 + \theta_A^2)}; \quad i = 0, 1, 2.\tag{7.31}$$

Note that for the second and higher Sakharov peaks $l_n^2/d_n^2 \gg 1$, and for the first peak only we have $l_n^2/d_1^2 \simeq 5$. For the analytical approximation of the integral in Eq. (7.31) the asymptotic behaviour $l_n^2/d_n^2 \gg 1$ is required for all peaks of the spectrum (7.28). Using this approximation, we arrive at the formulas for the spectrum parameters σ_i^2, as follows:

$$\sigma_0^2 = \frac{1}{2}\left[2\ln\frac{r}{\xi} - C + \sqrt{\frac{\pi}{2}}\sum_n A_n \frac{d_n}{l_n}\exp\left(\frac{-l_n^2\xi^2}{1 + 2d_n^2\xi^2}\right)\cdot\left(1 + 2d_n^2\right)^{1/2}\right],\tag{7.32}$$

$$\sigma_1^2 = \frac{1}{2\xi^2} + \sqrt{\frac{\pi}{2}}\sum_n \frac{A_n l_n d_n \exp\left(\frac{-l_n^2\xi^2}{1+2d_n^2\xi^2}\right)}{\left(1 + 2d_n^2\xi^2\right)^{3/2}}\left[1 + \Phi\left(\frac{l_n}{d_n\sqrt{2(1 + 2d_n^2\xi^2)}}\right)\right],\tag{7.33}$$

$$\sigma_2^2 = \frac{1}{2\xi^4} + \sqrt{\frac{\pi}{2}}\sum_n \frac{A_n l_n^3 d_n \exp\left(\frac{-l_n^2\xi^2}{1+2d_n^2\xi^2}\right)}{\left(1 + 2d_n^2\xi^2\right)^{7/2}}\left[1 + \Phi\left(\frac{l_n}{d_n\sqrt{2(1 + 2d_n^2\xi^2)}}\right)\right],\tag{7.34}$$

where C is the Euler constant, $\xi^2 = \theta_A^2 + s^2$, and $\Phi(x) = 2/\sqrt{\pi}\int_0^x dx\, e^{-x^2}$ is the probability integral. We see from Eq. (7.32) that only the first Sakharov peak is important in the calculation of the variance σ_0^2. The effects of the second and subsequent peaks are practically negligible because of the drop in the amplitude A_n and due to the relation d_n/l_n. However, these peaks dictate the topological structure of the maps $\Delta T/T$ (see Eqs (7.35) and (7.36)) – for example, the number of maxima and minima for different thresholds $\nu_n\sigma_0 = \Delta T/T$. Equations (7.33) and (7.34) describe a realistic model with $d_n^2\xi^2 \ll 1$ and $l_n^2\xi^2 \le 1$. In this model the density of all peaks for $\nu(-\infty, \infty)$ is given by an especially simple expression,

$$N_{\text{pk}}^+ = N_{\text{pk}}^- = \frac{1}{8\pi\sqrt{3}}\frac{\sigma_2^2}{\sigma_1^2} \text{ ster}^{-1},\tag{7.35}$$

where N_{pk}^+ and N_{pk}^+ are the densities of all maxima and all minima, respectively.

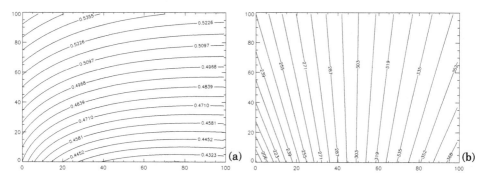

Figure 7.3 (a) $\gamma(x, y)$ and (b) $N_{\text{pk}}(x, y)$ as functions of the parameters $x = 10^2 A_2/A_1$ (horizontal axes) and $y = 10^2 A_3/A_1$ (vertical axes). The numbers on the curves correspond to the values $\gamma(x, y)$ and $N_{\text{pk}}(x, y)$. Note that the point $x = 26$, $y = 46$ corresponds to the amplitudes of the first and two subsequent accoustic peaks according to the MAXIMA-1 and BOOMERANG experiments.

Let us now consider a model situation in which all Sakharov peaks are smoothed ($A_n = 0$). For this model, the spectral parameters θ_* and γ have the following form:

$$\theta_*^2 = 2\xi^2; \qquad \gamma = \left(2 \ln \frac{R}{\xi} - C\right)^{-1/2}, \tag{7.36}$$

and the densities of all maxima and minima of arbitrary height are given by $N_{\text{pk}}^+ = N_{\text{pk}}^- = (1/8\pi\sqrt{3})\xi^{-3}$. If we denote the degree of sky coverage in a certain $\Delta T/T$ experiment as f_{sky} (for example, $f_{\text{sky}} \simeq 0.3\%$ for the MAXIMA-1 experiment), then the number of maxima (or minima) on the observational map will be given by

$$N_{\text{max}} \simeq 16 \left(\frac{f_{\text{sky}}}{0.003}\right) \left(\frac{\theta_{\text{FWHM}}}{1°}\right)^{-2}. \tag{7.37}$$

According to De Bernardis *et al.* (2000) and Hanany *et al.* (2000a,b) the antennas of the MAXIMA-1 and BOOMERANG experiments have the equivalent FWHM of $\simeq 10'$. This means that in the absence of Sakharov peaks in the spectrum we could detect 576 maxima on the appropriate maps. However, Eqs (7.32)–(7.34) imply that the presence of Sakharov peaks in the primordial spectrum decreases the number of peaks on the map to 271. Therefore we must add to our analysis the understanding that the effect of the Sakharov peaks is to reduce the number of hot and cold spots on the map by a factor of approximately 2.

The next question of interest is: to what extent is the topology of the $\Delta T/T$ map sensitive to the amplitudes of the second, A_2, and the third, A_3, peaks if we assume that the amplitude and position of the first peak are known? In order to answer this question, we compare $\gamma(A_2, A_3)$ and $N_{\text{pk}}(A_2, A_3)$ for the following models. In model 1 we choose the amplitude A_1 corresponding to the data of Hanany *et al.* (2000a,b), and we assume the coordinates and widths of the peaks to be as follows: $l_1 = 210$ for width $d_1 = 95$, $l_2 = 580$ for $d_2 = 110$ and $l_3 = 950$ for $d_3 = 130$. The corresponding plots are shown in Fig. 7.3.

The next model (Fig.7.4) corresponds to a hypothetical situation when the amplitude of the first Doppler peak is only half of that in the case above. As we see from Eqs (7.32)–(7.34) and Figs 7.3 and 7.4, the structure of the spectral parameters $\gamma(A_2, A_3)$ and $N_{\text{pk}}(A_2, A_3)$ in this second ('toy') model changes dramatically. The number of maxima increases to above

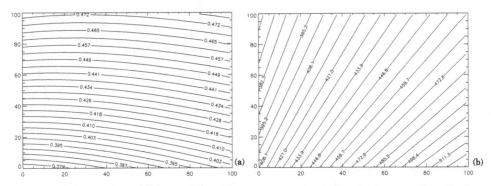

Figure 7.4 Same as in Fig. 7.3 but for a toy model with the first Sakharov peak amplitude one-half that in Fig. 7.3.

420, while the parameter γ retains almost the same value $\gamma \simeq 0.4$–0.47. This result is also important for analysing the global and local map topology. The position and amplitude of the first Sakharov peak in the $C(l)$ spectrum in the BOOMERANG and MAXIMA-1 experiments were measured to an accuracy of 10%. Because of this 10% error, the theoretically predicted number of CMB peaks on the observation maps can vary from 263 to 279. This difference of 16 peaks corresponds to a statistical fluctuation $\delta N/N \sim N^{1/2}$ of the number of peaks, N, on a map, with hardly any changes in the parameter γ. On the whole, starting with the results given above, we can state that the peak distribution on MAXIMA-1 maps corresponds to the hypothesis of the Gaussian nature of the signal.

However, another question arises: how will the local topology on a CMB map change if the resolving power of the receiving instrument increases and the noise level decreases? Will the future measurements be capable of revealing the internal structure of the peaks that were identified using the BOOMERANG and MAXIMA-1 experimental data? For example, will they be able to detect new peaks in the fine structure within the area $0 \leq \nu \leq 2$? If the answer is yes, what will the typical height of these peaks be? Answers to these questions depend on the peak-to-peak correlation on high-resolution maps. The spread of the number of peaks from one realization to another on an Ω_p pixel grid is related to the peak-to-peak correlation function $C_{\text{pk}-\text{pk}}$ by the following formula:

$$\langle (\Delta N_{\text{pk}}^+)^2 \rangle / \langle N_{\text{pk}}^+ \rangle^2 = \langle N_{\text{pk}}^+ \rangle^{-1} + \int \frac{d\Omega_{\vec{q}} \, d\Omega_{\vec{q}'}}{\Omega_p^2} C_{\text{pk}-\text{pk}}(\vec{q} - \vec{q}'), \qquad (7.38)$$

where $\langle N_{\text{pk}}^+ \rangle = n_{\text{pk}}^+(\nu_t)\Omega_p$, $n_{\text{pk}}^+(\nu_t)$ is the concentration of maxima having height ν above a certain threshold ν_t. Note that the first term in Eq. (7.38) corresponds to the Poisson peak distribution. Heavens and Sheth (1999) have recently conducted analytical and numerical calculations of the peak-to-peak correlation function and showed that $C_{\text{pk}-\text{pk}}$ tends to zero for $\theta < \theta_*$ and reaches a negative value $C_{\text{pk}-\text{pk}} = -1$ for $\theta = 0$. This result reflects the fact that various high peaks cannot be located close to one another. For instance, two high peaks with amplitudes $\nu_1 \sim \nu_2 \sim 2$–2.5σ must be separated by a distance $\theta \gg \theta_*$. According to Heavens and Sheth (1999), a typical angular scale θ_* for the favourite cosmological ΛCDM model is close to $20'$. This scale is twice as large as the FWHM in the BOOMERANG and MAXIMA-1 experiments and four times as large as it is in the PLANCK experiment. It is useful to remark, nevertheless, that about ten low-amplitude $\nu \leq 1$ peaks can exist around a high peak in the above-mentioned region on a high resolution map. Therefore, we can conclude that isolated 2–2.5σ peaks found on poorly resolved BOOMERANG and MAXIMA-1 maps manifest themselves as isolated peaks on PLANCK maps.

Let us return to considering a high peak with $\delta = 58.6°$, $RA = 15^h35'$ on the MAXIMA-1 map. The location of this peak is almost independent of high angular resolution of the future PLANCK experiment, and its amplitude can be described as follows. Imagine that the ideal experiment with a δ-function antenna revealed the highest peak with coordinates $\overline{\delta}$ and $\overline{RA}$. The amplitude of this peak, measured in units of variance, can be written as

$$\nu_{in} = \Delta T/\sigma_{0(in)}, \tag{7.39}$$

where $\sigma_{0(in)}$ corresponds to Eq. (7.33) for $\theta = 0$ and $\xi = s$. We assume for simplicity that the distribution $\Delta(x, y)$ around the point of maximum is Gaussian, with characteristic scales a and $b = ka$, k a constant, such that

$$\Delta T(x, y) = \nu_{in}\sigma_{0(in)} \exp\left(-\frac{x^2}{2a^2} - \frac{y^2}{2b^2}\right), \tag{7.40}$$

and the parameter a is proportional to a typical correlational scale of the primary signal. Following Bond and Efstathiou (1987), we can describe the local shape of the peak of height ν by measuring the radial curvature Γ and 'ellipticity' ϵ in polar coordinates θ and φ:

$$\delta(\overline{\theta}, \overline{\varphi}) = \sigma_{0(in)}\left[\nu_{in} - \frac{1}{2}\gamma\Gamma\left(\frac{\overline{\theta}}{\theta_c}\right)^2(1 + 2e_1 \cos 2\overline{\varphi})\right]. \tag{7.41}$$

Let $\theta^2 = x^2 + y^2$ and $\cos 2\varphi = (x^2 - y^2)/(x^2 + y^2)$. We now find

$$a^2 = \frac{\nu_{in}\theta_c^2}{(1 + 2e)\gamma\Gamma}; \qquad b^2 = \frac{\nu_{in}\theta_c^2}{(1 - 2e)\gamma\Gamma}; \qquad \kappa^2 = \frac{1 + 2e_1}{1 - 2e_1}. \tag{7.42}$$

Taking the eccentricity into account, we rewrite this last relation in the following form:

$$\kappa^2 = 1/(1 - \epsilon^2). \tag{7.43}$$

We will compare two experiments that would investigate the same part of the sky in the neighbourhood of the peak but with different resolutions θ_1 and θ_2. We assume that θ_1 corresponds to the MAXIMA-1 experiment and θ_2 to the PLANCK experiment ($\theta_1 \simeq 2\theta_2$). We denote the amplitude of the maximum in the poor-resolution experiment by ν_{MAXIMA} and that in the high-resolution experiment by ν_{PLANCK}. In such models, the amplitude is given by the equation

$$\widetilde{\Delta T}_j(x, y) = \frac{1}{2\pi\theta_j^2} \int dx'dy' \Delta T(x', y') \exp\left[-\frac{(\vec{r} - \vec{r}')^2}{2\theta_j^2}\right], \tag{7.44}$$

where the index $j = 1, 2$ corresponds to θ_1 and θ_2, and $\vec{r}(x, y)$ and $\vec{r}(x', y')$ are vectors in a Cartesian reference frame, with the origin at the central point of the maximum. The form of the function $\widetilde{\Delta T}_j(x, y)$ of Eq. (7.40) is described for the poor- and high-resolution experiments by the equation

$$\widetilde{\Delta T}_j(x, y) = \frac{\nu_{in}\sigma_{0(in)}ab}{[(a^2 + \theta_j^2)(b^2 + \theta_j^2)]^{1/2}} \exp\left[-\frac{x^2}{2(a^2 + \theta_j^2)} - \frac{y^2}{2(b^2 + \theta_j^2)}\right]. \tag{7.45}$$

This curve determines the parameter $\xi^2 = (b^2 + \theta^2)/(a^2 + \theta^2)$, which can be measured in the neighbourhood of the peak at a certain threshold $\nu_t \sigma_0^{(1)}$, where $\sigma_0^{(1)}$ is the variance of perturbations in the poor-resolution experiment. As a result, Eq. (7.41) yields the following peak amplitude:

$$\nu_{\mathrm{MAXIMA}} \sigma_0^{(1)} = \frac{\nu_{\mathrm{in}} \sigma_0^{(in)} \kappa}{\xi_1 \left(1 + \theta_1^2/a^2\right)}. \tag{7.46}$$

In high-resolution experiments, Eq. (7.45) yields

$$\nu_{\mathrm{PLANCK}} \sigma_0^{(2)} \simeq \frac{\nu_{\mathrm{in}} \sigma_0^{(in)} \kappa}{\left[\left(1 + \theta_2^2/a^2\right)\left(\kappa^2 + \theta_2^2/a^2\right)\right]^{1/2}}, \tag{7.47}$$

and since the quantities $\sigma^{(in)}$, $\sigma^{(1)}$ and $\sigma^{(2)}$ differ only logarithmically, we obtain

$$\nu_{\mathrm{PLANCK}} \simeq \nu_{\mathrm{MAXIMA}} \frac{\xi_1 (1 + 4\mu^2)}{[(1 + \mu^2)(\kappa^2 + \mu^2)]^{1/2}}. \tag{7.48}$$

For instance, this ratio for the peak of MAXIMA-1 with coordinates $\delta = 58.6°, RA = 15^{\mathrm{h}}35^{\mathrm{m}}$ is ~ 1.2–1.4. Taking this result into account, we can transform the peak with $\nu_1 \simeq 2$–3 on a map based on MAXIMA-1 into a peak on the anticipated future map provided by PLANCK. This means that the peak we consider, given by $\delta \simeq 58.6°$ and RA=$15^{\mathrm{h}}35^{\mathrm{m}}$, corresponds to a maximum in the distribution of the primordial signal at the level $\nu \simeq 4$. It is quite clear that a similar prediction can also be made regarding the high-amplitude peaks on the map of the radio skies obtained by BOOMERANG.

7.6 Clusterization of peaks on anisotropy maps

When analysing the structure of ΔT in the neighbourhood of extrema, we used the results of the theory of clusterization of peaks in random Gaussian fields whose astrophysical applications were studied in detail in Bardeen *et al.* (1986), Bond and Efstathiou (1987), Heavens and Sheth (1999) and Novikov and Jørgensen (1996a,b). Obviously, an analysis of peak clusterization, added to peak statistics depending on peak height, is an additional test of the Gaussian nature of the signal or of possible deviations of its characteristics from the normal distribution.

First, we need to give a definition of what we call a cluster of peaks. A map of anisotropy ΔT obtained in the framework of the MAXIMA-1 project clearly shows lighter and darker zones corresponding to isolated regions with enormously high (compared to the mean level) signal and enormously low values of the field ΔT. Obviously, this separation into zones is purely relative and depends on the height at which the section is made of the field $\Delta T(x, y)$ by a plane $\Delta T^* = \nu_t \cdot \sigma_0$. If $\Delta T^*(x, y) = \nu_t \sigma_0$, then the formal solution of this equation implies a set of points $\{\overline{x}, \overline{y}\}$ that form the level 'contour' $L\{\overline{x}, \overline{y}\}$. To be precise, this level contour is also a random function for a random field whose properties obviously depend on the spectral parameters σ_i. For high-level sections $\nu_t \gg 3$–5, the corresponding lines resemble a set of close contours at a considerable distance from one another. This effect has a very simple explanation that is totally based on the statistics of high peaks of Gaussian fields. As we saw in Section 7.2, the concentration of peaks for high ν_t decreases as ν_t increases. Correspondingly, the mean distance between peaks increases as well. Then the corresponding level contours for each isolated peak of ΔT automatically approaches elliptical shape (see

Section 7.3). The ellipticity of level contours in the neighbourhood of a point of extremum is therefore a sign of it being isolated from other peaks.

Let us take a closer look at the mathematical aspect of the problem of peak clusterization in the approximation of small size of the region to be analysed in comparison with the entire sphere. Assume now that we cut the surface $\Delta T(x, y)$ by a plane $\Delta T^*(x, y) = v_t \sigma_0$ and calculate the number of clusters N_k,

$$\sum_{k=1}^{\infty} N_k \cdot k = n_{\max}(v_t), \tag{7.49}$$

which corresponds to the total number of maxima above the level v_t on the map in question. We need to point out that in addition to ΔT peaks (the maxima) each cluster may also contain minima and saddle points. If, for instance, a cluster of dimension k contains $k = k_{\max}$ maxima and $k_{\min}$ minima, it as a rule contains k_s saddle points, such that $k_s = k_{\max} + k_{\min} - 1$.

Following Novikov and Jørgensen (1996a,b), we introduce normalization of the concentration of clusters of dimension k, as follows:

$$\sum_{k=1}^{\infty} n_k = n_{\max}(v_t) + n_{\min}(v_t) - n_s(v_t), \tag{7.50}$$

where $n_s(v_t)$ is the total number of saddle points above the section v_t. Then, making use of Eqs (7.49) and (7.50), we can determine the mean cluster length at the level v_t:

$$\langle k \rangle = \frac{\sum k N_k}{\sum N_k} = \frac{n_{\max}(v_t)}{n_{\max}(v_t) + n_{\min}(v_t) - n_s(v_t)}. \tag{7.51}$$

We described in Section 7.3 the general properties of extrema and we have used the expression for $n_{\max}(v_t)$ and $n_{\min}(v_t)$ in Eqs (7.23) and (7.24). For the concentration of saddle points, we will use the expression obtained by Novikov and Jørgensen (1996a,b):

$$n_s(v_t) = \frac{1}{8\pi\sqrt{3}} \frac{\sigma_2^2}{\sigma_1^2} \left\{ 1 - \Phi \left[\frac{\sqrt{3}v_t}{\sqrt{2(3 - 2\gamma^2)}} \right] \right\}, \tag{7.52}$$

where $\Phi(x)$ is the probability integral (Gradstein and Ryzhik, 1994). Equation (7.51) shows that the mean cluster length at the level v_t is a function of a single parameter γ. In its turn, this parameter is determined by a combination of spectral parameters and characterizes the general topology of anisotropy maps. As $\gamma \to 1$, the clusterization of peaks is largely suppressed and the signal on a map looks like an ensemble of a large number of isolated peaks. In the opposite asymptotics, when $\gamma \to 0$, the clusterization of maxima is extremely high and the ΔT map should look very fuzzy.

As for any type of statistics of Gaussian fields, the distribution and clusterization of maxima depending on the section height v_t can be used to test the potential non-Gaussian noises on the map $\Delta T(x, y)$. Let us consider as a noise of this sort, the background of non-resolved pointlike sources that create an excess signal $\Delta T_{ps} > 0$ at points $\{x_{ps}, y_{ps}\}$. On the whole, the presence of this noise results in biasing the level $\Delta T_{CMB}(x, y) + \Delta T_{ps} = 0$, where ΔT_{CMB} is the primordial anisotropy. What will the qualitative change be in the rate of the peak clusterization in the presence of this type of noise? The answer to this question will be easily obtained by analysing the behaviour of the function $\langle k \rangle(v_t, \gamma)$ for a 'bare' signal or a 'signal+pointlike source' combination.

On a map of infinite size, the function $\langle k \rangle (v_t, \gamma)$ formally tends to infinity, whereas on a finite map the value of $\langle k \rangle (v_t, \gamma)$ is high but finite. This sharp increase in the mean cluster dimension results from the percolation effect investigated by Naselsky and Novikov (1995). The effect consists in the following: as $v_t \to 0$ 'from above', i.e. while $v_t \geq 0$, the zones with $v > v_t$ gather into clusters of higher dimensionality but are nevertheless separated from one another by regions with negative values of ΔT. A similar pattern is observed when we move in the direction of $v_t = 0$ from the zone of $v_t < 0$. In this case the pattern is absolutely symmetric, as it is for $v_t > 0$, but now for minima of the field ΔT. The surface $v_t = 0$ is therefore peculiar. To be precise, any deviation upwards ($v_t > 0$) or downwards ($v_t < 0$) immediately transforms the 'maxima–minima' system to one of the two states listed above.

We have already remarked that the presence of noise on a map shifts the percolation level to the region $v_t < 0$ and breaks the symmetry in the distribution of maxima and minima of the field ΔT.

To conclude this section, we look in more detail at the field distribution $\Delta T(x, y)$ in the neighbourhood of two peaks that form a cluster of dimension 2 at the section level v_t. We invite the reader to pay attention again to the fact that different statistics of the Gaussian field respond in a different way to the presence of non-Gaussian noises or to the non-Gaussian nature of the signal itself. In this sense, the distribution of the field in a dimension-2 cluster is, literally, a local characteristic of the signal topology, and this test can be used to single out localized noises that manifest themselves in the shape of additional ΔT peaks.

A detailed analysis of the structure of field distribution in dimension-2 clusters as applied to the Gaussian field ΔT was given in Novikov and Jørgensen (1996a,b). For isolated peaks the structure of the signal in the vicinity of each peak was investigated in Bond and Efstathiou (1987) (see Section 7.3). The main conclusion is that the shape of the signal in the neighbourhood of a peak is elliptical, with the ellipticity parameters depending on the peak height v and section level v_t. For clusters of dimension 2, this condition will naturally remain valid as the section height v_t is increased because a dimension-2 cluster then automatically splits into two clusters of dimension $k = 1$. However, as the height v_t decreases, peak-to-peak correlation results in phasing of the orientation of ellipses corresponding to any section v_t^*, $v_t \geq v_t^* < v_{\max}$, where $v_{\max}$ is the height of the lowest of the peaks and v_t is the level of the section that singles out the cluster with $k = 2$.

It is also quite interesting that the ellipses corresponding to the field distribution around each peak lose their shape in response to peak-to-peak correlations in the direction of the major semi-axes.[4] Therefore, if a $\Delta T(x, y)$ map reveals dimension-2 clusters within which the field level contours intersect along the minor semi-axes of the ellipses, this would signify that one of the peaks in the cluster is definitely of a noise origin and is not related to the Gaussian signal. Another important feature of the local signal topology in the neighbourhood of a $k = 2$ cluster must be pointed out. This topology is stable and independent of the spectrum of primordial perturbations. Therefore, any disruptions would point to a non-Gaussian nature of the signal.

---

[4] The corresponding analytical expressions for the field distribution probabilities in dimension-2 clusters are given in Novikov and Jørgensen (1996a). These expressions are quite cumbersome, so we refer the reader to the original publication.

To conclude this section, note that the methods of analysing the clusters of anisotropy fields as described above are readily generalizable to the polarization field. A detailed analysis of clusterization of the Q and U components of the Stokes vector is given in Arbuzov *et al.* (1997a,b).

7.7   Minkowski functionals

'In reality', that is in the context of the differential and integral geometry, the Minkowski functionals (MFs) (Minkowski, 1903) were introduced into cosmology (Mecke, Buchert and Wagner, 1994) as three-dimensional statistics for distributions of objects in the Universe, and then for the isodensity contours of continuous random fields (Schmalzing and Buchert, 1997).

The idea of testing the statistical nature of the signal on the CMB anisotropy maps using the geometrical characteristics was developed in Doroshkevich (1970), Gott *et al.* (1990), Naselsky and Novikov (1995), Novikov *et al.* (2001), Schmalzing and Gorski (1998) and Winitzki and Kosowsky (1997). An analysis of this approach, and especially of its applications oriented aspect, to CMB anisotropy maps for the already implemented experiments (COBE, MAXIMA-1, BOOMERANG) demonstrated the high efficiency of the Minkowski functional techniques for verifying the Gaussian nature of a signal. This progress is largely caused by the accumulation in Minkowski functionals of a number of characteristics that were already discussed when studying the statistics of the CMB peaks. Following Novikov *et al.* (2001), we single out the global Minkowski functionals: $A = \sum a_i$ is the entire investigated area within the isotherms, L is the total length of the contour that encompasses the area at the section level ν_t, and G is the genus for the number of isolated maxima minus the number of isolated minima at a given level ν_t that are identifiable on the entire ΔT map. We also construct local (partial) Minkowski functionals that are used for one or several selected parts of the map. The need for this separation is obvious. If one of the regions on the map ΔT of area S contains a clearly pronounced non-Gaussian noise but its area remains small, it is clear that the distortion of global Minkowski functionals will be of the order $S/S_{map} \ll 1$, where S_{map} is the area of the entire map.

We therefore consider a simply connected region R_i of the anisotropy map with $\nu(\theta, \varphi) \equiv \Delta T(\theta, \varphi)/\sigma_0 > \nu_t$. To characterize its topology, we consider three parameters: area a_i of the region, the contour length l_i and the number of holes n_{hi} in it. These are exactly the three Minkowski functionals. In order to generate the global Minkowski functionals, we calculate the numerical values of all these quantities for all separated regions of the map; that is, we generate the sums $A = \sum a_i$ and $L = \sum l_i$; $G = \sum g_i$ is the number of isolated regions with $\nu > \nu_t$ minus the number of isolated areas with $\nu < \nu_t$. Clearly the total area $A(\nu_t)$ is proportional to the cumulative distribution function of the random field.

Minkowski functionals possess mathematical properties which make them rather unique among all other geometric characteristics. They are translationally and rotationally invariant, are additive[5] and have a simple (and intuitive) geometrical representation. Moreover, as was shown in Hadwiger (1957), all global morphological properties (that satisfy the property of invariance under movements and the additivity) of any D-dimensional space can be completely described using $D + 1$ Minkowski functionals.

---

[5] 'Additivity' means, among other things, that the MFs of a conglomerate of several disconnected regions can be easily obtained if we know the MFs for each individual region.

The global Minkowski functionals of a Gaussian field are known analytically; in two-dimensional space they take the following form:

$$A(v) = \frac{1}{2} - \frac{1}{2}\Phi\left(\frac{v}{\sqrt{2}}\right),$$

$$L(v) = \frac{1}{8\theta_c}\exp\left(-\frac{v^2}{2}\right), \qquad (7.53)$$

$$G(v) = \frac{1}{(2\pi)^{3/2}}\frac{1}{\theta_c^2}v\exp\left(-\frac{v^2}{2}\right),$$

where $\Phi(x) = (2/\sqrt{\pi})\int_0^x e^{-x^2}\,dx$ is the error function. The way these functionals depend on the spectrum can be expressed via the field correlation length $\theta_c = \sigma_0/\sigma_1$, where σ_0 and σ_1 can be calculated using the spectrum $C(l)$:

$$\sigma_0^2 = 1/4\pi\sum_l(2l+1)C(l),$$

$$\sigma_1^2 = 1/4\pi\sum_l(2l+1)(l+1)lC(l). \qquad (7.54)$$

Unfortunately, there are no analytical formulae for partial Minkowski functionals even for Gaussian fields. However, this is not an insurmountable obstacle to their application because they can be obtained by numerical calculations. We need to emphasize that to apply them in practical work, one has to know not only the mean value of a quantity, but also its variance. In most cases the variance cannot be found analytically, even if it is possible to find its mean value. For example, it is possible to calculate analytically the mean value of hot/cold spots, but the variance of this value can only be evaluated numerically.

Let us discuss the application of Minkowski functionals to two-dimensional maps. We refer to all unconnected areas above the threshold ($v > v_t$) as positive peaks and to those below the threshold ($v < v_t$) as negative ones. In each region R_i we can calculate three Minkowski functionals: area $v_1^i = a_i$, perimeter $v_2^i = l_i$ (that is, the length of the boundary), the number of holes (an equivalent of genus) $v_3^i = g_i$, and the number of maxima within the area $v_4^i = n_{mi}$. Now we need to analyse the cumulative function $F(v_t, v^k)(k = 1, 2, 3, 4)$ for these quantities.

As an example of a CMB map, we can look at COBE data from which all the radiation of the Galaxy (the entire galactic background) has been removed. This removal of the galactic background from the cosmic signal was achieved by using two independent techniques. The construction of the maps was described in detail in Bennet et al. (1992, 1994), published in COBE DMR ASDS. We will describe both techniques. The first of them is the so-called combination method (map 1 given in Fig. 7.5(a)). Here the galactic background was removed by a linear combination of all DMR maps followed by removing the free–free radiation, and then by normalizing the cosmic signal to the thermodynamical temperature.

The second technique is that of subtraction (map 2, Fig. 7.5(b)). In this case one constructs a map of synchrotron radiation and one of emission by dust, which is then subtracted from the DMR data. Then the galactic free–free emission is subtracted. In Section 7.7.1 we follow Novikov, Feldman and Shandarin (1999) and analyse both maps. We also describe the numerical algorithm required to calculate the distribution of partial Minkowski functionals on a sphere and the application of this algorithm to the COBE data.

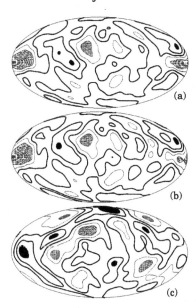

Figure 7.5 (a) COBE map 1 constructed using the combination technique (see text). Thick solid curves correspond to 0σ, 1σ and 2σ thresholds (areas in black are those within the 2σ contour), thin solid curves correspond to 1σ and 2σ (areas within the contour lines 2σ are shaded). (b) COBE map 2 constructed by the subtraction technique. The notation is the same as in (a). (c) Example of a Gaussian map with the same amplitudes as in (a).

7.7.1 Computation of maps

For a fixed position of pixels on a sphere, we choose a spherical system of coordinates. We consider the temperature distribution on a pixel map as a function of two variables in the following reference frame: $-\pi/2 < \theta < \pi/2$ and $-\pi < \varphi < \pi$. This function is in fact defined only at points (θ_k, φ_k) so that

$$\nu_{k_1, k_2} = \nu(\theta_{k_1}, \varphi_{k_2}), \qquad \theta_{k_1} = k_1 k_\theta, \qquad \varphi_{k_2} = k_1 k_\varphi. \tag{7.55}$$

We also assume that $h_\theta = h_\varphi = h = 2\pi/M$, where M is the number of pixels on the φ coordinate axis. Then the total number of pixels equals $M^2/2$. The original COBE maps were recalculated according to this pixelization as

$$\Delta T_{\text{data}}(\theta, \varphi) = B \int \Delta T_{\text{COBE}}(\theta', \varphi') \, e^{-\frac{\gamma^2}{2\gamma_0^2}} \, d\cos(\theta') \, d\varphi', \tag{7.56}$$

where ΔT_{COBE} and ΔT_{data} are the temperature at the points of COBE pixels and at the points defined by Eqs (7.55), respectively, γ is the angle between pixels, $\gamma = 7°$ is the angle of smoothing and B is the normalization coefficient. Temperature perturbations are completely described by the coefficients of the spectrum C_k^m. Using this description, we can write an expression for the CMB temperature as follows:

$$\Delta T_{\text{data}}(\theta, \varphi) = \sum_{l=2}^{\infty} \sum_{m=-1}^{m=l} C_l^m Y_l^m(\theta, \varphi),$$

$$\nu_{\text{data}}(\theta, \varphi) = \Delta T_{\text{data}}(\theta, \varphi) / \left(\Delta T_{\text{data}}^2\right)^{1/2}, \tag{7.57}$$

where Y_l^m are spherical harmonics. The summation in Eqs (7.57) begins with $l = 2$.

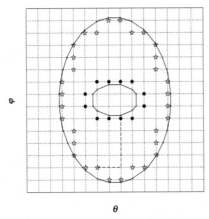

θ

Figure 7.6 Area bounded by the level contour. Closed polygons are approximations for the boundaries based on a linear interpolation. Stars and circles are two rows of internal boundary pixels corresponding to boundaries. The dashed line corresponds to a possible path on a grid, connecting a pair of internal pixels which may belong to one and the same area or two different areas. If this path intersects both boundaries (i.e. the boundaries corresponding to pixels) an even number of times, then both pixels belong to the same (simply connected) region; otherwise they belong to different regions. In this particular case both numbers are zeros (i.e. are even) and the pixels belong to the same area.

Novikov *et al.* (1999) numerically calculated 1000 different Gaussian realizations of temperature distribution on a sphere and compared the distribution of partial Minkowski functionals in observational data with a random Gaussian field. Following this paper, we introduce a section of the map at a height v_t. Each isolated hot spot (region with $v > v_t$) can be regarded as a cluster that can be described using the area, length of boundary and the Euler characteristic (this last is equivalent to the genus and both are related directly to the number of non-connected boundaries). For instance, the total area of a map for which $v > v_t$ is the sum of areas of all isolated hot spots with $v > v_t$. The global Minkowski functionals, that is the total area, the total boundary length and the total genus, can be found by summing up their partial values over all clusters of a map. A calculation of partial Minkowski functionals for a pixelized map requires that the algorithm satisfies the convergence properties

$$\frac{\left(v_k^i\big|_p - v_k^i\right)}{v_k^i} \sim O(h^m), \quad k = 1, 2, \tag{7.58}$$

where $v_k^i\big|_p$ is the kth MF in the ith cluster calculated over a pixelized map, v_k^i is the exact value of this functional on the continuous field and m is the interpolation index. Further on we will use for our algorithm a linear interpolation with $m = 1$.

The pixels (k_1, k_2) inside an area of $v > v_t$ satisfy the condition $v_{k_1 k_2} > v_t$. Let us define a pixel (k_1, k_2) inside this area as an inner boundary pixel if the value of the field is below the threshold v_t in at least one of its four neighbours, $((k_1 + 1, k_2), (k_1 - 1, k_2), (k_1, k_2 + 1), (k_1, k_2 - 1))$; that is, $v_{k_1+1, k_2} < v_t$. We now approximate the smooth boundary curve by a broken line using a linear interpolation of the field between the inner and outer boundary pixels (Fig. 7.6). After this we find the intersection of the boundary curve by the lines of the grid, as follows:

$$\theta_b = k_1 h + h \frac{v_t - v_{k_1, k_2}}{v_{k_1+1, k2} - v_{k_1 k_2}}, \qquad \varphi_b = k_2 h, \tag{7.59}$$

for the lines of the φ grid and

$$\theta_b = k_1 h, \qquad \varphi_b = k_2 h + h \frac{v_t - v_{k_1,k_2}}{v_{k_1+1,k2} - v_{k_1 k_2}}, \qquad (7.60)$$

for the lines of the θ grid. Here θ_b and φ_b are the coordinates of the boundary points on the broken line $\vec{X}_m = (\theta_b, \varphi_b)'$. Obviously this broken line converges to the smooth boundary curve as $h \to 0$. Now the cluster analysis algorithm consists of two steps. We describe these steps below.

Determination of the boundaries and calculation of their length

First closed boundary curves are sought at a height $v = v_t$. Then each row of boundary points is ordered under the assumption that $\vec{X}_{m+1}$ is the boundary point closest to the point $\vec{X}_m$. The length of the closest boundary is then given by

$$l_n = \sum_{m=1}^{m=M_n+1} |\vec{X}_{m+1}^n - \vec{X}_m^n|, \qquad (7.61)$$

where M_n is the total number of boundary points on the nth closed line at a height $\vec{X}_{M_{m+1}}^n = \vec{X}_1^n$ and the norm is given by

$$|\vec{X}_{m+1}^n - \vec{X}_m^n| = \left[(\theta_{m+1} - \theta_m)^2 + \sin\left(\frac{\theta_{m+1} + \theta_m}{2}\right) (\varphi_{m+1} - \varphi_m)^2 \right]^{1/2}.$$

The first point, X_1, is arbitrary. Different boundary curves on the map correspond to arrays of boundary points $\vec{X}_{X_m}^n$ and to inner boundary pixels $\vec{Y}_{X_m}^n$. The total boundary of isolated regions with $v > v_t$ can consist of a set of closed lines (two lines in Fig. 7.6).

Determination of cluster boundaries and calculation of the total boundary and genus

Let us combine all closed lines that form boundaries of the same cluster, using the arrays of inner boundary pixels $\vec{Y}_m^n$. We assume that we need to test whether two different lines are boundaries of the same cluster. These lines correspond to two rows of inner boundary pixels $\vec{Y}_m^{n_1}$ and $\vec{Y}_m^{n_2}$. If we take two arbitrary inner pixels, one from each row, and connect them by a segment along the grid lines (see Fig. 7.6), then this segment may intersect the boundaries N_{int}^i times ($i = 1, 2$), where $N_{int}^i \geq 0$. If two numbers N_{int}^1 and N_{int}^2 are even, then inner boundary pixels belong to the same cluster; otherwise they belong to two different clusters. Consequently, all boundary lines belonging to one cluster form its boundary, whose total length equals the sum of the lengths of each line. The number of closed lines for each cluster is equivalent to the genus of this cluster. Therefore, we find the total number of clusters and two partial MFs for each one of them: the length and the genus.

7.7.2 Calculation of the cluster area

All pixels located between the inner boundaries of the cluster belong to this cluster. The area of a cluster can be crudely approximated by the total area of all these pixels, including the inner boundary pixels. After this we can calculate total and partial Minkowski functionals.

Figures 7.7 and 7.8 show cumulative distribution functions $F(v_t, nu_k)$ (from Novikov *et al.* (1999), where $nv_t = \Sigma_i nv_t^i$ for two COBE maps described above. The mean value and variance were obtained from 1000 random realizations of the Gaussian field, having the same amplitudes as the one shown but with different phases. Both Figs 7.7 and 7.8

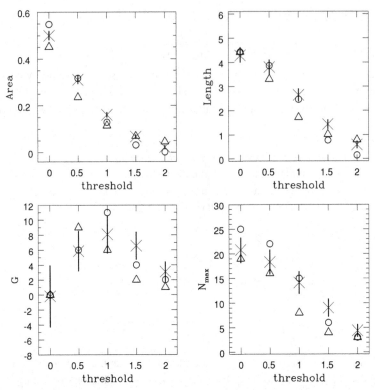

Figure 7.7 Joint distribution function of global Minkowski functionals and the number of maxima/minima as a function of temperature threshold (in units of σ for the COBE map in Fig.7.5(a)). Circles and triangles mark the values of the positive and negative thresholds, respectively. Measurement errors correspond to σ calculated over 1000 Gaussian realizations.

reveal considerable deviation from the Gaussian behaviour. It is interesting to point out that each set of statistics demonstrates small deviations from the Gaussian behaviour for different thresholds $F(A)$ at $\nu_t = 0.5$, $F(L)$ at $\nu_t = -1$, $F(G)$ at $\nu_t = \pm 1$ and $F(N_{\max})$ at $\nu_t = 0, 0.5, -1$. Roughly speaking, these deviations are identical for both maps and are based on the assumption that each of the four statistics carries its own specific statistical information.

It can be expected that partial Minkowski functionals will provide more detailed information. Figures 7.9–7.13 present partial Minkowski functionals for ten thresholds[6] $\nu_t = \pm 2, \pm 1.5, \pm 1, \pm 0.5$ and ± 0. Plotted in each figure are two curves, one for the positive threshold $\nu > \nu_t$ (solid curve) and another for the negative threshold, $-\nu < \nu_t$ (dashed curve). The threshold ν_t has the same absolute value $|\nu_t|$ for each map. Thick and thin solid curves correspond to the COBE maps 1 and 2, respectively (see Figs 7.5(a),(b)). The mean Gaussian curve independent of the sign of the threshold is plotted by a dotted curve. The hatched region is the Gaussian variance 1σ.

The main features implied by Figs 7.9–7.13 are as follows.

Figure 7.9: $\nu_t = 2$; the functions $F(a)$ and $F(b)$ show strong non-Gaussian signal while the functions $F(g)$ and $F(N_{\max})$ are roughly in agreement with the Gaussian behaviour.

---

[6] The thresholds $\nu_t = +0$ and $\nu_t = -0$ correspond to regions with $\nu > 0$ and $\nu < 0$, respectively.

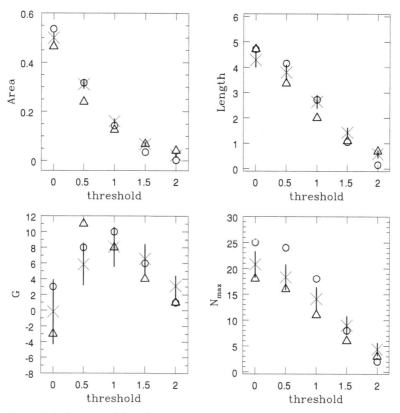

Figure 7.8 Same as in Fig. 7.7 but for the COBE map in Fig. 7.5(b).

Figure 7.10: $v_t = 1.5$; all statistics point to non-Gaussian behaviour.

Figure 7.11: $v_t = 1$; the strongest non-Gaussian signal gives the distribution of maxima $F(N_{max})$, and other statistics are roughly in agreement with the Gaussian behaviour.

Figure 7.12: $v_t = 0.5$; all statistics point to strong deviations from the Gaussian behaviour.

Figure 7.13: $v_t = 0$; all statistics are roughly in agreement with the Gaussian behaviour.

What are the conclusions we can draw from these results? First of all, we note that owing to low angular resolution, the spectrum $C(l)$ for COBE data stretches to 30–40 multipoles; furthermore, which is very important, we are dealing here with a single realization of a random process on a sphere. Moreover, we see that the main source of a non-Gaussian signal is the emission of the Galaxy. The various methods of eliminating this signal, given above, demonstrate that despite the subtraction of the galactic background, the result nevertheless contains a remnant non-Gaussian component. To a certain extent this conclusion is confirmed by an analysis of statistical features of the signal in near-polar zones that are free of the influence of the galactic emission. For example, Colley, Gott and Park (1996) studied the genus curve and failed to detect substantial deviations from the Gaussian behaviour. Schmalzing and Gorski (1998) arrived at similar conclusions and so did Ferreira, Magueijo and Gorski (1998), who also analysed the properties of the COBE signal in near-polar regions. Note nevertheless that this method proved to be adequately efficient for the COBE data with their characteristic low resolution and low signal-to-noise ratio, ~ 1.

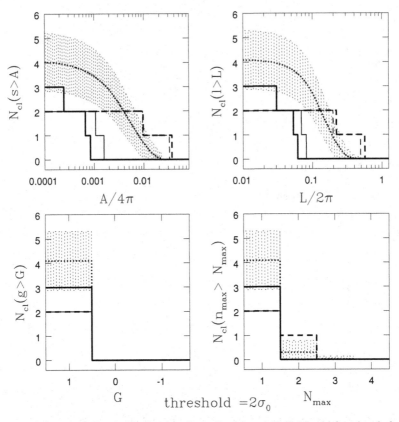

Figure 7.9 The joint distribution functions, N_{cl}, of partial Minkowski functionals for both COBE maps (Figs 7.5(a),(b)). The hatched areas are the ($\pm\sigma$ regions of the Gaussian realizations. Solid lines represent the cumulative distribution function for positive ($\nu > \nu_t$) thresholds and the dashed lines that for negative thresholds. Thick solid lines correspond to the COBE map 1 and thin solid lines to the COBE map 2. Threshold $\nu_t = 2\sigma$.

The problem of extracting the Galaxy's contribution will be very acute in future high-resolution experiments, not only for the central region of the map where it dominates, but also in other areas. This effect is largely created by the 'leaking' of the galactic signal through side lobes of the antenna diagrams. Therefore, an analysis of the statistical properties of the COBE data again highlights the importance of developing methods to eliminate the strong non-Gaussian character of the CMB anisotropy maps caused by various types of non-cosmological noise.

7.8   Statistical nature of the signal in the BOOMERANG and MAXIMA-1 data

We need to point out that along with the COBE experiment that covers the entire area of the sky, an important role for testing the Gaussian nature of primordial anisotropy of the CMB radiation is played by the analysis of the statistical nature of the signal in experiments with low coverage of the sky, for example BOOMERANG and MAXIMA-1. We have already mentioned that these experiments are special in that they cover a relatively small area of the

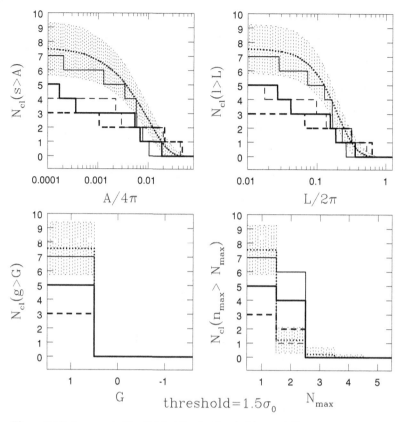

Figure 7.10 Same as in Fig. 7.9 but for the threshold $v_t = 1.5\sigma$.

sky. However, in contrast to COBE, an analysis of the properties of the CMB anisotropy selects those areas of the sky that contain no contribution from the emission of the Galaxy. In analysing the properties of the signal obtained by the BOOMERANG mission, we follow Polenta *et al.* (2002) who made use of the following tests:

(a) skewness and kurtosis (third and fourth moments of the distribution function);
(b) three Minkowski functionals – the area, length and genus.

The area of the sky that contains the signal occupies 1.19% of the entire area of the sphere and covers a zone with coordinates $70° < RA < 105°$, $-55° < \delta < -35°$, where no defects are present on the map and the time of signal gathering is maximal. We will now give the main characteristics of the signal in the B150A frequency channel of the BOOMERANG mission (Polenta *et al.*, 2002).

The third moment of the distribution function S_3 (skewness) and the fourth moment S_4 (kurtosis) are defined in a standard manner, as follows:

$$\sigma_0^2 = \sum_i (T_i - \langle T \rangle)^2 / (N - 1),$$

$$\mu_3 = \sum_i (T_i - \langle T \rangle)^3 / N, \qquad (7.62)$$

$$\mu_4 = \sum_i (T_i - \langle T \rangle)^4 / N,$$

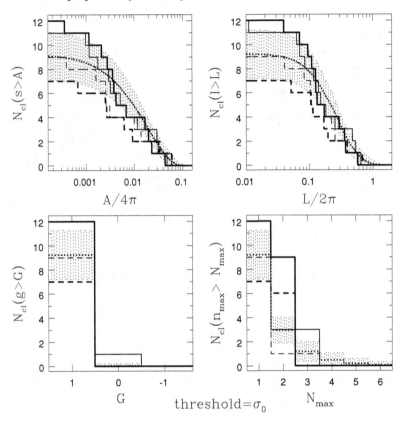

Figure 7.11 Same as in Fig. 7.9 but for the threshold $\nu_t = \sigma$.

and are based on local values of T_i and $\langle T \rangle = \frac{1}{N} = \sum_i T_i$, where N is the number of pixels on the map.

The normalized values of the third and fourth moments, $S_3 = \mu_3/\sigma_0^3$ and $S_4 = \mu_4/\sigma_0^4$, after data processing are correspondingly given by $S_3 = -0.03$ and $S_4 = 0.19$. Owing to the effect of noise and effects of systematics, these values are, of course, non-vanishing. When testing the hypothesis of the Gaussian nature of the signal, Polenta *et al.* (2002) analysed the model maps of CMB anisotropy, simulated for a given type of spectrum $C(l)$ that was obtained by analysing the actual maps of the BOOMERANG experiment. The difference between the models S_3 and S_4 and those obtained directly from the map do not go beyond the errors of simulation. Similar conclusions follow also for the Minkowski functionals calculated for the maps far from the galactic plane. These conclusions are identical to those reported by the MAXIMA-1 collaboration (Wu *et al.*, 2001a,b).

Figure 7.14 plots three Minkowski functionals for two variants of map filtration. As we see from this figure, all Minkowski functionals are in perfect agreement with the Gaussian statistics.

Does this mean that we can be completely sure of the statistical (Gaussian) nature of the ΔT signal, or is the jury still out and does the problem need more detailed experimental and theoretical investigation? We make an attempt to analyse the possible pitfalls of the Gaussian

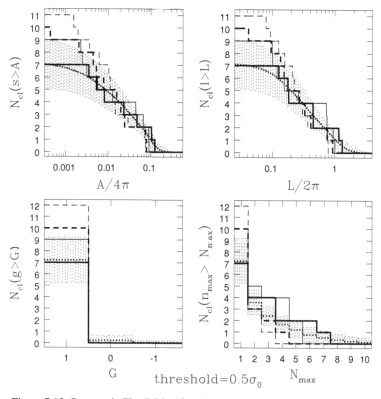

Figure 7.12 Same as in Fig. 7.9 but for the threshold $\nu_t = 0.5\sigma$.

test using a simple model suggested by Novikov *et al.* (2000); this work forms the basis of the following section.

7.9   Simplest model of a non-Gaussian signal and its manifestation in Minkowski functionals

Following one of the most important predictions of the inflation theory, we have considered in the preceding sections of this chapter the primordial perturbations of the metric, density and velocity of the plasma as driven by quantum fluctuations of the vacuum physical fields at the earliest stages in the expansion of the Universe. In the framework of this paradigm, the Gaussian nature of quantum noise is transferred automatically to the angular distribution of the CMB anisotropy in view of the linear mechanism of its generation. By virtue of this paradigm, all possible deviations from the normal distribution of ΔT were interpreted as manifestations either of non-cosmological noise or of systematic errors of the experiment. However, the following question arises: could there be a 'fallacy' in this logic of analysis? Can the experiment itself, possessing certain peculiar characteristics (for example finite resolution, characteristic relaxation time of receiving electronics, etc.) destroy the initial non-Gaussian behaviour of the signal and transform it into a 'Gaussian look-alike' signal? In other words, what is hiding within the error limits that characterize, for instance, Minkowski functionals in the real and the smoothed MAXIMA-1 maps?

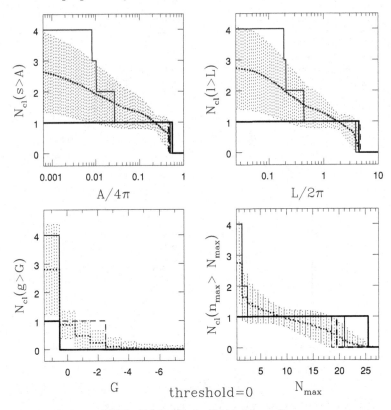

Figure 7.13 Same as in Fig. 7.9 but for the threshold $\nu_t = 0$.

Answers to these questions were recently given in Novikov *et al.* (2001), where one of the simplest models of inflation was analysed; it predicted the so-called χ^2 distribution instead of the normal law of fluctuation distribution. Obviously, this model can be used as one of the possible models for the global non-Gaussian behaviour of the signal.

We will consider a temperature perturbation of the CMB radiation $\Delta T/T$ on the sky in spherical coordinates ϑ and φ. Normalizing $\Delta T/T$ to the variance $\sigma = \langle (\Delta T/T)^2 \rangle^{1/2}$, we obtain a random field $u(\vartheta, \varphi)$ with zero mean value $\langle u \rangle = 0$ and unit variance $\langle u^2 \rangle = 1$. In what follows we consider two models of the random field that are postulated as normalized temperature perturbations on the sky.

The standard approach is to simulate u as a random Gaussian field. The properties of this random Gaussian field are well known (see, for example, Adler (1981)); we have discussed it in the preceding sections of this chapter. Furthermore, we use the field Φ^2 with a single degree of freedom, as suggested by Linde and Mukhanov (1997). To retain the zero mean value and unit variance, we use the expression

$$\psi = \frac{1 - \Phi^2}{\sqrt{2}}. \tag{7.63}$$

Therefore, the two fields u and ψ simulate the Gaussian and non-Gaussian distributions ΔT (Novikov *et al.*, 2001).

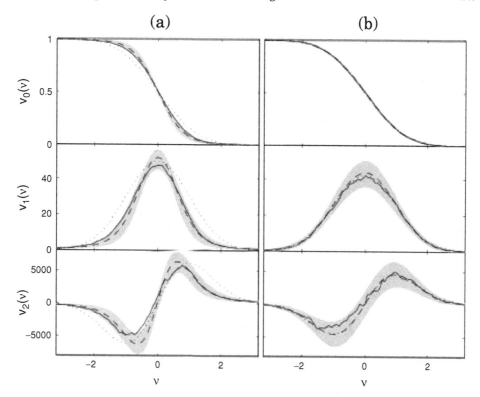

Figure 7.14 Minkowski functionals of the MAXIMA-1 maps. (a) Minkowski functionals for the original signal maps. (b) MFs on a Wiener-filtered map. Adapted from Wu *et al.* (2001a,b).

We simulate the non-Gaussian component of ΔT by a field ψ^2, which is a quadratic function of the Gaussian field Φ. Thus, in the expansion of the anisotropy field in spherical harmonics,

$$\Delta T = \sum_{l} \sum_{l=0}^{l} a_{lm} C(l) Y_{lm},$$

of the field u, the distribution a_{lm} is Gaussian, but a_{lm} for the field ψ are distributed according to the χ^2_{2l+1} law. Note that the variance of a_{lm} both for Gaussian and non-Gaussian fields is normalized to unity and $C(l)$ is chosen to be identical for the two fields. For the Gaussian field, these functions were given earlier in the chapter and for the χ^2 distribution of a_{lm} they are given in Table 7.3.

Obviously the properties both of the Minkowski functionals and of the distribution of extrema for the Gaussian and non-Gaussian fields differ drastically. On the whole, the situation appears to be fairly predictable: the differences in properties of the fields generate differences in statistics, and this is confirmed by calculation results. Assume that the properties of the receiving instruments are not ideal. For instance, that an antenna characterized by quite specific angular resolution possesses nonideal properties. How can this affect the properties of the signal? It is quite clear that the answer depends on the relation between the characteristic

Table 7.3.

	Gaussian field	ψ field
Area ν_0	$\frac{1}{2}\left(1 - \Phi_0\frac{\nu_t}{\sqrt{2}}\right)$	$\Phi\left(\sqrt{\frac{1}{2}} - \frac{\nu_t}{\sqrt{2}}\right)$
Length ν_1	$\frac{\sqrt{\tau}}{8}\exp(-\nu_t^2/2)$	$\frac{\sqrt{\tau}}{4\sqrt{2}}\exp\left(-\frac{1}{2} + \frac{\nu_t}{\sqrt{2}}\right)$
Genus ν_2	$\frac{\tau}{\sqrt{8\pi^3}}\nu_t\exp(-\nu_t^2/2)$	$\frac{\tau}{4\pi^{3/2}}\sqrt{\frac{1}{2} - \frac{\nu_t}{\sqrt{2}}}\exp\left(-\frac{1}{2} + \frac{\nu_t}{\sqrt{2}}\right)$

angular scales of the signal in which the non-Gaussian component is concentrated (if it is concentrated) and the angular resolution of the antenna. We will transform the problem from a qualitative to a quantitative one by modelling the effect of the antenna as a linear filter that acts on the transform of the original signal.

We concentrate our attention on a one-point field distribution function and introduce a smoothing filter $g(\mathbf{x}, t)$, where t is the smoothing scale and $g(\mathbf{x}, 0) = \delta(\mathbf{x})$; then the smoothed field $u(\mathbf{x}, t)$ is defined as

$$u(\mathbf{x}, t) = \mathcal{N}(t)\int d^2y\, g(\mathbf{x} - \mathbf{y}, t)u(\mathbf{y}), \tag{7.64}$$

where the constant $N(t)$ is chosen in such a way that the smoothed field remains normalized to unit variance. If the filter g is Gaussian,[7] it satisfies the diffusion equation:

$$\frac{\partial g(\mathbf{x}, t)}{\partial t} = t\Delta g(\mathbf{x}, t), \tag{7.65}$$

where Δ is the Laplacian in the space of the angular variables $\vec{x}$. By combining Eqs (7.64) and (7.65), we can obtain an 'equation of evolution' for the field u in response to changing the scale t:

$$\frac{\partial u(\mathbf{x}, t)}{\partial t} = t\left(\Delta + r_{\mathrm{corr}}^{-2}\right)u(\mathbf{x}, t). \tag{7.66}$$

The second term r_{corr}^{-2} in Eq. (7.66) appears because of the scale dependence on the normalizing factor $N(t)$. This equation makes it possible to study the one-point probability, $P(u, t)$, of the distribution of the smoothed field $u(\mathbf{x}, t)$ in response to a changed smoothing scale. By writing this probability density as

$$P(u, t) = \langle\delta(u(\mathbf{x}, t - u))\rangle, \tag{7.67}$$

and taking the partial derivative with respect to t, we obtain

$$\frac{\partial(u, t)}{\partial t} = -t\frac{\partial}{\partial u}\left[\left(\langle\Delta u\rangle_u + \frac{u}{r_{\mathrm{corr}}^2}\right)P(u, t)\right]. \tag{7.68}$$

The quantity $\langle\Delta u\rangle_u$ is the averaged value of the field Laplacian $u(x, t)$, provided the value

---

[7] Note that the Gaussian approximation of the antenna diagram is widely used for processing the observational data of CMB anisotropy.

of the field u is fixed. Incidentally, this equation is written in a conservative form, that is its integral over du vanishes.

The conditional mean of the Laplacian for the set of random fields can be calculated analytically. The most interesting relation for the two fields, the Gaussian field and the χ^2 field, is $\langle \Delta u \rangle_u = -(u/r^2_{\text{corr}})$. In such cases the right-hand side of Eq. (7.68) simply vanishes.

Note that $P(u) \propto \exp(-u^2)$ is a stationary solution of Eq. (7.68). This is a reflection of the well known fact that a Gaussian random field remains random and Gaussian after smoothing, as it does after any linear filtration. However, the probability distribution for the χ^2 field differs from zero only if $u \ll 1/\sqrt{2}$ (see Eq. (7.1)) and is not differentiable at the outer field boundary. Consequently, smoothing makes the field evolve away from the χ^2 distribution so it ultimately approaches a distribution close to the stationary Gaussian solution for large smoothing lengths.

Therefore, the effect of the antenna (acting as a Gaussian filter) may result in the fact that the primary non-Gaussian signal, localized on scales smaller than the antenna diagram width, will test as a Gaussian signal. This is why experiments with the highest possible angular resolution, which makes it feasible (in principle) to restrict the angular scale θ_* of the possible initial non-Gaussian system, become especially important.

7.10 Topological features of the polarization field

In contrast to the analysis of the statistical properties of the anisotropy field, procedures needed to test the nature of the CMB polarization are not as well elaborated. This is mostly caused by a greater complexity of the properties of the polarization field, which in contrast to anisotropy is not scalar. Moreover, as we saw in Chapter 6, polarization is described by using both local and non-local approaches, and the choice between the two is dictated, in our opinion, by the properties of the noise present in polarization maps alongside the primary signal. The first question to which we want to attract attention is: which of the characteristics of polarization carry information on the statistical nature of perturbations on the surface of the last scattering of quanta? We have already discussed the geometric characteristics of CMB polarization in Section 3.5, and, among other things, pointed to the emergence of field anomalies in the neighbourhood of the points Q, where both the U and Q Stokes components vanish simultaneously. In what follows we analyse the results of statistical calculations of the genus of $p = |\overline{P}|/\sigma_0$, where σ_0^2 is the variance and $|P|$ is the polarization vector magnitude (Naselsky and Novikov, 1998).

Let us split the CMB polarization map up into two types of regions: regions with relatively strong polarization $p > p_0$ ('strongly polarized zones') and regions with relatively weak polarization $p < p_0$ ('weakly polarized zones') (Figs 7.15 and 7.16). Assume now that we can measure the signal of polarization $p \geq p_t$, where p_t is the threshold which reflects the instrument sensitivity. If it is possible to measure only the 'strongly polarized' signal $p_t > 0$, then only some polarized spots will be observed and no percolation between spots will be possible. Therefore, percolation via polarized zones is only possible if the instrument sensitivity $p_t \leq p_0$.

The value of p_0 can be found analytically. We assume that $|\overline{P}|$ is a random two-dimensional scalar field with the Rayleigh distribution (Coles and Barrow, 1987). This field can be presented as a two-dimensional surface in three-dimensional space. This surface contains extremal points: minima, maxima, saddle points and singular points. The densities of maxima,

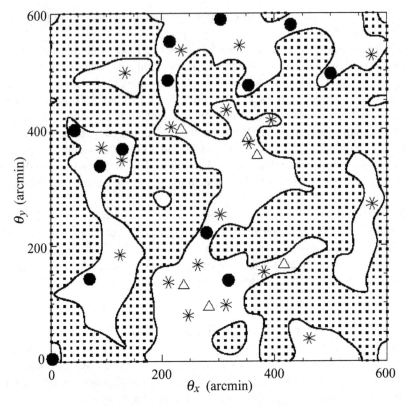

Figure 7.15 A $10° \times 10°$ map of the CMB polarization field for the CDM model. The dotted area corresponds to regions with polarization degree $p > p_t$. Solid curves plot the boundaries between the regions of $p > p_t$ and $p < p_t$. Circles, triangles and stars mark the foci, nodes and saddle points, respectively. This map comprises 13 foci, 6 nodes and 19 saddles.

minima and saddle points for the Rayleigh field P are found as follows:

$$N_{max}(p) = \int_p^\infty n_{max}(p')\,dp',$$

$$N_{min}(p) = \int_p^\infty n_{min}(p')\,dp',$$ (7.69)

$$N_{sad}(p) = \int_p^\infty n_{sad}(p')\,dp'.$$

Here $n_{max}(p)$, $n_{min}(p)$ and $n_{sad}(p)$ are the concentrations of maxima, minima and saddle points, respectively, inside a certain interval $(p, p + dp)$, and $N_{max}(p)$, $N_{min}(p)$ and $N_{sad}(p)$ are the concentrations of maxima, minima and saddle points above a certain level p. Note that in this case saddle points are saddle points on the two-dimensional surface $p(x, y)$. As in the case of the anisotropy field, we define the genus for the polarization field by

$$g(p) = n_{max}(p) + n_{min}(p) - n_{sad}(p).$$ (7.70)

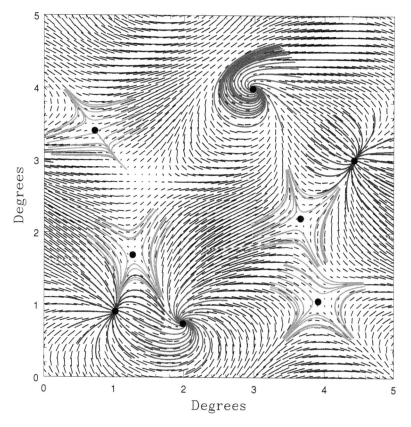

Figure 7.16 A $2° \times 2°$ map of the CMB polarization field for the scale-invariant adiabatic model of cold dark matter with $\Omega_b = 0.03$ and $h = 0.75$, with the smoothing angle of $5'$ (FWHM). Calculation techniques for small regions of the sky and the spectrum were borrowed from Bond and Efstathiou (1987). The length of each vector is proportional to the degree of polarization and the orientation represents the polarization direction. To simplify the visual perception, only 50 vectors were used. Only the orientation of the polarization in the neighbourhood of non-polarized points has been marked (solid curves). This map contains seven non-polarized points: two foci, one node and four saddle points.

Integrating Eq.(7.70) from a certain section height P to ∞, we obtain

$$G(p) = N_{\max}(p) + N_{\min}(p) - N_{\text{sad}}(p) = \int_p^\infty g(p')\,\mathrm{d}p'. \qquad (7.71)$$

First of all, we find the percolation height for the polarization field using the condition $G(p_0) = 0$ and taking into account the definition (7.70). Note that this condition does not mean that p_0 is automatically the percolation level for any scalar field. It is well known that the level of percolation for the random Gaussian field corresponds to the level at which the genus curve intersects the zero. In the case of the Rayleigh approximation, this condition also signifies that the level p_0 corresponds to the percolation contour.

As for the anisotropy field, we introduce a combination of independent random quantities q and u and their first and second derivatives q_i, u_i, q_{ij}, u_{ij}, ($q_{ij} = Q_{ij}/\sigma_2$, $u_{ij} = U_{ij}/\sigma_2$, $i = 1, 2$), where σ_2 is the spectral parameter defined as $\sigma_2^2 = \langle Q_{ii}^2 \rangle = \langle U_{ii}^2 \rangle$. These quantities obey the following conditions:

$$p^2 = q^2 + u^2,$$

$$p_i = qq_i + uu_i,$$

$$\gamma p_i p_j + p_{ij} = \gamma(q_i q_j + u_i u_j) + qq_{ij} + uu_{ij},$$

$$\langle qu \rangle = \langle q_i u_j \rangle = \langle q_{ij} u_{kl} \rangle = \langle qu_i \rangle - \langle q_i u \rangle = 0,$$

$$\langle qq_{ij} \rangle = \langle uu_{ij} \rangle = -\frac{\gamma}{2}\delta_{ij}, \tag{7.72}$$

$$\langle q_i q_j \rangle = \langle u_i u_j \rangle = -\frac{1}{2}\delta_{ij},$$

$$\langle q_{ij} u_{kl} \rangle = \frac{1}{8}(\delta_{ik}\delta_{jl} + \delta_{il}\delta_{jk} + \delta_{ij}\delta_{kl}),$$

$$\gamma = \frac{\sigma_1^2}{\sigma_0 \sigma_2}.$$

The joint distribution function F for the quantities q, q_i, q_{ij}, u, u_i, u_{ij} is chosen to be Gaussian in accordance with the hypothesis of the normal distribution of perturbations of metric, velocity and density of matter, as follows:

$$F dq \, du \, dq_i \, du_i \, dq_{ij} \, du_{ij} = \frac{1}{\sqrt{(2\pi)^{12} \det M}} e^{-\frac{A}{2}} dq \, du \, dq_i \, du_i \, dq_{ij} \, du_{ij}, \tag{7.73}$$

$$A = v \times m^{-1} \times v^{\mathrm{T}},$$

where M is the covariant matrix and A is a quadratic form of the 12-dimensional vector $v(q, q_i, q_{ij}, u, u_i, u_{ij})$. A substitution of p, p_i, p_{ij} into Eq. (7.73) from Eqs (7.72) and integration over six variables yields the joint probability $f dp \, dp_i \, dp_{ij}$ of the quantities p, p_i, p_{ij} falling within the interval from p, p_i, p_{ij} to $p + dp$, $p_i + dp_i$, $p_{ij} + dp_{ij}$.

By analogy to Bond and Efstathiou (1987), the differential density of extremal points obeys the following equation:

$$n_{\text{ext}}(p) = \frac{\sigma_2^2}{\sigma_1^2} \int |\det(p_{ij})| f \delta(p_1)\delta(p_2) \, dp_{ij}, \tag{7.74}$$

where n_{ext} is the density of extrema. An extremum may be a maximum, a minimum or a saddle point, depending on the limits of integration over du_{ij}. These limits are dictated by the quantities $\mathrm{tr}(p_{ij})$ and $\det(p_{ij})$ of the matrix of second derivatives (p_{ij}).

We find from the definitions (7.71) and (7.74) that the genus curve is described by the following equation:

$$g(p) = n_{\max}(p) + n_{\min}(p) - n_{\text{sad}}(p) = \frac{\sigma_2^2}{\sigma_1^2} \int \det(p_{ij}) f(p, p_i = 0, p_{ij}) \, dp_{ij}. \tag{7.75}$$

Integrating this equation, we obtain

$$g(p) = \frac{1}{4\pi} \left(\frac{\sigma_1^2}{\sigma_0^2} \right)^2 (p^2 - 3) e^{-\frac{p^2}{2}}. \tag{7.76}$$

Then the genus curve has the form

$$G(p) = \frac{1}{4\pi r_c^2}(p^2 - 1)e^{-\frac{p^2}{2}}. \tag{7.77}$$

The condition $G(p) = 0$ yields the value p_0:

$$p_0 = 1. \tag{7.78}$$

Therefore, the principal and most important difference between the field of polarization modulus and the anisotropy field is the bias of the percolation level from $p_0 = 0$ (typical of a Gaussian field) to $p_0 = 1$, which reflects specific features of the Rayleigh distribution. We turn now to analysing Minkowski functionals for maps of the modulus of polarization 'vector'.

A geometrical interpretation of Minkowski functionals on a two-dimensional map is very simple. We consider the polarization intensity as a two-dimensional surface in the three-dimensional space, as we did in the preceding section. If we cut this surface at a certain height p_t, then the map area splits into two parts: one where polarization is above a certain threshold p_t, and the other where polarization is below this threshold, $p < p_t$. The Minkowski functionals in the case of a two-dimensional distribution correspond to the following quantities.

   (1) A is the part of the map area where $p > p_t$.
   (2) L is the length of the boundary between the parts where $p < p_t$ and $p > p_t$, per unit area.
   (3) $G = N_{max} + N_{min} - N_{sad}$ are the Euler characteristics per unit area (an equivalent of genus).

The threshold is therefore an independent variable on which all these functionals depend. In fact, the third functional has already been discussed above where we analysed the percolation level. The first functional is $\exp(-\nu/2)^2$. The second functional has the following form (Naselsky and Novikov, 1998):

$$L = \frac{1}{r_c} p_t e^{-\frac{\nu_t^2}{2}}. \tag{7.79}$$

Functionals for the Rayleigh distribution are zero at $p_t < 0$. These functionals can be used to describe the morphology of the CMB polarization field in the same manner as they are used for the CMB anisotropy (Winitzki and Kosowsky, 1997).

8

The Wilkinson Microwave Anisotropy Probe (WMAP)

8.1   Mission and instrument

The successful launch of the WMAP[1] (Wilkinson Microwave Anisotropy Probe) mission in June 2001 signalled a new epoch in the investigation of the cosmic microwave background. This experiment differs from all previous satellite-, balloon- and ground-based experiments by unprecedented precision and sensitivity.

The WMAP mission has been designed to determine the power spectrum of the CMB anisotropy and polarization and subsequently to estimate such cosmological parameters as the Hubble constant H_0, the baryonic fraction of dark matter Ω_b, the geometry Ω_K of the Universe, etc. These observations provide an independent check on the COBE results, determine whether the anisotropy obeys Gaussian statistics, and verify whether the predicted temperature–polarization correlation is present.

The high-level features of the WMAP mission can be briefly described as follows. The mission is designed to produce an almost full (>95% of the entire sky) map of the CMB temperature fluctuations with $\simeq 0.2°$ angular resolution, accuracy on all angular scales $>0.2°$, accurate calibration (<0.5% uncertainty), an overall sensitivity level of $\Delta T_{\mathrm{rms}} < 20\ \mu K$ per pixel (for 393 216 sky pixels, 3.2×10^{-5} ster per pixel) and systematic errors limited to <5% of the random variance on all angular scales.

We have taken information from Page (2000) to describe the instrument. The instrument measures temperature differences from two regions of the sky separated by ~140°. It is composed of ten symmetric, passively cooled, dual-polarization differential microwave receivers. There are four receivers in W band, two receivers in V band, two in Q band, one in Ka band and one in K band. The receivers are fed by two back-to-back Gregorian telescopes. The primary mirrors are 1.4 m ×1.6 m. The secondaries are roughly one metre across. For computing the CMB angular spectrum it is very important to have precise knowledge of the antenna beams. Using Jupiter as a source, the satellite team measured the beam to less than 30 dB of its peak value. The beams are not symmetric, neither are they Gaussian. Fortunately, the scan strategy symmetrizes the beam, greatly facilitating the analysis.

The instrument is passively cooled. There are no cryogenic or mechanical refrigerators and thus no inboard sources of thermal variation. The WMAP spins around its axis with a period of 2 min and precesses around a 22°.5 cone every hour. Consequently, ~30% of the sky is covered in one hour. The axis of this combined rotation–precession approximately sweeps

---

[1] WMAP was proposed in 1995, building work began in 1996, and the major push started in 1997 after the confirmation review.

out a great circle as the Earth orbits the Sun. In six months the whole sky is mapped. With this scan, the WMAP is continuously calibrated on the CMB dipole.

Systematic errors in the sky maps can originate from several different sources: external emission sources, calibration errors, internal emission sources, onboard electronics, striping and map-making errors and uncertainty in the beam shape. In order to reduce the systematic errors, WMAP has a symmetric differential design, rapid large-sky-area scans, a highly interconnected and redundant set of differential observations, an L_2 (the second Lagrangian point of the Sun–Earth system) orbit to minimize contamination from the Sun, Earth and Moon emission, and to allow for thermal stability, and five independent frequency channels to enable a separation of galactic and cosmic signals.

8.2 Scientific results

The scientific result of the mission is a set of maps of the microwave sky at different frequency bands. These maps may be used both to analyse Galactic and extragalactic emission and to analyse the CMB cosmological anisotropy to address the most basic cosmological questions and extract information on fundamental cosmological parameters.

Here we will focus on the cosmological results. Special regions of the sky have been selected for cosmological analysis. The regions on the sky which are significantly contaminated by diffuse emission from our Galaxy and by pointlike source emission have been masked (Bennett *et al.*, 1996). After applying the diffuse and point source masks followed by additional cleaning up (Finkbeiner, 2003), the residual signal was assumed to represent the cosmological CMB signal and became the subject of the subsequent analysis. If the fluctuations in the CMB are Gaussian, then all the information in the CMB is contained in the angular power spectrum. We will discuss the possible non-Gaussianity at the end of this chapter.

Under the assumption of Gaussianity of the primordial cosmological signal, the best-fit parameters can be determined from the peak of the N-dimensional likelihood surface. For this purpose, Spergel *et al.* (2003a,b) used a basic cosmological model, which is a flat Universe with radiation, baryons, cold dark matter, cosmological constant, and a power-law spectrum of adiabatic primordial fluctuations. Such a model describes both TT (temperature–temperature) and TE (temperature–E-component of polarization) CMB power spectra with the following parameters: the Hubble constant h (in units of 100 km s^{-1} Mpc^{-1}), the physical matter and baryon densities $w_m \equiv \Omega_m h^2$ and $w_b \equiv \Omega_b h^2$, the optical depth to the decoupling surface τ, and the scalar spectral index n_s.

This simple model provides an acceptable fit to both the WMAP TT and TE data (see Figs 8.1 and 8.2).

Table 8.1 summarizes the results obtained by Spergel *et al.* (2003a,b).

It is essential to compare the best-fit power law ΛCDM model with other cosmological observations and produce the best-fit model to the full data set. In particular, Spergel *et al.* (2003a,b) considered determinations of the local expansion rate (i.e. the Hubble constant), the amplitude of fluctuations on galaxy scales, the baryon abundance, ages of the oldest stars, large-scale structure data and supernova Ia data.

8.2.1 Hubble constant

The Hubble Key Project (Freedman *et al.*, 2001) used Cepheids to calibrate several secondary distance indicators (Type Ia supernovae, Tully–Fisher, Type II supernovae,

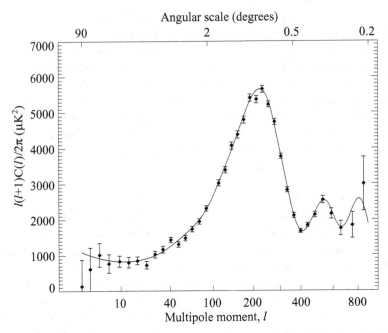

Figure 8.1 This figure compares the best-fit power law ΛCDM model to the WMAP temperature angular power spectrum. The grey dots are the unbinned data. Adapted from Spergel *et al.* (2003a,b).

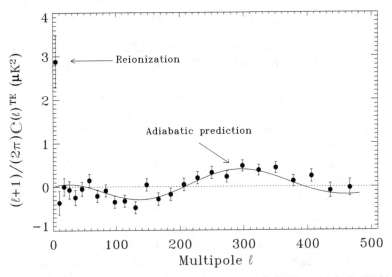

Figure 8.2 This figure compares the best-fit power law ΛCDM model to the WMAP TE angular power spectrum. Adapted from Spergel *et al.* (2003a,b).

and surface brightness fluctuations). Their estimate for the Hubble constant is $H_0 = 72 \pm 3$(statistical) ± 7(systematic) km s^{-1} Mpc^{-1} (see Chapter 1). The agreement between this estimated value and the value $h = 0.72 \pm 0.05$ given by WMAP, is very impressive, taking into account the complete independence of these two approaches.

Table 8.1. *Derived cosmological parameters*

Parameter	Mean (68% confidence range)
Amplitude of density/fluctuations	$\sigma_8 = 0.9 \pm 0.1$
Characteristic amplitude of velocity fluctuations	$\sigma_8 \Omega_m^{0.6} = 0.44 \pm 0.10$
Baryon density/critical density	$\Omega_b = 0.047 \pm 0.006$
Matter density/critical density	$\Omega_m = 0.29 \pm 0.07$
Age of the Universe	$t_0 = 13.4 \pm 0.3$ Gyr
Redshift of reionization	$z_r = 17 \pm 5$
Redshift at decoupling	$z_{dec} = 1088^{+1}_{-2}$
Age of the Universe at decoupling	$t_{dec} = 372 \pm 14$ kyr
Thickness of surface of last scatter	$\Delta z_{dec} = 194 \pm 2$
Age of Universe at Last Scatter	$\Delta t_{dec} = 115 \pm 5$ kyr
Redshift at matter/radiation equality	$z_{eq} = 3454^{+385}_{-392}$
Sound horizon at decoupling	$r_s = 144 \pm 4$ Mpc
Angular diameter distance to the decoupling surface	$d_A = 13.7 \pm 0.5$ Gpc
Acoustic angular scale	$\ell_A = 299 \pm 2$
Current density of baryons	$n_b = (2.7 \pm 0.1) \times 10^{-7}$ cm^{-3}
Baryon/photon ratio	$\eta = \left(6.5^{+0.4}_{-0.3}\right) \times 10^{-10}$

8.2.2 Weak lensing and galaxy velocity fields

Weak lensing and galaxy velocity fields are potentially powerful tools for measuring mass fluctuations. These techniques directly probe density fluctuations of dark matter, and therefore can be compared with the CMB model predictions. Several groups have reported different measurements within the past years. There is still a significant scatter in the reported amplitude of fluctuations (0.72–0.98), but the best-fit model to the WMAP data lies in the middle of the reported range: 0.9 ± 0.1, where σ_8 measurements have been normalized to $\Omega_m = 0.283$ (the best fit for WMAP data). Improvement in the measurements of weak lensing and the CMB can usefully complement these independent observations.

8.2.3 Cluster number counts

Detection of the contribution of the Sunyaev–Zeldovich effect to the CMB power spectrum on small scales is in fact a probe of the number density of high-redshift clusters. The recent cosmic background image (CBI) detection (Mason, Myers and Readhead, 2001, Bond *et al.*, 2002) at $\ell > 1500$ gives $\sigma_8 = 1.04 \pm 0.12$ (Komatsu and Seljak, 2002).

The result of the Bahcall and Bode (2003) analysis of the abundance of massive clusters at $z = 0.5{-}0.8$ yields $\sigma_8 = 0.95 \pm 0.1$ for $\Omega_m = 0.25$. Other cluster analyses yield different values; for example, the Borgani *et al.* (2001) best-fit values for a large sample of x-ray clusters are $\sigma_8 = 0.66^{+0.05}_{-0.05}$ and $\Omega_m = 0.35^{+0.13}_{-0.10}$. On the other hand, Reiprich and Böhringer (2002) find very different values: $\sigma_8 = 0.96^{+0.15}_{-0.12}$ and $\Omega_m = 0.12^{+0.06}_{-0.04}$. Pierpaoli *et al.* (2003) discuss the wide range of values that different x-ray analyses find for σ_8. With the larger REFLEX sample, Schuecker *et al.* (2003) find $\sigma_8 = 0.711^{+0.039}_{-0.031}{}^{+0.120}_{-0.162}$ and $\Omega_m = 0.341^{+0.031}_{-0.029}{}^{+0.087}_{-0.071}$, where the second set of errors includes the systematic uncertainties.

One can see that the best-fit WMAP value lies in the middle of the relevant range.

8.2.4 Baryon abundance

Acoustic peaks in the spectrum of CMB are a natural manifestation of the evolution of the baryonic dark matter during recombination. At the same time, primordial abundance of deuterium (Boesgaard and Steigman, 1985) is a very sensitive function of the cosmological density of baryons. The best-fit baryon abundance based on WMAP is $\Omega_b h^2 = 0.024 \pm 0.001$; it gives us a baryon/photon ratio of $\eta = (6.5^{+0.4}_{-0.3}) \times 10^{-10}$. For this abundance, standard Big Bang nucleosynthesis (Burles, Nollett and Turner, 2001) implies a primordial deuterium abundance relative to hydrogen of $[D]/[H] = 2.37^{+0.19}_{-0.21} \times 10^{-5}$.

The Kirkman *et al.* (2003) analysis of QSO HS 243+3057 yields a D/H ratio of $2.42^{+0.35}_{-0.25} \times 10^{-5}$. They combine this measurement with four other D/H measurements (Q0130 − 4021: D/H $< 6.8 \times 10^{-5}$, Q1009 + 2956: $3.98 \pm 0.70 \times 10^{-5}$, PKS 1937 − 1009: $3.25 \pm 0.28 \times 10^{-5}$, and QSO HS0105+1619: $2.5 \pm 0.25 \times 10^{-5}$), to obtain their current best D/H ratio of $2.78^{+0.44}_{-0.38} \times 10^{-5}$, implying $\Omega_b h^2 = 0.0214 \pm 0.0020$. D'Odorico, Dessauges-Zavadsky and Molaro (2001) find $2.24 \pm 0.67 \times 10^{-5}$ from their observations of $Q0347 − 3819$ (although a reanalysis of the system by Levshakov *et al.* (2003) reports a higher D/H value of 3.75 ± 0.25). Pettini and Bowen (2001) report a D/H abundance of $1.65 \pm 0.35 \times 10^{-5}$ from STIS measurements of QSO 2206 − 199, a low metallicity ($Z \sim 1/200$) damped Lyman α system. The WMAP value lies between this estimate, $\Omega_b h^2 = 0.025 \pm 0.001$, and that by Kirkman *et al.* (2003), $\Omega_b h^2 = 0.0214 \pm 0.0020$.

It is worth noting that such good agreement between independent measurements of the D/H ratio from different physical aspects is important for the basic Big Bang model.

8.2.5 Large-scale structure and supernova data

The large-scale structure observations, the Ly-α forest data and CMB measurements deal with similar physical scales during completely different epochs and therefore can be considered independent and complementary to each other. According to an analysis conducted by the Anglo-Australian Telescope Two Degree Field Galaxy Redshift Survey (2dFGRS) (Colless *et al.*, 2001) and the Sloan Digital Sky Survey[2] (SDSS) large-scale structure data, one can conclude that the ΛCDM model obtained from the WMAP data alone provides a fairly good fit to the 2dFGRS power spectrum. The best fit has $\beta = 0.45$, consistent with the (Peacock *et al.* (2001) measured value of $\beta = 0.43 \pm 0.07$.

Systematic studies by the Supernova Cosmology Project (Perlmutter *et al.* 1999) and by the High z Supernova Search Team (Riess *et al.*, 1998) provide evidence for an accelerating Universe. The combination of the large-scale structure and the CMB and supernova data provide strong evidence for a flat Universe dominated by a cosmological constant (Bahcall *et al.*, 1999). Since the supernova data probes the luminosity distance versus redshift relationship at moderate redshift $z < 2$, and the CMB data probes the angular diameter distance relationship to high redshift ($z \sim 1089$), the two data sets are complementary. The supernova constraint on the cosmological parameters (Knop *et al.*, 2003; Melchiori *et al.*, 2003; Tonry *et al.*, 2003) are consistent with the ΛCDM WMAP model. The SNIa likelihood surface in the $\Omega_m - \Omega_\Lambda$ and in the $\Omega_m - w$ planes provides useful additional constraints on cosmological parameters (Spergel *et al.*, 2003b).

---

[2] See www.sdss.org.

Table 8.2. *Basic and derived cosmological parameters: Running spectral index model*

Amplitude of fluctuations	$A = 0.83^{+0.09}_{-0.08}$
Spectral index at $k = 0.05$ Mpc^{-1}	$n_\mathrm{s} = 0.93 \pm 0.03$
Derivative of spectral index	$dn_\mathrm{s}/d\ln k = -0.031^{+0.016}_{-0.018}$
Hubble constant	$h = 0.71^{+0.04}_{-0.03}$
Baryon density	$\Omega_\mathrm{b} h^2 = 0.0224 \pm 0.0009$
Matter density	$\Omega_\mathrm{m} h^2 = 0.135^{+0.008}_{-0.009}$
Optical depth	$\tau = 0.17 \pm 0.06$
Matter power spectrum normalization	$\sigma_8 = 0.84 \pm 0.04$
Characteristic amplitude of velocity fluctuations	$\sigma_8 \Omega_\mathrm{m}^{0.6} = 0.38^{+0.04}_{-0.05}$
Baryon density/critical density	$\Omega_\mathrm{b} = 0.044 \pm 0.004$
Matter density/critical density	$\Omega_\mathrm{m} = 0.27 \pm 0.04$
Age of the Universe	$t_0 = 13.7 \pm 0.2$ Gyr
Reionization redshift	$z_\mathrm{r} = 17 \pm 4$
Decoupling redshift	$z_\mathrm{dec} = 1089 \pm 1$
Age of the Universe at decoupling	$t_\mathrm{dec} = 379^{+8}_{-7}$ kyr
Thickness of surface of last scatter	$\Delta z_\mathrm{dec} = 195 \pm 2$
Age of Universe at last scatter	$\Delta t_\mathrm{dec} = 118^{+3}_{-2}$ kyr
Redshift of matter/radiation equality	$z_\mathrm{eq} = 3233^{+194}_{-210}$
Sound horizon at decoupling	$r_\mathrm{s} = 147 \pm 2$ Mpc
Angular size distance to the decoupling surface	$d_\mathrm{A} = 14.0^{+0.2}_{-0.3}$ Gpc
Acoustic angular scale	$\ell_\mathrm{A} = 301 \pm 1$
Current density of baryons	$n_\mathrm{b} = (2.5 \pm 0.1) \times 10^{-7}$ cm^{-3}
Baryon/photon ratio	$\eta = (6.1^{+0.3}_{-0.2}) \times 10^{-10}$

8.2.6 *Basic results of WMAP data analysis*

The WMAP mission has provided cosmology with a standard model: a flat Universe composed of matter, baryons and vacuum energy with a nearly scale-invariant spectrum of primordial fluctuations. This cosmological model is a result of best possible fitting to many observational data that include star formation, small-scale CMB data, large-scale structure data and supernova data. This model is also consistent with the baryon/photon ratio inferred from observations of D/H in distant quasars, the HST Key Project measurement of the Hubble constant, stellar ages and the amplitude of mass fluctuations inferred from clusters and from gravitational lensing. Table 8.2 (Spergel *et al.*, 2003a) lists the best-fit parameters for this model.

While there have been a host of papers on cosmological parameters, WMAP has brought this program to a new stage: WMAP's more accurate determination of the angular power spectrum has significantly reduced parameter uncertainties, its detection of TE fluctuations has confirmed the basic model, and its detection of the reionization signature has reduced the n_s–τ degeneracy. Most importantly, the rigorous propagation of errors and uncertainties in the WMAP data has strengthened the significance of the inferred parameter values.

In spite of the great success of the WMAP experiment, there are still some uncertainties to be solved and explained. Many different investigations have been performed to establish Gaussianity or non-Gaussianity of the WMAP data (Komatsu *et al.*, 2003). Since the inflation predicts a Gaussian random distribution of initial perturbations in the Universe, the expectation of the Gaussian CMB temperature fluctuations on the sky seems to be well justified. This is a directly testable prediction since WMAP's high-resolution, all-sky data set of the has become available.

The sophisticated non-Gaussianity test by Chiang *et al.* (2003) on derived maps from the first-year WMAP data by Tegmark, de Oliveria-Costa and Hamilton (2003) shows significant non-Gaussianity features. This test was based on a phase (of the spherical harmonic coefficients) mapping technique, which has the advantage of testing non-Gaussianity in separate multipole bands. It has been shown that the foreground-cleaned WMAP map contradicts the random-phase hypothesis in all four multipole bands, which points to non-Gaussianity. The evidence of non-Gaussianity for $l > 350$ has been discovered and is yet to be explained. Explanation of such a deviation from Gaussianity is crucial because most of the data analysis technique used for the analysis of the WMAP data has been adopted only for the primordial Gaussianity of the cosmological signal.

To detect Gaussianity or non-Gaussianity, Larson and Wandelt (2005) performed an analysis of one- and two-point statistics of the hot and cold spots in the CMB data obtained by WMAP. Investigating the pattern of cold and hot spots on the sky, they found an anomaly in the full-sky minima–minima temperature–temperature two-point correlation function in the form of an excessively large fluctuation, which is unlikely at the 3 sigma level. To obtain this result they approached the problem numerically: they found some way to reduce the entire CMB sky to a single number – a single statistic computed on the hot and cold spots. They compared the statistic for the measured CMB sky to the distribution of statistics for the simulated CMB skies. If the measured statistic falls significantly above or below the others, then large statistical fluctuations were assumed to exist, which they quantified. It is then, to paraphrase Larson and Wandelt, up to the reader to determine if this should be interpreted as merely an unlikely statistical fluctuation, an indication of non-Gaussianity of a cosmological CMB signal, a residual foreground, or a systematic effect. It should be mentioned that Larson and Wandelt observed the anomaly in the minima–minima temperature–temperature two-point function in the form of over large fluctuations only on the full sky. This suggests that this effect is distinct from those that led to recent claims of global anisotropy in the CMB (see the following).

Eriksen *et al.* (2004) also investigated the three Minkowski functionals (see Chapter 7) and the so-called skeleton of the two-dimensional WMAP data in order to check for Gaussianity or non-Gaussianity. The skeleton length was introduced as a diagnostic for Gaussianity by Novikov, Colombi and Dore (2003). The skeleton of the field is a set of lines that extend from extremum to extremum along the lines of maximum or minimum gradient. The results of statistics of the WMAP data are compared with 5000 Monte Carlo simulations, based on Gaussian fluctuations with the *a priori* best-fit running-index power-spectrum and WMAP-like beam and noise properties. Several power-spectrum-dependent quantities, such as the number of stationary points, the total length of the skeleton and the spectral parameter, γ, are also estimated. While the area and length of the Minkowski functionals and the length of the skeleton show no evidence of departure from the Gaussian hypothesis, the northern hemisphere genus has a χ^2 that is large at the 95% level for all scales. For the particular smoothing scale of 3.40 degrees FWHM it is larger than that found in 99.5% of the simulations. In addition, the WMAP genus for negative thresholds in the northern hemisphere has an amplitude that is larger than in the simulations with a significance of more than 3 sigma. On the smallest angular scales considered, the number of extrema in the WMAP data is high at the

3 sigma level. However, this can probably be attributed to the effect of point sources. Finally, the spectral parameter γ is high at the 99% level in the northern Galactic hemisphere, while perfectly acceptable in the southern hemisphere. The results provide strong evidence for the presence of both non-Gaussian behaviour and an unexpected power asymmetry between the northern and southern hemispheres in the WMAP data.

Land and Magueijo (2005) made an analysis of the WMAP data based on an orthonormal frame for each multipole and on a set of invariants to investigate the statistical isotropy and Gaussianity of the WMAP data. This sophisticated analysis did not show any evidence of non-Gaussianity and also showed that the signal is statistically isotropic. However, as the authors emphasize, their method is very limited by noise and overlooks subtle features in the data.

Another new method for analysing non-Gaussianity of CMB maps has been presented by Naselsky *et al.* (2005). The purpose of this paper was to show that any correlations of phases of the spherical harmonic coefficients of the CMB signal determine the morphology in the space of phases. The authors generalize the method of phase correlations of Naselsky, Doroshkevich and Verkhodanov (2003) for testing the phase coupling in the CMB maps. In particular, they introduce the mean angle Θ_l for each multipole l, averaged over all m-modes in order to check whether the distribution of Θ_l is uniform (as it should be for Gaussian signals); if it is not, they check the possible preferred directions for each multipole.

To illustrate the sensitivity of the method, Naselsky *et al.* (2003) use the foreground-cleaned map (FCM) and the Wiener filtered map (WFM) derived by Tegmark *et al.* (2003), the ILC map derived by the WMAP science team[3] and the ILC map reproduced by Eriksen *et al.* (2004). All these maps contain some features of the foreground residues, non-uniformity of the noise, Galactic plane substraction, etc. The main task is to show how any detected non-Gaussian features relate to known properties of the non-Gaussian components of the signal. All these maps include Galactic plane contamination, which is excluded by different types of masking in order to estimate the power spectrum of CMB anisotropies.

Naselsky *et al.* (2003) apply these analyses to the CMB maps derived from one-year WMAP data. These CMB maps are generated using different foreground cleaning methods; hence their morphologies are somewhat different. As phases are closely related to morphology, the analysis of phases not only demonstrates the existence of non-Gaussian residuals among these CMB maps, but also reveals the differences between morphologies of these maps.

It is worth mentioning that application of different statistical methods to the investigation of the temperature fluctuations on the sky can sometimes lead to completely different results that can 'contradict' each other. But this contradiction occurs only because different techniques and different methods are sensitive to different non-Gaussian properties of the signal. Another problem is related to anomalies in the low-multipole range of the WMAP CMB anisotropy (Hinshaw, Barnes and Bennett, 2003; Spergel *et al.*, 2003b, Tegmark *et al.*, 2003). The deficit of power for quadrupole, planarity of octupole and alignment between the quadrupole and octupole were discussed in numerous papers, but questions about their nature are still unanswered.

At the time of writing (2006) the WMAP mission continues to gather information and send data. Many important questions on the structure and the evolution of the Universe are yet to be investigated. In the meantime, we are waiting for a new space mission, the

---

[3] http://lambda.gsfc.nasa.gov/product/map/m_products.cfm.

PLANCK mission, with its higher sensitivity, greater angular resolution and larger frequency range.

Measurement of the polarization in the CMB by PLANCK opens a new window onto an exciting future for the study of the CMB. This new coming era in cosmology will be discussed in Chapter 9.[4]

---

[4] More information about WMAP and the data are available at http:// lambda.gsfc.nasa.gov/

9

The 'Planckian era' in the study of anisotropy and polarization of the CMB

9.1   Introduction

The completion of the recent balloon experiments BOOMERANG, MAXIMA-1, TOP-HAT and ground-based experiments DASI, CBI and VSA ushered in a new phase in the experimental investigation of anisotropy and polarization of the cosmic microwave background. The satellite WMAP, launched successfully in June 2001, opened a new era in the study of the CMB anisotropy and polarization, which differs from previous experiments in its greater accuracy of finding the characteristics of the cosmological signal, and hence of the cosmological parameters. We mean here the Hubble constant H_0, the density Ω_b of the baryonic fraction of matter in the Universe, the hidden mass density Ω_{dm}, the dark energy density Ω_Λ (the vacuum?), the exponent n in the spectrum of adiabatic perturbations, and a number of others.

An international satellite, PLANCK, whose aim is to measure anisotropy and polarization of the primordial radiation with unprecedented accuracy is scheduled for launch in 2007. The accepted opinion is that PLANCK will summarize more than 35 years of progress in the theoretical and experimental study of polarization and anisotropy of the CMB and will allow us to come very close to building a realistic model of the Universe.

Before discussing the main features contributed by WMAP and PLANCK, we briefly sum up the results achieved by radioastronomy and cosmology in studying the CMB anisotropy. Figure 9.1 brings together the the the main results of observing the anisotropy spectra of the CMB, indicating measurement errors (for references and descriptions of experiments, see Wang, Tegmark and Zaldarriaga (2001) and http://space.mit.edu/home/tegmark/index.html) and a comparison of the results of BOOMERANG, MAXIMA-1, WMAP, VSA, DASI, ACBAR and CBI data. As we see from Fig. 9.1, the spectra $C(l)$ obtained in the above experiments are all in quite satisfactory agreement. On the basis of this data, Efstathiou (2002, 2003a,b) obtained parameters of the CDM model that best describe the available set of observational data (see Fig. 9.2). According to Efstathiou (2003a,b), the best agreement between the predictions of the theory and experimental data is obtained in the standard CDM model with adiabatic perturbations and the following set of parameters: $\Omega_b h^2 \simeq 0.021$, $\Omega_{dm} h^2 = 0.12$, $\Omega_\Lambda = 0.7$, $\Omega_K = 1 - \Omega_\Lambda - \Omega_b - \Omega_{dm} = 0$ (curvature parameter) and $n_s = 1$ (the Harrison–Zeldovich spectrum).

First, we can be quite certain that the detected small-scale anisotropy (for angles $\theta < 1°$) proved to be in complete agreement with the theoretical predictions. The first Sakharov peak ($l \sim 200$) is clearly seen in both Figs 9.1 and 9.2, and theoretical predictions and experimental results appear to be in good agreement in the area of the second, and possibly the third, peaks. This agreement in the shape of the theoretical curve and experimental data

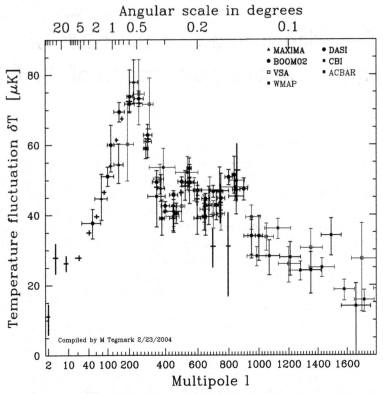

Figure 9.1 $\delta T(l)$ as a function of multipole number using the data of seven experiments completed by mid 2004. Adapted from http://space.mit.edu/home/tegmark/index.html.

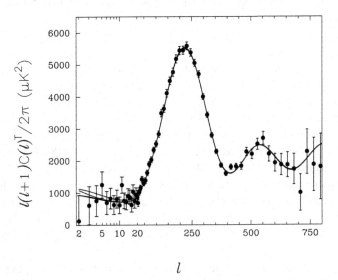

Figure 9.2 Anisotropy spectrum for the CDM model with the set of parameters $\omega_b = 0.022$, $\Omega_{\mathrm{dm}} = 0.12$, $\Omega_K = 0$, $\Omega_\Lambda = 0.7$ (Efstathiou, 2003a,b). Full circles: data of observations; solid curve: most suitable CDM model; thin solid lines: models listed in Table 9.1. These models have identical values of ω_b and ω_c fixed to the best-fit values from WMAP. Adapted from Efstathiou (2003a,b).

Table 9.1. *Parameters for degenerate models in Fig. 9.2*

Ω_k	Ω_b	Ω_c	Ω_Λ	h
0.00	0.0463	0.2237	0.73	0.720
−0.05	0.0806	0.3894	0.58	0.546
−0.10	0.1114	0.5386	0.45	0.446
−0.20	0.1714	0.8286	0.20	0.374

indicates that our understanding of the dynamics of the processes that were taking place in the pre-galactic plasma in the epoch of redshifts $z \leq 10^3$ is clearly correct. Less obvious is the answer to this question: which of the cosmological models is the best one? The answer is obviously dependent on what we use as a criterion of a 'good' or 'bad' model. We intuitively feel that the smaller the deviation of the theoretical anisotropy spectrum $C(l)^t$ from the experimental spectrum $C(l)^{exp}$, the better the chosen theoretical model is in comparison with models not providing this minimum. However, current and future experiments cannot measure the spectrum $C(l)^{exp}$ without errors, $\Delta C(l)$. Therefore, two most important tasks arise: to minimize the error (increase the accuracy of the theoretical prediction of $C(l)^t$ and reduce theoretical $\Delta C(l)$) and at the same time to analyse the nature of the experimental error $\Delta C(l)$ (systematics+random spread of values). As an illustration of the most important role played by systematic and random errors, we invite the reader to look at Fig. 9.2. As we see from this figure, the entire range of large-scale anisotropy measured by COBE ($l < 30$) happens to lie outside the optimal curve that mostly approximates the data of BOOMERANG, MAXIMA-1, DASI, CBI, WMAP, etc. that we discussed in the preceding chapters. This means that the array of observational data selected for processing contains systematic errors of various experiments; this may affect the value of the principal cosmological parameters. Following the generally accepted terminology for most important cosmological parameters, we will keep to the following notation:

(a) $\omega_b = \Omega_b h^2$ is the density of the baryonic fraction of matter in units of critical density ρ_{cr};

(b) $\omega_c = \Omega_{dm} h^2$ is the density of the CDM fraction in units of ρ_{cr};

(c) Q_{10} is the amplitude of the adiabatic mode normalized to $C(10)^{1/2}$ for $l = 10$ relative to the COBE data;

(d) n_s and n_t are the exponents of the spectra of adiabatic perturbations and gravitational waves, respectively;

(e) $r_2 = C(2)^t / C(2)^s$ is the ratio of the spectra of gravitational waves and adiabatic perturbations for $l = 2$;

(f) $\Omega_k = 1 - \Omega_b - \Omega_c - \Omega_\Lambda$ is the curvature parameter of the Universe.

We assume that systematic experimental errors are completely eliminated[1] and that the errors $\Delta C(l)$ are statistical in nature.

---

[1] Naturally, this assumption is far from being realistic. Presumably, the role of 'systematics' will be decisive in evaluating the noise level both in the WMAP and in the PLANCK experiments. At the same time, this simplification makes it possible to compare errors of various experiments in determining $C(l)$ at least at an academic level.

Table 9.2.

Parameter	WMAP			PLANCK		
	no const.	$r_2 = -7n_t$	$r_2 = 0$	no const.	$r_2 = -7n_t$	$r_2 = 0$
$\delta\omega_b/\omega_b$	0.052	0.028	0.030	0.0064	0.0056	0.0056
$\delta\omega_c/\omega_c$	0.097	0.028	0.031	0.0042	0.0042	0.0039
δQ	0.0066	0.0047	0.0050	0.0013	0.0010	0.0011
δr	0.49	0.043	—	0.33	0.023	—
δn_s	0.030	0.0061	0.0098	0.0049	0.0032	0.0042
δn_t	0.56	0.0061	—	0.40	0.0032	—
$\delta h/h$	0.082	0.020	0.028	0.0045	0.0045	0.0041
$\delta\Omega_\Lambda$	0.16	0.049	0.068	0.012	0.012	0.011

Which of the ongoing WMAP or future PLANCK experiments, given a chosen model of random $\Delta C(l)$, will be able to determine the cosmological parameters with higher accuracy, corresponding best to the observational data? To answer this, we see from Table 9.2 that the expected level of systematic and statistical errors for the PLANCK instruments must be almost five times lower than for WMAP.

In other words, PLANCK will open up new possibilities for both theoretical and experimental studies of anisotropy and polarization of the CMB radiation, thus creating for scientists new problems that lie beyond the sensitivity at the level of several per cent. We will briefly outline the scope of this quest. We assume, as before, that the level of systematic noise does not exceed the level of random errors. Then the uncertainty in the values of the main cosmological parameters are as in Table 9.2. In this table, column $r_2 = 0$ corresponds to the situation of no gravitational waves, column $r_2 = -7n_t$ corresponds to the inflation model discussed in Chapter 5 based on the slow roll approximation, and the column 'no const.' corresponds to the ability of the experiment to measure cosmological parameters without any additional assumptions concerning the characteristics of gravitational waves. The potential results of PLANCK compared to those of WMAP, let alone those of balloon experiments, are impressive, even taking due account of the importance of correcting for systematic errors. The logical question now is: what factors are responsible for this high sensitivity of the PLANCK instruments and what methods will be used to achieve it? To answer this, we give below the formal description of the project; the details can be found on the PLANCK web page.[2] First of all, the unique capabilities of the PLANCK mission are based on two instruments (low-(LFI) and high-(HFI) frequency instruments) that comprise 54 detectors in the range 30–70 GHz and 56 detectors in the range 100–857 GHz, respectively. The choice of ten frequency channels is predicated on the need for multifrequency filtration of galactic and extragalactic noise. Table 9.3 shows the variance of the CMB anisotropy, the variance of the pixel noise, σ_{noise}, and the size of the pixels and half-widths of the antenna diagram for all PLANCK's frequency channels (see Vielva *et al.* (2001) for reference). The three HFI channels, 143, 217 and 545 GHz, will measure the CMB polarization. It is expected that the level of galactic and extragalactic noise in the first two ranges will be minimal compared with other ranges (see the following section).

---

[2] http://astro.estee.esa.nl/PLANCK

Table 9.3.

Frequency (GHz)	σ_{CMB} (10^{-5})	σ_{noise} (10^{-5})	FWHM (arcmin)	Pixel size (arcmin)
857	4.47	2221.11	5.0	1.5
545	4.47	48.951	5.0	1.5
353	4.48	4.795	5.0	1.5
217	4.43	1.578	5.5	1.5
143	4.27	1.066	8.0	1.5
100(HFI)	4.07	0.607	10.7	3.0
100(LFI)	4.10	1.432	10.0	3.0
70	3.88	1.681	14.0	3.0
44	3.43	0.679	23.0	6.0
30	3.03	0.880	33.0	6.0

Launch of the PLANCK mission is planned for February 2007; it is intended that the satellite should be placed in orbit around the second Lagrange point L2. The observation of the radio sky is to begin in July 2007. The preliminary scenario of the strategy for scanning the sky is shown schematically in Fig. 9.3. The general rotation of the satellite is maintained in the plane of the ecliptic, at constant Sun–Earth–satellite orientation. The rotation axis of the satellite remains constant in the selected coordinate system for an hour, during which time the optical axis performs 60 or 120 revolutions (with fixed orientation of the rotation axis).[3] Then the orientation of the rotation axis is changed by 2.5 arcmin; this operation is repeated once every hour. As the satellite changes its position relative to L2 and its rotation, the orbit slowly deviates from the plane of the ecliptic by an angle $\leq 10°$. It is anticipated that the entire radio sky will be covered during the first year of observation. The special distinction of the PLANCK project from those already implemented in measuring the anisotropy and polarization of primordial radiation lies in a detailed preliminary analysis, simulation and development of techniques for eliminating systematical effects that include the calibration of antenna's profile and elimination of low-frequency noise (the $1/f$ noise), etc. (for details see the PLANCK web page given in the footnote on p. 228).

Even though the problem of the elimination of systematics is extremely important and timely, it is natural that maximum interest is concentrated on the new horizons of the physics of primordial radiation; we will explore this in more detail in the following sections.

9.2 Secondary anisotropy and polarization of the CMB during the reionization epoch

This section presents one of the new concepts in CMB physics devised by the scientific community in recent decades in response to the impressive achievements of experimental research. As recently as the 1980s the hypothesis of a more complicated ionization history of the cosmic plasma than that predicted by the standard model of recombination (Peebles, 1968; Zeldovich *et al.*, 1969) was discussed in the literature with reference to the problem of the absence of anisotropy in primordial radiation at the level $\Delta T/T \sim 10^{-2}$–10^{-3}. It is very

---

[3] Precisely these two options are discussed in various versions of the scanning strategy.

unlikely that anyone at the beginning of the 1980s could seriously foresee that progress in optical and radio astronomy would virtually overturn our concepts concerning the structure and evolution of the Universe and open not only theoretical, but also (more importantly) experimental possibilities of studying the process of the birth of structures in the Universe. Nevertheless, after the discovery of objects with redshifts of $z \geq 6$ and of large- and small-scale anisotropy of primordial radiation, the time has come to analyse the new possibilities in the structure and evolution of the Universe, in which the physics plays a very important role. The immensity of the problem dictates that certain requirements should be imposed on both theory and experiment, but also that it should stimulate progress in both. In the CMB field, this makes us reexamine the processes of formation of the CMB anisotropy and polarization, taking into account the fact that this radiation propagates toward us not through vacuum but through space filled with mature galaxies or galaxies and clusters in the process of formation.

The evolution of these objects transforms the gravitational energy of matter into radiation, acting as a gigantic reactor. This radiation lies in various ranges of the spectrum and affects the properties of the medium through which the primordial radiation reaches the 'last scattering' surface (at $z \sim 10^3$) as quanta propagate toward the observer ($z = 0$). The term 'last scattering' must now be used quite justifiably in quotes since the quanta of the CMB do undergo some – fortunately slight – scattering on the way.

A reservation must be made here: the moment we 'move' to the level of signals generated in the epoch of the secondary plasma ionization, we come very close to the limit of sensitivity and angular resolution of today's experiments and of those planned for the future. Therefore, a reasonable lower bound on the level of the signal must be introduced into the theoretical studies of the effects of secondary anisotropy generation that would automatically put constraints on the types of processes that result in the formation of anisotropy and polarization above the prescribed limit. For such a reasonable threshold, it is natural to set the limit $\Delta T \sim 1-3\,\mu\text{K}$ on the area $1.5' \times 1.5'$ that corresponds to the sensitivity limit both in amplitude and in the angular scale of the PLANCK experiment. Then we automatically narrow the scope of physical processes that we consider to be responsible for the formation of the CMB anisotropy to the following sequence: the effect of attenuation of primordial anisotropy, the linear and quadratic Doppler effects, and lensing effects.

These effects reflect the emergence of both the gravitational influence of the evolving perturbation of CMB and the effects of additional scattering that arises at relatively small redshifts, $z \leq 1-20$.

9.2.1   *Attenuation of primordial anisotropy and generation of polarization*

In this subsection we wish most of all to emphasize the fact that the main role in the formation of the secondary anisotropy of the CMB in the reionization epoch is played by the background of the cosmological hidden mass, whose particles do not interact electromagnetically with the photons of the primordial radiation. As a result, perturbations in the hidden mass evolve independently from the CMB. The second most important conclusion that follows from the comparison of theoretical predictions and the BOOMERANG and MAXIMA-1 data, and having the most important predictive power for WMAP and PLANCK, is that the reionization of cosmic plasma was not accompanied by any significant increase of the optical depth τ_T relative to the Thomson scattering; the most probable constraints on τ_T give $\tau_T \leq 0.1-0.3$ (Tegmark and Zaldarriaga, 2000).

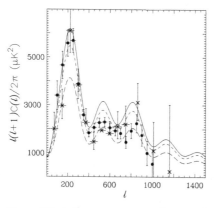

Figure 9.3 Anisotropy spectrum for a number of models. The solid curve corresponds to the standard model (see Fig. 9.2). Long-dash curves correspond to the model with delayed recombination; short-dash curves to the spectrum with $n = 0.95$, and the dotted curve represents the standard model with the optical depth of reionization, $\tau = 0.1$.

A point of principal importance for analysing the secondary anisotropy of primordial radiation generated by scattering of quanta by electrons in the reionization epoch is the emergence of this 'secondary contact' which is immediately followed by a Doppler scattering. This additional scattering at relatively late stages of evolution results first of all in attenuation of the level of primary anisotropy by a factor $e^{-\tau}$ which does not exceed 10% at $\tau \sim 0.1$ and 40% at $\tau \simeq 0.3$. At the same time, if we take into account the expected sensitivity of the PLANCK mission, this attenuation should be detectable, especially as far as polarization effects are concerned (Naselsky *et al.*, 2001). To illustrate this argument, we consider several modifications of the extensively used cosmological model ΛCDM with the following set of parameters: $\omega_c = 0.127$, $\omega_b = 0.019$, $\Omega_\Lambda = 0.7$, $h = 0.65$, $\Omega_K = \Omega_c = 0$.

Figure 9.4 compares the results of the simulation of the CMB anisotropy spectrum with the data of the BOOMERANG, MAXIMA-1 and CBI experiments for various modifications of the initial cosmological model. The value $z_{rec} = 15$ was chosen as the redshift of the reionization epoch, providing the optical depth $\tau_T \simeq 0.2$. For comparison, Fig. 9.3 gives the spectra of the CMB anisotropy in models that are close to the standard one and include gravitational waves ($r_2 = 0.15$), a slight slope of the power spectrum for adiabatic modes relative to the Harrison–Zeldovich spectrum ($n = 0.95$), and a model with 'delayed' recombination ($\varepsilon_\alpha = 7$).

We see from Fig. 9.3 that these models are practically indistinguishable if we stay within the observational data array. However, they will all be significantly different from one another at the sensitivity level of the PLANCK mission (see Table 9.2).

A point of special interest is the response of the spectrum of the E mode of CMB polarization to the secondary ionization epoch. In contrast to anisotropy, which is mostly decreasing in the range $l \leq 10^3$–2×10^3, substantial anomalies arise in the power spectrum of polarization that reflect specific features of the ionization history in this period (see Figs 9.4 and 9.5).

The physics of generation of these anomalies was studied in detail (for references, see Zaldarriaga (1997)). The hypothesis that underlies this analysis is that hydrogen reionization in the epoch of redshift z_{rec} completes over a characteristic time $\Delta z \ll z_{rec}$. In fact, we

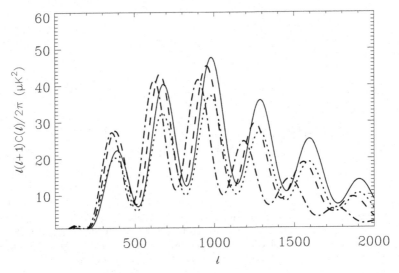

Figure 9.4 As Fig. 9.3, but for polarization. The notation is the same as in Fig. 9.3.

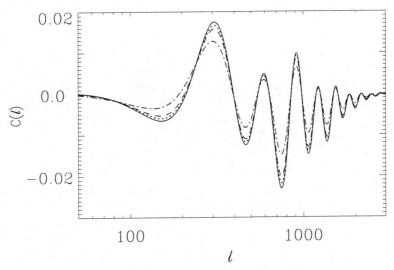

Figure 9.5 Cross-correlation of anisotropy and polarization. The notation is the same as in Fig. 9.3.

can assume as a zeroth approximation that the degree of ionization of hydrogen $x_H = 0$ for $z > z_{rec}$ while $x_H = 1$ at $z < z_{rec}$; that is, hydrogen becomes completely ionized. The level of the additionally generated polarization can then be readily evaluated as follows. We remarked in Chapter 6 that the main source of polarization of the background radiation is the scattering of quanta by electrons that move in the field of this radiation. The important factor for the generation of polarization in scattering is the quadrupole component $\Delta T/T \propto v_b$, where v_b is the peculiar velocity of electrons. Therefore, the level of the effect is on the order of $\Delta_p(k) \propto k\tau_T \Delta T$, where k is the wave factor and $\Delta_p(k)$ is the polarization amplitude at each

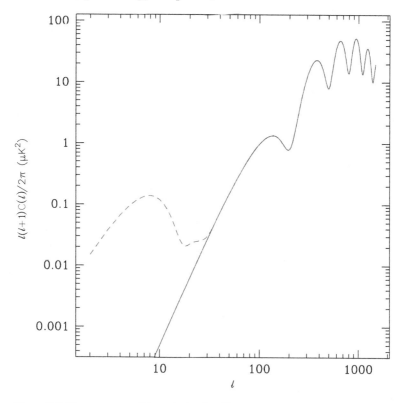

Figure 9.6 The spectrum of the polarization E component in the standard CDM model with the optical depth of reionization $\tau = 0.1$ (dashed curve). The solid curve corresponds to the SCDM model. Adapted from Zaldarriaga (1996).

point k (Zaldarriaga, 1997). A detailed numerical analysis of the dynamics of polarization generation can be carried out using the previously mentioned software package CMBFAST.

Figure 9.6 illustrates the behaviour of the CMB polarization in the standard CDM model with reionization taken into account ($\tau \simeq 0.1$) (Zaldarriaga, 1997). We see from this figure that the main specific feature of the spectrum is the emergence of a peak in the $l \leq 10$ area and the practically unchanged shape of the spectrum in the $l > 40$ range, as we had in the model with $\tau_T = 0$. Note that the generation of such peaks in the polarization power spectra is typical of the cosmological models that include a virtually instantaneous change in the degree of ionization in hydrogen in the z_{rec} epoch. The question arises of how the predictions of the theory are changed if hydrogen reionization occurs not jumpwise but gradually, beginning with redshifts $z \gg 10^2$ up to $z \sim 5$–10. One of the possible versions of this ionization solution was analysed in Doroshkevich and Naselsky (2002) for the case of the model of decay of supermassive particles that generate the flux of superhigh-energy cosmic rays at the surface of the Earth. Decays of these particles, from the hydrogen recombination epoch and up to the present time, result in gradual ionization of hydrogen, the dynamics of which is plotted in Fig. 9.8 (for notation, see Section 3.11).

Figure 9.9 plots the power spectrum of polarization for ionizer models shown in Fig. 9.8. We see from this power spectrum that no peak is formed in this case; however, a considerable

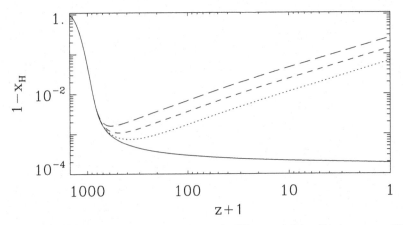

Figure 9.7 The fraction of ionized hydrogen in different models of ionizer power. The dotted curve corresponds to $\varepsilon_i = 1/(1 + z)$; the short-dashed curve corresponds to $\varepsilon_i = 2/(1 + z)$; the long-dashed curve corresponds to $\varepsilon_i = 3/(1 + z)$; and the solid curve corresponds to the standard model.

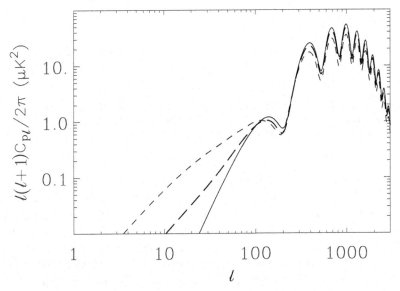

Figure 9.8 Polarization power $C_P(l)$ corresponding to the regimes shown in Fig. 9.7. Solid curve: standard model; short-dashed curve: $\varepsilon_i = 3/(1 + z)$; long-dashed curve: $\varepsilon_i = 3/(1 + z)$ (see the text).

rise in power takes place by more than an order of magnitude in the range $l \leq 20$. A legitimate question is: are the WMAP and PLANCK missions capable of detecting these sort of peculiarities of the ionization history that reveal themselves in the polarization spectrum? Leaving aside the role played by galactic and extragalactic noise in polarization experiments till the following section, we give in Fig. 9.9 the comparative sensitivity of polarization measurements in WMAP and PLANCK. This figure clearly shows that the sensitivity of PLANCK at $20 < l < 100$ greatly exceeds the potential of WMAP and is mostly limited by the presence

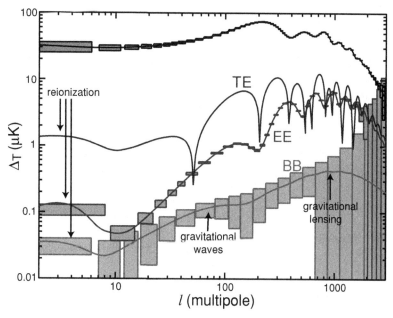

Figure 9.9 A comparison of the possibility for PLANCK to measure the polarization components E and B. Adapted from Delabruille (2004).

of galactic and extragalactic noise and the previously mentioned 'cosmic variance' effect (see Chapter 5). For this effect, the error level $\Delta C(l)/C(l) \sim (l \cdot f_{\text{sky}})^{-1/2}$ at $l \leq 200$ is found to be on the order of 10%, $(l/100)^{-1/2}$, with total sky coverage ($f_{\text{sky}} = 1$).

We see that even if $l \sim 10\%$, with the relative accuracy of measuring $C(l)$ expected to be about several tens of per cent, PLANCK's instruments will safely detect the signal. Therefore, one of the most important elements of the CMB physics involved in launching this mission is a unique possibility of testing the $z < 10^3$ epoch. Hence, polarization measurements may bring us closer to the current age of the Universe.

9.2.2 Linear and quadratic Doppler effects

In addition to attenuating the primordial anisotropy in the course of reionization, scattering of quanta by free electrons generated secondary anisotropy both in the v/c-linear and in the non-linear approximations. The physics of the linear Doppler effect does not differ from the case of generation of primary anisotropy but possesses a number of specific features. First, we need to take into account the gravitational growth of perturbations of density and peculiar velocities of dark matter whose distribution correlates with that of the baryonic fraction of matter, at least on scales accessible to PLANCK observations. Secondly, the growth of perturbations in the baryonic fraction of matter results in an additional modulation of electronic density $n_e = \bar{n}_e(1 + \delta_b)$. This modulation produces fluctuations of optical depth $\Delta \tau_T \sim \delta_b$ and hence the anisotropy response to the velocity field of electrons contains a non-linear term $\sim \delta_b \cdot v_b$. Taking this term into account in the generation of anisotropy reproduces excess power in high multipoles, which is known as the Vishniak effect (Vishniak, 1987).

Figure 9.10 presents the aggregate picture of anticipated spectra for the linear Doppler effect in comparison with the primordial anisotropy and the instrumental noise level in the

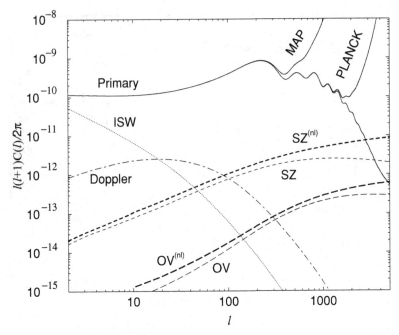

Figure 9.10 A comparison of the contributions of the primary and secondary effects to the CMB anisotropy spectrum. The dotted curve corresponds to the Sachs–Wolfe effect (ICV), and the dash–dot curve corresponds to the linear Doppler effect; $OV^{(nl)}$ and OV stand for the Vishniak effect in the non-linear and linear approximations, respectively, and SZ is the Zeldovich–Sunyaev effect in the same approximations. Adapted from Cooray and Hu (2000).

WMAP and PLANCK experiments (Cooray and Hu, 2000). Note that both these effects are sensitive to the rate of plasma reionization (see Haiman and Knox (1999) and references therein). At the same time, we need to emphasize that both the linear and the quadratic Doppler effects in the range $l < 2000$ are lower than the primordial anisotropy by almost 2–3 orders of magnitude, and their experimental investigation constitutes a serious problem for the PLANCK experiment.

Figure 9.10 (taken from Cooray and Hu (2000)) also shows contributions from the Zeldovich–Sunyaev effect described in detail in Chapter 3, with the spatial distribution of galaxy clusters in the epoch of their formation taken into account. As we saw for the Doppler effect, the contribution of the Zeldovich–Sunyaev effect is found to be 1 to 2 orders of magnitude lower than both the level of primordial anisotropy and that of the instrument noise in the $l < 2000$ range. In the $l > 2000$ range, the role played by the Zeldovich–Sunyaev effect becomes better pronounced. Moreover, the frequency dependence of $C(l)$ warrants the hope that this effect will be studied in detail in the framework of the PLANCK project. Our optimism is based to a large extent on the recent results of the land-based CBI experiment (Bond *et al.*, 2002; Mason *et al.*, 2003) in which the first indications of the presence of the signal in the range $l \simeq 2000$–3500 were obtained. We would like to point out that the statistical reliability of this result at high l is not large; nevertheless, the attempt to 'peek' into the part of the spectrum where secondary anisotropy effects predominate is of great importance, both for progress in the theory and for future experiments.

9.3   Secondary anisotropy generated by gravitational effects

In this section we wish to take a closer look at the analysis of gravitational effects that distort the spectrum of primordial CMB anisotropy. We must immediately make the reservation that the possibility of detecting these effects is predicated on the unique parameters of the PLANCK mission.

Gravitational effects include the previously discussed Sachs–Wolfe effect (Sachs and Wolfe, 1967) for that part of the perturbation spectrum which is still evolving in the linear mode, and the effect of 'scattering' of quanta on perturbations of the gravitational potential produced by non-linear structures (Rees and Sciama, 1968). This latter effect, known as the Rees–Sciama effect, is found to be extremely small for the CDM model with adiabatic inhomogeneities – for clearly understood reasons. Even if the density contrast $\delta \gg 1$ on the inhomogeneity scale $\lambda \ll ct$, non-linear additions to the potential cannot be larger than $\delta\varphi \sim (\lambda/ct)^2 \delta$. For typical scales of structures $\lambda \geq 1$ Mpc and $ct \sim 10^4$ Mpc, potential perturbations will be $\Delta\varphi \sim 10^{-8}\delta$. Even if on these scales $\delta \sim 10$–10^2, $\Delta\varphi$ should not exceed the 10^6 level, and density perturbations for realistic values of the mean-square level should be even smaller (Seljak, 1996a,b).

One of the most important and non-trivial gravitational effects capable of significantly distorting the spectra of primordial anisotropy and polarization is the effect of lensing of the primordial radiation background in the course of propagation through space. The gravitational lensing of quanta on objects 'clustered' into large-scale structures results in distortion of the spectra of CMB anisotropy and polarization. Obviously, if there is no primordial anisotropy, the lensing effect cannot generate it all by itself. At the same time, primordial anisotropy may be significantly distorted in the process of propagation of quanta from the surface of last scattering ($z = 10^3$) toward the observer or owing to additional scattering if reionization is taken into account ($z \sim 10$–20).

In the approximation of weak lensing, the angle by which rays are deflected is related to the projection of the gravitational potential by the following formula (Kaiser, 1992):

$$\vec{\Theta}_{\rm L}(m) = -2 \int_0^{r_0} dr\, \frac{d_{\rm A}(r_0 - r)}{d_{\rm A}(r)d_{\rm A}(r_0)} \Phi(r, \hat{m}r), \tag{9.1}$$

where $r(z) = \int_0^z dz'/H(z')$; $H(z')$ is the Hubble parameter, $r_0 = r(z = \infty)$, $d_{\rm A}(r) = H_0^{-1}\Omega_K^{-1/2}\sinh(H_0\Omega_K^{1/2}r)$ is the angular diameter, $\Omega_k = 1 - \sum_i \Omega_i$, Ω_i is the density of the ith component in units of critical density, and Φ is the gravitational potential. As we see from Eq. (9.1), the angle $\vec{\Theta}_{\rm L}$ is related to the projection of the gravitational potential gradient on the direction $\vec{m}$ as follows:

$$\vec{\Theta}_{\rm L}(\vec{m}) = \nabla\Phi(\vec{m}).$$

As a result of the deflection of rays from the initial direction in the course of lensing, the biased value of the the CMB anisotropy in the $\vec{n}$ direction is now related to the initial values (without lensing) by

$$\Delta T_{\rm L}(\vec{n}) = \Delta T(\vec{n} + \vec{\Theta}_{\rm L}). \tag{9.2}$$

Owing to the non-Gaussian type of potential distribution, the distribution of $\vec{\Theta}_{\rm L}$ around $\vec{n}$ will also be non-Gaussian, and therefore the statistical properties of the lensed signal $\Delta T_{\rm L}$ will be different from the properties of the pre-lensing signal. The lensing effect then

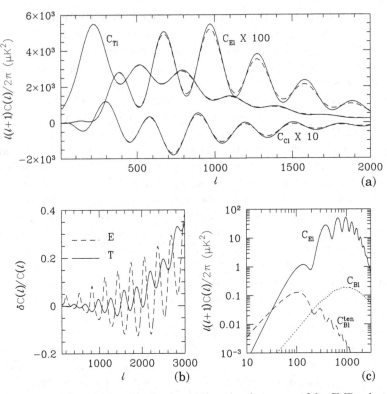

Figure 9.11 The role played by lensing in distorting the spectra of the CMB anisotropy and polarization. (a) C_{Tl} is the anisotropy spectrum, C_{El} and C_{Cl} are the spectra of lensed E component and $T - E$ correlations. Solid curves correspond to models without lensing, dashed curves trace distortions when lensing is taken into account. The difference between the lensed and non-lensed spectra are shown for temperature (E) and polarization (c). Dotted curve: corresponds to the contribution of the B mode, and C_{Bl}^{ten} corresponds to the contribution of gravitational waves. Adapted from Zaldarriaga and Seljak (1998).

perturbs the spectrum of the CMB anisotropy and polarization (see, for example, Hu (2002) and Zaldarriaga and Seljak (1998)).

Figure 9.11 plots the anticipated levels of distortion of the CMB anisotropy and polarization spectra caused by lensing (Zaldarriaga and Seljak, 1998). As expected, this effect leaves the anisotropy spectrum practically undistorted around the first three peaks and becomes significant (at the level $\delta C(l)/C(l) \sim 5\%$) only with $l \geq 1200$ (see Fig. 9.11(b)). For polarization, the same effect results in 5% deviations at $l \sim 200$, which increase to 10–20% at $l > 1000$.

As a result, taking lensing effects into account acquires maximum importance for the analysis of future experimental data of the PLANCK mission because, on the one hand, lensing distorts the anisotropy and polarization spectra and, on the other hand, it makes the signal non-Gaussian. Both manifestations of lensing may lead to incorrect values of the most important cosmological parameters (listed in the introduction to this chapter) if they are extracted from the PLANCK data ignoring lensing. What we face here is another extension of the horizon of applications of the primordial background radiation as a kind of probe of the state of matter in the Universe; we can achieve this due to the unique characteristics of the PLANCK mission.

9.4 Galactic and extragalactic noise

When we discussed the data that were potentially achievable from the PLANCK mission and compared them with the sensitivity of ongoing experiments earlier in this chapter, we emphasized that observational data are not free from noise. It was often sufficient for a theoretical simulation of various effects to use the simplest model of instrumental noise, considering it not to be correlated with either the signal or with itself. This model of 'white noise' is frequently sufficient for halting the flight of imagination at a reasonable level (as far as experimenters are concerned) and to narrow down the set of effects that need to be taken into account in analysing observational data. However, in addition to instrumental errors that include the effects of systematics, there exists another class of noise, one that cannot be eliminated in principle and which reflects the simple fact that together with the Earth and the Sun we exist on the periphery of the Galaxy and are surrounded by magnetic fields, cosmic rays, clouds of dust and gas, etc., and also stars, whose radiation activates the components mentioned above. In other words, the experimental conditions are far from ideal, especially if we take into account that if other galaxies differ from ours, they are typically even worse as far as noise is concerned.

The factors given above are usually separated into galactic and extragalactic sources of noise; their radiation covers a broad range of wavelength bands from the long-wavelength radio background ($\lambda \geq 0.5$ m) up to superhigh-energy cosmic rays (see Chapter 1). From the practical point of view, we will now concentrate on that range of radiation frequencies that are generated by galactic and extragalactic sources of noise and stretches from 30 to 900 GHz, in correspondence with the possibilities of the LFI+HFI instruments of the PLANCK mission. Note that this frequency range covers all bands – including the current ground-based experiments and balloon missions – and in this sense is the most general for analysing various sorts of noise.

Therefore, to predict the contribution of galactic and extragalactic noise to CMB anisotropy and polarization maps, one inevitably has to extrapolate the known results obtained in other experiments or those belonging to adjacent frequency ranges. The same is true in the case of high uncertainty of spectral characteristics of the noise. In this connection the results of the WMAP and PLANCK missions appear to be also important in the context of studying the radio background of the Universe, including its cosmological component.

10

Conclusion

This book is devoted to a systematic discussion of the physics of the cosmic microwave background. This field of physics is very complex and involves an enormous diversity of processes that occur in the course of expansion of the Universe. The CMB proved to be a true 'goldmine' for extracting scientific information on these processes; it has in fact grown into the central branch of modern cosmology. Comparing this highly perfected theory with observational data makes it possible to obtain essential information on the early Universe and on the physical parameters of the Universe as a whole.

Combining these results with results from other branches of astrophysics, such as the large-scale structure of the Universe, supernovae, etc., provides a robust basis of modern cosmology.

Results from BOOMERANG, MAXIMA-1, ARCHEOPS, CBI, DASI and WMAP were so important that they have taken the field into the era of 'precision cosmology'. As we have discussed, these studies have produced impressive constraints on many fundamental cosmic parameters and have led to a very definite picture of the structure and evolution of the Universe. As a result, one could become overexcited and declare that almost everything in cosmology is now known. However, we want to emphasize that even after those remarkable projects, the study of CMB physics is not coming to an end. There are still many unsolved problems in cosmology, and another generation of satellite experiments, as well as ground-based and balloon-borne experiments, is needed. The deeper science penetrates and the more mysteries it solves, the more problems it discovers, each more daunting and less predictable than the last.

Among the fundamental problems which arose after the recent success of cosmology, we mention the following.

(a) What is the nature and origin of 'dark energy'? How did it evolve?
(b) Why does $\Omega_\Lambda \simeq 0.73$ in the current state of the Universe?
(c) What is the nature of 'dark matter'?
(d) Why does $\Omega_{dm} \simeq 0.3$? What theory could provide this value?

These problems present a tremendous challenge. Thus, serious work on these and other cosmological problems continues.

There are also problems and uncertainties directly related to the interpretation of recent observational data. Recent experiments focused much attention on the separation of the primordial cosmological signal from noise and the foreground and on the statistical properties of the signal.

240

As we discussed in Chapter 8, sophisticated non-Gaussianity testing on derived maps from first-year WMAP data shows significant non-Gaussian features. What are these features?

(a) Could they be foreground residuals? If yes, then what is wrong with the methods of separating the primordial signal and the models of the foreground?

(b) Could they be systematic effects? If yes, then what kind?

(c) Or could they exhibit primordial non-Gaussianity? If yes, then this is a fact of great importance. The physics of their origin would probably be related to the origins of 'dark energy' and 'dark matter'.

In addition, we want to emphasize that the assumption that the statistical properties of the primordial CMB signal are Gaussian is the crucial requirement for deriving cosmological parameters from temperature and polarization power spectra.

Should the primordial CMB signal possess a non-Gaussian origin in the form of a quadratic non-linearity in the gravitational potential, the connection between $C_T(l)$, $Cp(l)$ and the cosmological parameters would need additional, probably non-trivial, investigation. The importance of the non-Gaussianity of the CMB signal can be illustrated by assuming that at some range of multipoles, say $l \sim 200$, the a_l coefficients of the spherical harmonics expansion of the anisotropy ΔT are highly correlated. Without comprehensive testing for non-Gaussianity in the map, these correlations can easily mimic the first acoustic peak of the $C_T(l)$, leading to the wrong conclusions about the properties of the CMB and cosmological parameters. Preparation and implementation of sensitive non-Gaussianity tests on the anisotropy and polarization maps is therefore pivotal for the PLANCK mission.

Another important problem scientists are working on now is the ionization history of the Universe. We discussed this problem in Chapters 3 and 9. Here we want to emphasize that the study of the reionization process is a crucial test of the correctness of our knowledge of the processes of the formation of structure in the Universe. It also tests our knowledge of the possible nature of the hypothetical unstable particles, the decays of which influenced the kinetics of hydrogen recombination.

We also want to mention the following important problems, which are related to CMB science and are under active study by cosmologists.

First of all, there is an open question regarding primordial gravitational waves. Polarization measurements of the CMB can serve as a detector of stochastic background of the primordial gravitational waves. As we discussed in Chapters 6 and 7, the pattern of polarization directions in the sky will be different if a stochastic primordial gravitational radiation exists. In this case the so-called pseudoscalar or 'magnetic' part of the polarization would not be equal to zero. It should be emphasized that the inflation model predicts the existence of such radiation. These measurements, then, with better precision than those of DASI and WMAP, will open a window on the early Universe.

These investigations are especially important because there is a huge project, the Laser Interferometer Space Antenna (LISA), which may allow direct detection of a continuous spectrum of primordial gravitational radiation. Comparison of the PLANCK and LISA results, obtained by these absolutely different methods of observation, is extremely important.

It should be mentioned that the possible existence of a primordial magnetic field can also be tested using CMB observations (Naselsky *et al.*, 2004). Another fundamental problem of modern cosmology is the possibility of other types of primordial perturbations in the early Universe, different from the adiabatic ones (for example, isocurvature perturbations).

We should remember that the impressive constraints on many fundamental cosmic parameters, produced by WMAP and other projects, reside within the framework of a definite cosmological model. If we take into account the possibility of a wider class of cosmological models, it may be that the actual uncertainty is much greater. We will have to wait for forthcoming observations to reduce the current uncertainties.

After the beginning of the era of 'precision cosmology', the number of questions affecting the basic fundamentals of cosmology increased significantly. And the show goes on! Both cosmological and physical communities are now working on future projects such as PLANCK, ALMA, LISA, etc., and there remains plenty of room for surprises.

References

Abbott, L. F. and Wise, M. B. (1984). *Nucl. Phys. B* **244**, 541.

Adler, R. J. (1981). *The Geometry of Random Fields* (Chichester: John Wiley and Sons).

Alcock, C., Allsman, R. A., Alves, D. *et al.* (1997). *Astrophys. J.* **486**, 697.

Alpher, R. and Herman, R. (1953). *Ann. Rev. Nucl. Sci.* **2**, 1.

Ambrosio, M., Antolini, R., Aramo, C. *et al.* (MACRO Collaboration) (1998). *Phys. Lett. B* **434**, 451.

Ambrosio, M., Antolini, R., Aramo, C. *et al.* (2001). *Astrophys. J.* **546**, 1038.

Andersen, R. C., Henry, R. C., Brune, W. H., Feldman, P. D. and Fastie, W. G. (1979). *Astrophys. J.* **234**, 415.

Arbuzov, P., Kotok, E., Naselsky, P. and Novikov, I. (1997a). *Int. J. Mod. Phys. D* **6**, 409.

Arbuzov, P., Kotok, E., Naselsky, P. and Novikov, I. (1997b). *Int. J. Mod. Phys. D* **6**, 515.

Athanassopoulos, C., Auerbach, L. B., Burman, R. L. *et al.* (1998). *Phys. Rev. Lett.* **81**, 1744.

Ave, M., Hinton, J. A., Vázquez, R. A., Watson, A. A. and Zas, E. (2000). *Phys. Rev. Lett.* **16**, 405.

Bahcall, N. A. and Bode, P. (2003). *Astrophys. J.* **588**, L1.

Bahcall, M. A. and Chen, R. (1993). *Astrophys. J.* **407**, L49.

Bahcall, M. A., Lubin, L. M. and Dorman, V. (1995). *Astrophys. J.* **447**, L81.

Bahcall, M. A., Krastev, P. I. and Smirnov, A.Yu. (1998). *Phys. Rev. D* **58**, 096016.

Bahcall, N. A., Ostriker, J. P., Perlmutter, S. and Steinhardt, P. J. (1999). *Science* **284**, 1481.

Balbi, A. M., Cabella, P., de Gasperis, G., Natoli, P. and Vittorio, N. (2002). In Cecchini, S., Cortiglioni, S., Sault, R. and Sbarra, C., eds, *Astrophysical Polarized Backgrounds*, AIP Conference Proceedings, vol. 609 (Melville, NY: American Institute of Physics), pp. 78–83.

Bardeen, J. M. (1980). *Phys. Rev. D* **22**, 1882.

Bardeen, J. M., Bond, J. R., Kaiser, N. and Szalay, A. S. (1986). *Astrophys. J.* **304**, 15.

Barkana, R. and Loeb, A. (2000). *Astrophys. J.* **539**, 20.

Barreiro, R. B., Sanz, J. L., Martinez-Gonzales, E., Cayon, L. and Silk, J. (1997). *Astrophys. J.* **478**, 1.

Basko, M. M. (1981). *Astrofiz.* **17**, 69.

Basko, M. M. and Polnarev, A. G. (1979). *Astron. Zh.* **24**, 3.

Bassani, L., Dean, A. J., Di Cocco, G. and Perotti, F. (1985). In Dyson, J. E., ed., *Active Galactic Nuclei* (Manchester: Manchester University Press), p. 252.

Bennett, C. L., Smoot, G. F., Hinshaw, G. *et al.* (1992). *Astrophys. J.* **396**, L7.

Bennett, C. L., Kogut, A., Hinshaw, G. *et al.* (1994). *Astrophys. J.* **436**, 423.

Bennett, C. L., Hill, R. S., Hinshaw, G. *et al.* (1996). *Astrophys. J.* **464**, L1.

Bennett, C. L., Banday, A., Gorski, K. M. *et al.* (1996). astro-ph/9601067.

Bensadoun, M., Bersanelli, M., De Amici, G. *et al.* (1993). *Ann. NY Acad. Sci.* **668**, 792.

Berlin, A. V., Bulaenko, E. V., Vitkovsky, V. V., Parijskij, Yu. N. and Petrov, Z. V. (1983). In Abell, G. O. and Chincarini, G., eds, *Early Evolution of the Universe and its Present Structure*. Proceedings of IAU Symposium, Kolumbari, Greece, 1982 (Dordrecht: D. Reidel), p. 121.

Bernstein, G. M., Fischer, M. L., Richards, P. L., Peterson, J. B. and Timusk, T. (1990). *Astrophys. J.* **362**, 107.

Bernstein, L. and Dodelson, S. (1990). *Phys. Rev. D* **41**, 354.

Bersanelli, M., Witebsky, C., Bensadoun, M. *et al.* (1989). *Astrophys. J.* **339**, 632.

Bersanelli, M., Bensadoun, M., De Amici, G. *et al.* (1994). *Astrophys. J.* **424**, 517.

Bhattacharjee, P. and Sigl, G. (2000). *Phys. Rev.* **327**, 109.

Biggs, A. D., Browne, I. W. A., Helbig, P., Koopmans, L. V. E., Wilkinson, P. N. and Perley, R. A. (1999). *Mon. Not. Royal Astron. Soc.* **304**, 349.

Birkinshaw, M. (1999). *Phys. Rep.* **310**, 97.

Birkinshaw, M. and Hughes, J. P. (1994). *Astrophys. J.* **420**, 331.

Bisnovatiy-Kogan, G. S. and Novikov, I. D. (1980). *Astron. Zh.* **57**, 899.

Bisnovatiy-Kogan, G. S., Lukash, V. N. and Novikov, I. D. (1983). In Abell, G. O. and Chincarini, G., eds, *Early Evolution of the Universe and its Present Structure*. Proceedings of IAU Symposium, Kolumbari, Greece, 1982, (Dordrecht: D. Reidel) p. 327.

Blasi, P. (1999). *Phys. Rev. D* **60**, 023514.

Boesgaard, A. M. and Steigman, G. (1985). *Ann. Rev. Astron. Astrophys.* **23**, 319.

Bond, J. R. and Efstathiou, G. (1984). *Astrophys. J.* **285**, L45.

Bond, J. R. and Efstathiou, G. (1987). *Mon. Not. Royal Astron. Soc.* **226**, 665.

Bond, J. R., Contaldi, C. R., Pen, U.-L. *et al.* (2002). astro-ph/0205386.

Bonnor, W. B. (1957). *Mon. Not. Royal Astron. Soc.* **117**, 104.

Borgani, S., Rosati, P., Tozzi, P. *et al.* (2001). *Astrophys. J.* **561**, 13.

Boschan, P. and Biltzinger, P. (1998). *Astron. Astrophys.* **336**, 1.

Boschan, P. and Biltzinger, P. (1999). astro-ph/9911032.

Boynton, R. A. and Stokes, R. A. (1974). *Nature* **247**, 528.

Boynton, R. A., Stokes, R. A. and Wilkinson, D. T. (1968). *Phys. Rev. Lett.* **B21,** 462.

Bronshtein, I. N. and Semendyaev, K. A. (1955). *Spravochnik po Matematike* (Moscow: Gostehizlat).

Bunn, E. F., Hoffman, Y. and Silk, J. (1995). *Astrophys. J.* **464**, 1.

Burles, S., Nollett, K. M. and Turner, M. S. (2001). *Astrophys. J.* **552**, L1.

Calberg, R. G., Yee, H. K. C., Ellingson, E. *et al.* (1996). *Astrophys. J.* **462**, 32.

Calberg, R. G., Yee, H. K. C. and Ellingson, E. (1997). *Astrophys. J.* **478**, 462.

Carlstrom, J. E., Joy, M. and Grew, L. (1996). *Astrophys. J.* **456**, 75.

Carlstrom, J. E., Joy, M. K., Greco, L. *et al.* (1999). astro-ph/9905255.

Carlstrom, J. E. *et al.* (DASI Collaboration) (2000). *Astron. Astrophys. Suppl.* **197**, 5501.

Carlstrom, J. E. *et al.* (2001). In Durret, F. and Gerbal, G., eds, IAP Conference, July 2000 (astro-ph/0103480).

Cayre, R., Spite, M., Spite, F., Vangioni-Flam, E., Casse, M. and Audouze, A. A. (1999). *Astron. Astrophys.* **343**, 923.

Challinor, A. (2000). *Class. Quant. Grav.* **17**, 871.

Challinor, A. and Lasenby, A. (1997). In Sanchez, N., ed., *Current Topics in Astrofundamental Physics* (Dordrecht: Kluwer Academic), p. 37.

Chandrasekhar, S. (1950). *Radiative Transfer* (Oxford: Clarendon Press), p. 17.

Chernin, A. D. (2001). *Sov. Phys. Uspekhi* **44**, 1099.

Chiang, L.-Y., Naselsky, P. D., Verkhodanov, O. V. and Way, M. J. (2003). *Astrophys. J. Lett.* **590**, L65.

Chibisov, G. V. (1972a). *Astron. Zh.* **49**, 74.

Chibisov, G. V. (1972b). *Astron. Zh.* **49**, 286.

Chiu, W. A., Gnedin, N. Y. and Ostriker, J. (2001). *Astrophys. J.* **563**, 21.

Clark, T. A., Brown, L. W. and Alexander, J. K. (1970). *Nature* **228**, 847.

Coles, P. and Barrow, J. D. (1987). *Mon. Not. Royal Astron. Soc.* **228**, 407.

Colless, M., Dalton, G., Maddox, S. *et al.* (2001). *Mon. Not. Royal Astron. Soc.* **328**, 1039.

Colley, W. N., Gott, J. R. III and Park, C. (1996). *Mon. Not. Royal Astron. Soc.* **281L**, 82.

Cooray, A. and Hu, W. (2000). *Astrophys. J.* **534**, 533.

Corbelli, E. and Salucci, P. (1999). astro-ph/9909252.

Crane, P., Hegyi, D. J., Mandolesi, N. and Danks, A. C. (1986). *Astrophys. J.* **309**, 12.

Crittenden, R. G., Coulson, D. and Turok, N. G. (1995). *Phys. Rev. D* **52**, 5402.

Davis, M. and Peebles, P. J. E. (1983). *Ann. Rev. Astron. Astrophys.* **21**, 109.

de Amici, G., Smoot, G. F., Friedman, S. D. and Witebsky, C. (1985). *Astrophys. J.* **298**, 710.

de Amici, G., Limon, M., Smoot, G. F. *et al.* (1991). *Astrophys. J.* **381**, 341.

de Bernardis, P., Ade, P. A. R., Bock, J. J. *et al.* (2000). *Nature* **404**, 955.

de Freitas-Pacheco, J. A. and Peirani, S. (2004). *Int. J. Mod. Phys. D* **13**, 1335.

de Oliveira-Costa, A. and Tegmark, M. (1999). *Microwave Foregrounds*, ASP Conference Series, vol. 181 (San Francisco: Astronomical Society of the Pacific).

de Vaucouleurs, G. (1982). *The Observations* **102**, 178.

Dekkel, A., Eldar, A., Kolatt, T. *et al.* (1999). *Astrophys. J.* **522**, 1.

Delabruille, J. (2004). *Astrophys. Space Sci.* **290**, 87.

Dell'Antonio, J. P. and Rybicki, G. B. (1993). In Chincarini, G., Iovino, A., Maccacaro, T. and Maccagni, D., eds, *Observational Cosmology*, ASF Conference Series, vol. 51 (San Francisco: Astronomical Society of the Pacific), p. 548.

Dicke, R. H., Peebles, P. J. E., Roll, P. G. and Wilkinson, D. T. (1965). *Astrophys. J.* **142**, 414.

Djorgovski, S. G., Castro, S. M., Stern, D. and Mahabal, A. (2001). *Astrophys. J.* **142**, 414.

D'Odorico, S., Dessauges-Zavadsky, M. and Molaro, P. (2001). *Astron. Astrophys.* **368**, L21.

Dolgov, A. D. and Sommer-Larsen, J. (2001). *Astrophys. J.* **551**, 608.

Dolgov, A. D., Doroshkevich, A. G., Novikov, D. I. and Novikov, I. D. (1999). *Int. J. Mod. Phys. D* **8** (2), 189.

Dolgov, A. D., Hansen, S. H., Pastor, S. and Semikoz, D. V. (2001). *Astrophys. J.* **559**, 123.

Doroshkevich, A.G. (1970). *Astrofiz.* **6**, 581.

Doroshkevich, A. G. (1985). *Pis'ma Astron. Zh.* **11**, 723.

Doroshkevich, A. G. and Naselsky, P. D. (2002). *Phys. Rev. D* **65**, 123517.

Doroshkevich, A. G. and Novikov, I. D. (1964). *Acad. Sci. USSR Doklady* **154**, 809.

Doroshkevich, A. G., Lukash, V. N. and Novikov, I. D. (1974). *Astron. Zh.* **51**, 554.

Doroshkevich, A. G., Novikov, I. D. and Polnarev, A. G. (1977). *Astron. Zh.* **54**, 932.

Doroshkevich, A. G., Zeldovich, Ya. B. and Sunyaev, R. A. (1978). *Astron. Zh.* **22**, 523.

Doroshkevich, A. G., Naselsky, I. P., Naselsky, P. D. and Novikov, I. D. (2003). *Astrophys. J.* **586**, 709.

Dwek, E. and Arendt, R. G. (1998). *Astrophys. J.* **508**, L9.

Efstathiou, G. (2002). *Mon. Not. Royal Astron. Soc.* **332**, 193.

Efstathiou, G. (2003a). astro-ph/0303127.

Efstathiou, G. (2003b). *Mon. Not. Royal Astron. Soc.* **343**, L95.

Eisenstein, D. J., Zehavi, I., Hogg, D. W. *et al.* (2005). *Astrophys. J.* **633**, 560.

Epstein, R., Lattimer, J. and Schramm, D. N. (1976). *Nature* **263**, 198.

Eriksen, H. K., Novikov, D. I., Lilje, P. B., Banday, A. J. and Gorski, K. M. (2004). *Astrophys. J.* **612**, 64.

Esposito, S. (1999). astro-ph/9904411.

Ewing, M. S., Burke, B. F., Staelin, D. M. *et al.* (1967). *Phys. Rev. Lett.* **19**, 1251.

Fabricant, D., Beers, T. C., Geller, M. J., Gorenstein, P., Huchra, J. P., Kurtz, M. J. (1986). *Astrophys. J.* **308**, 530.

Fan, X., White, R. L., Dawis, M. *et al.* (2000). astro-ph/0005414.

Fan X., Narayan, V. K., Strauss, M. A. *et al.* (2002). *Astron. J.* **123**, 1247.

Ferrarese, L., Mould, J. R., Kennicutt, R. C. *et al.* (1999). astro-ph/9908192.

Ferreira, P. G., Magueijo J. C. R. and Gorski, K. M. (1998). *Astrophys. J.* **503**, L1.

Fich, M., Blitz, L. and Stark, A. A. (1989). *Astrophys. J.* **342**, 272.

Field, G. B. and Hitchcock, J. L. (1966). *Phys. Rev. Lett.* **16**, 817.

Finkbeiner, D. P. (2003). *Astrophys. J. Suppl.* **146**, 407.

Fix, J. D., Craven, J. D. and Frank, L. A. (1989). *Astrophys. J.* **345**, 203.

Fixsen, D. J., Cheng, E. S., Gales, J. M., Mather, J. C., Shafer, R. A. and Wright, E. L. (1996). *Astrophys. J.* **515**, 512.

Fixsen, D. J., Hinshaw, G., Bennet, K. and Mather, J. (1997). *Astrophys. J.* **486**, 623.

Franx, M. and Tonry, J. (1999). *Astrophys. J.* **506**, 1778.

Freedman, W. L., ed. (2004). *Measuring and Modelling the Universe*, Carnegie Astrophysics Series, vol. 2 (Cambridge: Cambridge University Press).

Freedman, W. L., Madore, B. F., Gibson, B. K. *et al.* (2001). *Astrophys. J.* **553**, 47.

Fukuda, Y., Hayakawa, T., Ichihara, E. *et al.* (Super-Kamiokande Collaboration) (1998). *Phys. Rev. Lett.* **81**, 1562.

Fukugita, M. (2000). astro-ph/0005069.

Fukugita, M., Hogan, C. J. and Peebles, P. J. E. (1998). *Astrophys. J.* **503**, 518.

Gamow, G. (1946). *Phys. Rev.* **70**, 527.

Gehrels, N. and Cheng, W. (1996). *Astron. Astrophys. Suppl.* **120**, 331.

Gibson, B. K., Stetson, P. B., Freedman, W. L. *et al.* (2000). *Astrophys. J.* **529**, 723.

Gispert, R., Lagache, G. and Puget, J. L. (2000). *Astron. Astrophys.* **360**, 1.

Gorski, K., Silk, J. and Vittorio, N. (1992). *Phys. Rev. D Lett.* **68**, 733.

Gott, J. R. III, Park, C., Juszkiewicz, R. *et al.* (1990). *Astrophys. J.* **352**, 1.

Gradstein, I. S. and Ryzhik, I. M. (1994). *Tables of Integrals, Series and Products* (New York: Academic Press Inc.).

Greisen, K. (1966). *Phys. Rev. Lett.* **16**, 748.

Grischuk, L. P. (1974). *Zh. Eksp. Teor. fiz.* **67**, 825.

Gunn J. E. and Peterson, B. A. (1965). *Astrophys. J.* **142**, 1633.

Gurvitz, L. I. and Mitrofanov, I. G. (1986). *Nature* **324**, 349.

Gush, H. P., Halpern, M. and Wishnow, E. H. (1990). *Phys. Rev. Lett.* **65**, 537.

Guth, A. H. (1981). *Phys. Rev. D* **23**, 347.

Hadwiger, H. (1957). *Vorlesungen über Inhalt, Oberfläche und Isoperimetrie* Berlin: Springer Verlag.

Haiman, S. and Knox, L. (1999). In de Oliviera, A. and Tegmark, M., eds, *Microwave Foregrounds*, ASP Conf. Series 181 (San Fransisco: ASP).

Halpern, M. and Scott, D. (1999). In de Oliviera, A. and Tegmark, M., eds, *Microwave Foregrounds* (San Francisco: ASP), P. 283.

Hamilton, A. J. S. (1998). In Hamilton, D., ed., *The Evolving Universe* (Dordrecht: Kluwer), p. 185.

Hamuy, M., Phillips, M. M., Suntzeff, N B., Schommer, R. A., Maza, J. and Aviles, R. (1996). *Astron. J.* **112**, 2391.

Hanany, S., Ade, P., Balbi, A. *et al.* (2000a). astro-ph/0005123.

Hanany, S., Ade, P., Balbi, A. *et al.* (2000b). *Astrophys. J.* **545**, L5.

Harrison, E. R. (1970). *Phys. Rev. D* **27**, 26.

Hasinger, G. and Zamorani, G. (1997). astro-ph/9712341.

Hauser, M. G. (1998). *Astron. Astrophys. Suppl.* **193**, 6202.

Hauser, M. G., Kelsall, T., Moseley, S. H. Jr *et al.* (1991). In Hoet, S., Bennett, C. L. and Trimble, V., eds, *After the First Three Minutes.* AIP Conf. Proc. 222 (New York: AIP), p. 61.

Hauser, M. G., Arendt, R. G., Kelsall, T. *et al.* (1998). *Astrophys. J.* **508**, 25.

Hawking, S. W. (1971). *Mon. Not. Royal Astron. Soc.* **152**, 75.

Hawking, S. W. (1974). *Nature* **248**, 30.

Hawking, S. W. (1982). *Phys. Lett. B* **115**, 195.

Hayashi, C. (1950). *Prog. Theor. Phys.* **5**, 224.

Hayashida, N., Honda, K., Honda, M. *et al.* (1994). *Phys. Rev. Lett.* **73**, 3491.

Heavens, A. and Sheth, R. (1999). *Mon. Not. Royal Astron. Soc.* **305**, 527.

Henry, R. C. and Murthy, J. (1996). In Calzetti, D., Livio, M. and Madau, P., eds, *Extragalactic Background Radiation*. Proceedings of the Extragalactic Background Radiation Meeting, Baltimore 1993 (Cambridge: Cambridge University Press), p. 51.

Hinshaw, G., Barnes, C. and Bennett, C. L. (2003). ApJS **148**, 63.

Howell, T. F. and Shakeshaft, J. R. (1966). *Nature* **210**, 1318.

Howell, T. F. and Shakeshaft, J. R. (1967). *Nature* **216**, 753.

Hu, W. (1995) Ph.D. thesis (UC Berkeley) astro-ph 9508126.

Hu, W. (2002). *Phys. Rev. D* **65**, 3003.

Hu, W. (2003). *Ann. Phys.* **303**, 203.

Hu, W. and Silk, J. (1993). *Phys. Rev. D* **48**, 2.

Hu, W. and Sugiyama, N. (1994). *Phys. Rev. D* **50**, 627.

Hu, W. and Sugiyama, N. (1995). *Astrophys. J.* **444**, 489.

Hu, W. and White, M. (1996). *Astrophys. J.* **471**, 30.

Hu, W. and White, M. (1997a). astro-ph/9706147.

Hu, W. and White, M. (1997b). *New Astronomy* **2**, 323.

Hu, W., Scott, D. and Silk, J. (1994). *Phys. Rev. D* **49**, 648.

Hu, W., Scott, D., Sugiyama, N. and White, M. (1995a). astro-ph/9505043.

Hu, W., Scott, D., Sugiyama, N. and White, M. (1995b). *Phys. Rev. D* **52**, 5498.

Hu, W., Sugiyama, N. and Silk, J. (1996). astro-ph/9604166.

Hu, W., Sugiyama, N. and Silk, J. (1997). *Nature* **386**, 37.

Hughes, Y. P. (1989). *Astrophys. J.* **337**, 21.

Hui, L., Haiman, Z., Zaldarriaga, M. and Alexander, T. (2002). *Astrophys. J.* **564**, 525.

Hummer, D. G. (1994). *Mon. Not. Royal Astron. Soc.* **268**, 109.

Hurwitz, M., Bowyer, S. and Martin, C. (1990). In Bowyer, S. and Leinet, C., eds, *The Galactic and Extragalactic Background Radiation*, Proc. IAU 139 (Dordrecht: Kluwer Academic), p. 229.

Illarionov, A. F. and Sunyaev, R. A. (1975a). *Astron. Zh.* **18**, 413.

Illarionov, A. F. and Sunyaev, R. A. (1975b). *Astron. Zh.* **18**, 691.

Itoh, N. Kawana, Y., Nozawa, S. and Kohyama, Y. (2001). *Mon. Not. Royal Astron. Soc.* **327**, 567.

Ivanchuk, A. V., Orlov, A. D. and Varshalovich, D. A. (2001). *Pis'ma Astron. Zh.* **27**, 615.

Ivanov, P., Naselsky, P. and Novikov, I. (1994). *Phys. Rev. D* **50**, 71731.

Izotov, Y. I. and Thuan, T. X. (1998). *Astrophys. J.* **500**, 188.

Izotov, Y. I., Thuan, T. X. and Lipovetsky, V. A. (1994). *Astrophys. J.* **435**, 647.

Jakobsen, P., Bowyer, S., Kimble, R. *et al.* (1984). *Astron. Astrophys.* **139**, 481.

Jansen, J., Tonry, J. and Blakeslee, J. (2004). In Freedman, W., ed., *Measuring and Modelling the Universe*, Carnegie Observatories Astrophysics Series, vol. 2 (Cambridge: Cambridge University Press), P. 99.

Jeans, J. H. (1902). *Phil. Trans.* **129**, 44.

Jha, S., Garnavich, P. M., Kirshner, R. P. *et al.* (1999). *Astrophys. J. Suppl.* **125**, 73.

Johnson, D. G. and Wilkinson, D. T. (1987). *Astrophys. J. Lett.* **313**, L1.

Jones, B. J. T. and Wyse, R. F.G (1985). *Astron. Astrophys.* **149**, 144.

Jørgensen, H., Kotok, E., Naselsky, P. and Novikov, I. (1993). *Mon. Not. Royal Astron. Soc.* **265**, 639.

Jørgensen, H., Kotok, E., Naselsky, P. and Novikov, I. (1995). *Astron. Astrophys.* **294**, 639.

Joubert, M. N., Mashou, J. L., Lequeux, J., Deharveng, J. M. and Cruvellier, P. (1983). *Astron. Astrophys.* **128**, 114.

Kaidanovsky, M. L. and Parijskij, Yu. N. (1987). *Istoriko Astronomicheskie Issledovaniya* (Moscow: Nauka), P. 59.

Kaiser, N. (1983). *Mon. Not. Royal Astron. Soc.* **202**, 1169.

Kaiser, N. (1992). *Astrophys. J.* **388**, 1272.

Kaiser, M. E. and Wright, E. L. (1990). *Astrophys. J. Lett.* **356**, L1

Kamionkowski, M. and Kosowsky, A. (1998). *Phys. Rev. D* **57**, 685.

Kamionkowski, M., Kosowsky, A. and Stebbins, A. (1997a). *Phys. Rev. Lett.* **78**, 2058.

Kamionkowski, M., Kosowsky, A. and Stebbins, A. (1997b). *Phys. Rev. D* **55**, 7368.

Kappadath, S. C., McConnell, M., Ryan, J. *et al.* (1999). *Bull. Am. Astron. Soc.* **31**, 737.

Kardashev, N. S. (1967). *Astron. Tsirkulyar* no. 430.

Karzas, W. J. and Latter, R. (1961). *Astrophys. J. Suppl.* **6**, 167.

Kendall, M. G. and Stuart, A. (1977). *The Advanced Theory of Statistics*, 4th edn (London: Charles Griffin).

Kelson, D. D., Illingworth, G. D., Tonry, J. L. *et al.* (2000). *Astrophys. J.* **529**, 768.

Kerr, F. N. and Lynden-Bell, D. (1986). *Mon. Not. Royal Astron. Soc.* **221**, 1023.

Kirkman, D., Tytler, D., Suzuki, N., O'Meara, J. and Lubin, D. (2003). *Astrophys. J. Suppl.* **149**, 1.

Klypin, A. A., Sazhin, M. M., Strukov, I. A. and Skulachev, D. P. (1987). *Pis'ma Astron. Zh.* **13**, 104.

Knop, R. A., Aldering, G., Amanullah, R. *et al.* (2003). *Astrophys. J.* **598**, 102.

Knox, L. (1995). *Phys. Rev. D* **52**, 4307.

Kodama, H. and Sasaki, M. (1984). *Prog. Theor. Phys.* **78**, 1.

Kofman, L. and Linde, A. (1987). *Nucl. Phys. B* **282**, 555.

Kofman, L. and Starobinsky, A. A. (1985). *Sov. Astron. Lett.* **9**, 643.

Kogut, A., Bensadoun, M., De Amici, G. *et al.* (1990). *Astrophys. J.* **355**, 102.

Kogut, A., Linewear, C., Smoot, G., Bennet, K. and Banday, A. (1993). astro-ph/9312056.

Kogut, A., Banday, A. J., Bennett, C. L. *et al.* (1996a). *Astrophys. J.* **464**, L29.

Kogut, A., Banday, A. J., Bennett, C. L. *et al.* (1996b). *Astrophys. J.* **470**, 653.

Kolb, E. W. and Turner, M. S. (1989). *The Early Universe* (Reading, MA: Addison-Wesley).

Komatsu, E. and Seljak, U. (2002). *Mon. Not. Royal Astron. Soc.* **336**, 1256.

Komatsu, E., Kogut, A., Nolta, M. R. *et al.* (2003). *Astrophys. J. Suppl.* **148**, 119.

Kompaneets, A. S. (1957). *Zh. Eksp. Teor. Fiz.* **4**, 730.

Kompaneets, D. A., Lukkash, V. N. and Novikov, I. D. (1982). *Astron. Zh.* **59**, 424.

Koopmans, L. V. E. and Fassnacht, C. D. (1999). *Astrophys. J.* **527**, 513.

Kosowsky, A. (1999). astro-ph/9904102.

Kotok, E. V., Naselsky, P. D., Novikov, D. I. (1995). *Mon. Not. Royal Astron. Soc.* **273**, 376.

Kotok, E. V., Novikov, D. I., Naselsky, P. D. and Novikov, D. I. (2001). *Int. J. Mod. Phys. D* **10**, 501.

Krolik, J. H. (1990). *Astrophys. J.* **353**, 21.

Lachiez-Rey, M. and Gunzig, E. (1999). *The Cosmological Background Radiation* (Cambridge: Cambridge University Press).

Lagache, G., Abergel, A., Bonlanger, F., Desert, F. X. and Puget, J.-L. (1998). *Astron. Astrophys.* **344**, 322.

Land, K. and Magueijo, J. (2005). astro-ph/0502574.

Landau, L. D. and Lifshits, E. M. (1962). *Teoriya Polya* (Moscow: Nauka).

Landau L. D. and Lifshits, E. M. (1984). *Fizicheskaya Kinetika* (Moscow: Nauka).

Landau, S., Harari, D. and Zaladarriaga, M. (2001). *Phys. Rev. D* **63**, 3505.

Landy, S. D. and Szalay, A. (1993). *Astrophys. J.* **412**, 64L.

Larson, D. L. and Wandelt, B. D. (2005). astro-ph/0505046.

Lawrence, M. A., Reid, R. J. O. and Watson, A. A. (1991). *J. Phys. G. Nucl. Part. Phys.* **17**, 733; for details, see http://ast.leeds.ac.uk/haverah/hav-home.html.

Lepp, S., Stancil, P. C. and Dalgarno, A. (1998). *Mem. Soc. Astron. It.* **69**, 331.

Levin, S. M., Witebsky, C., Bensadoun, M. *et al.* (1988). *Astrophys. J.* **334**, 14.

Levin, S. M., Bensadoun, M. Bersanelli, M. *et al.* (1992) *Astrophys. J.* **396**, 3.

Levshakov, S. A., Agafonova, I. I., D'Odorico, S., Wolfe, A. M. and Dessauges-Zavadsky, M. (2003). *Astrophys. J.* **582**, 596.

Liddle, A. (2003). *An Introduction to Modern Cosmology*, 2nd edn (Chichester: John Wiley and Sons).

Lifshits, E. M. (1946). *Zh. Eksp. Teor. Fiz.* **16**, 587.

Lifshits, E. M. and Khalatnikov, I. M. (1960). *Zh. Eksp. Teor. Fiz.* **39**, 149.

Lightman, A. P. (1981). ApJ **244**, 392.

Linde, A. D. (1984). *JETP* **40**, 1333L.

Linde, A. D. (1990). *Fizika Elementarnyh Chastits and Inflyatsionnaya Kosmologiya* (Moscow: Nauka).

Linde, A. D. and Mukhanov, V. F. (1997). *Phys. Rev. D* **56**, 535.

Linsky, J. L. (1998). *Space Sci. Rev.* **84**, 285.

Longair, M. S. (1993). In Calzetti, D., Livio, M. and madau, P., eds, *Extragalactic Background Radiation.* Proceedings of the Extragalactic Background Radiation Meeting, Baltimore 1993. (Cambridge: Cambridge University Press), pp. 223–234.

Longair, M. S. and Sunyaev, R. A. (1969). *Nature* **223**, 719.

Lubarsky, Y. E. and Sunyaev, R. A. (1983). *Astrophys. & Space Sci.* **123**, 171.

Lukash, V. N. (1980). *Zh. Eksp. Teor. Fiz.* **79**, 1601.

Lukash, V. N. and Mikheeva, (1998). In Mueller, V., Gottloeber, S., Muecket, J.P. and Wambsganss, J., eds, *Large Scale Structure: Tracks and Traces.* Proceedings of the 12th Potsdam Cosmology Workshop, 15–19 September, 1997 (Singapore: World Scientific), p. 381.

Lyubimov, V. A., Novikov, E. G., Nozik, V. Z., Tret'yakov, E. F. and Kozik, V. S. (1980). (Moscow: Preprint ITEF-62).

Ma, C.-P. and Bertschinger, E. (1995). *Astrophys. J.* **455**, 7.

Mahaffy, P. R., Donahue, T. M., Atreya, S. K., Owen, T. C. and Niemann, H. B. (1998). *Space Sci. Rev.* **84**, 251.

Mandolesi, N., Calzolare, P., Cortiglioni, S. *et al.* (1986). *Astrophys. J.* **310**, 561.

Martin, C. and Bowyer, S. (1990). *Astrophys. J.* **350**, 242.

Mason, B. S., Myers, S. T. and Readhead, A. C. S. (2001). *Astrophys. J.* **555**, L11.

Mason, B. S., Pearson, T. J., Readhead, A. C. S. *et al.* (2003). *Astrophys. J.* **591**, 540.

Mayers, S. T., Baker, Y. E., Readhead, A. C. S., Leitch, E. M. and Herbig, T. (1997). *Astrophys. J.* **485**, 1.

Mecke, K. R., Buchert, T. and Wagner, H. (1994). *Astron. Astrophys.* **288**, 697.

Melchiori, A. and Vittorio, N. (1996). astro-ph/9610029.

Melchiori, A., Sazhin, M. V., Shulga, V. V. and Vittorio, N. (1999). astro-ph/9901220.

Melchiori, A., Mersini, L., Ödman, C. J. and Trodden, M. (2003). *Phys. Rev. D* **68**, 43509.

Melchiori, B. and Melchiori, F. (1994). *Riv. Nuova Chim.* **17** (1), 1.

Meyer, D. M. and Jura, M. (1985). *Astrophys. J.* **297**, 119.

Meyer, S. S., Cheng, E. S. and Page L. A. (1989). *Astrophys. J. Lett.* **343**, L1.

Mihalas, D. M. (1978). *Stellar Atmospheres* (San Francisco: W. H. Freeman & Co.).

Millea, R., McColl, M., Pedersen, R. J. and Vernon, F. L. (1971). *Phys. Rev. Lett.* **26**, 919.

Minkowski, H. (1903). *Math. Annal.* **57**, 443.

Miralda-Escode, J. and Ostriker, J. (1990). *Astrophys. J.* **350**, 1.

Miyaji, T., Ishisaki, Y., Ogasaka, Y. *et al.* (1998). *Astron. Astrophys.* **334**, L13.

Molaro, P., Primas, F. and Bonifacio, A. (1995). *Astron. Astrophys.* **295**, L47.

Mukhanov, V. F. (2003). astro-ph/0303072.

Mukhanov, V. F. and Chibisov, G. V. (1981). *JETP Lett.* **33**, 523.

Mukhanov, V. F. and Chibisov, G. V. (1982). *Sov. Phys. JETP* **56**, 258.

Mukhanov, V. F. and Steinhard, P. (1998). *Phys. Lett. B* **422**, 52.

Mukhanov, V. F., Feldman, H. A. and Branderberger, R. H. (1992). *Phys. Rev.* **215**, 203.

Mulchaey, J. S., Davis, D. S., Mushotzky, R. F. and Burstein, D. (1996). *Astrophys. J.* **456**, 80.

Naselsky, P. D. (1978). *Pis'ma Astron. Zh.* **4**, 387.

Naselsky, P. D. and Chiang, L.-Y. (2004). *Phys. Rev. D* **69**, 123518.

Naselsky, P. D. and Novikov, D. (1995). *Astrophys. J. Lett.* **444**, L1.

Naselsky, P. D. and Novikov, D. (1998). *Astrophys. J.* **507**, 31.

Naselsky, P. and Novikov, I. (1993). *Astrophys. J.* **413**, 14.

Naselsky, P. and Novikov, I. (2002). *Mon. Not. Royal Astron. Soc.* **334**, 137.

Naselsky, P. D. and Polnarev, A. G. (1987). *Astrofiz.* **26**, 543.

Naselsky, P., Schmalzing, J., Sommer-Larsen, J. and Hannestad, S. (2001). astro-ph/0102378.

Naselsky, P. D., Doroshkevich, A. G. and Verkhodanov, O. V. (2003). *Astrophys. J. Lett.* **599**, L53.

Naselsky, P. D., Chiang, L.-Y., Olesen, P. and Verkhodanov, O. V. (2004). *Astrophys. J.* **615**, 45.

Naselsky, P. D., Chiang, L.-Y., Olesen, P. and Novikov, I. (2005). *Phys. Rev. D* **72**, 3512.

Ng, K. L. and Ng, K. W. (1995). *Phys. Rev. D* **51**, 364.

Ng, K. L. and Ng, K. W. (1996). *Astrophys. J. Lett.* **456**, L1.

Nordberg, E. and Smoot, G. (1998). astro-ph/9805123.

Novikov, D. I. and Jørgensen, H. E. (1996a). *Int. J. Mod. Phys. D* **5**, 319.

Novikov, D. I. and Jørgensen, H. E. (1996b). *Astrophys. J.* **471**, 521.

Novikov, D. I., Feldman, H. and Shandarin, S. (1999). *Int. J. Mod. Phys. D* **8**, 291.

Novikov, D. I., Schmalzing, J. and Mukhanov, V. (2000). *Astron. Astrophys.* **364**, 17.

Novikov, D. I., Naselsky, P. D., Jorgensen, H. E., Christensen, P. R., Novikov, I. D. and Norgaarrd-Nielsen, H. U. (2001). *Int. J. Mod. Phys. D* **10**, 245.

Novikov, D. I., Colombi, S. and Dore, O. (2003). (astro-ph/0307003).

Novikov, I. D. (1964). *Zh. Exsp. Teor. Fiz.* **46**, 686.

Novikov, I. D. (1968). *Astron. Zh.* **45**, 538.

Novikov, I. D. (2001). In Martinez, V. J., Trimble, V. and Pons-Bordea, M. J., eds, *Historical Development of Modern Cosmology*, ASP Conference Proceedings, vol. 252 (San Francisco: Astronomical Society of the Pacific), p. 43.

Ohm, L. A. (1961). *Bell Syst. Tech. J.* **40**, 1065.

Olive, K. A. (2000). astro-ph/0009475.

Olive, K. A. and Steigman, G. (1995). *Astrophys. J. Suppl.* **97**, 49.

Olive, K. A., Steigman, G. and Walker, T. P. (2000). *Phys. Rev. D* **333**, 389.

O'Meara, J. M., Tytler, D., Krikman, D., Suzuki, N., Prodaska, J. X., Lubin, D. and Wolf, A. M. (2001). *Astrophys. J.* **552**, 718.

Onaka, T. (1990). In Bowyer, S. and Leinet, C., eds, *The Galactic and Extragalactic Background Radiation*, Proc. IAU 139 (Dordrecht: Kluwer Academic), p. 379.

Padmanabhan, T. (1996). *Cosmology and Astrophysics Through Problems* (Cambridge: Cambridge University Press).

Page, L. A. (2000). The Wilkinson Microwave Anisotropy Probe. In Freedman, W. L., *Measuring and Modelling the Universe* (Cambridge: Cambridge University Press), p. 330.

Pagel, B. E. J., Simonson, E. A., Terlevich, R. J. and Edminds, M. (1992). *Mon. Not. Royal Astron. Soc.* **255**, 325.

Palazzi, E., Mandolesi, N., Crane, P., Kutner, M. L., Blades, J. C. and Hegyi, D. J. (1990). *Astrophys. J.* **357**, 14.

Palazzi, E., Mandolesi, N. and Crane, P. (1992). *Astrophys. J.* **398**, 53.

Parese, F., Margon, B., Bowyer, S. and Lampton, M. (1979). *Astrophys. J.* **230**, 304.

Partridge, R. B. (1995). *3K: The Cosmic Microwave Background Radiation* (Cambridge: Cambridge University Press).

Peacock, J. A. (1997). *Mon. Not. Royal Astron. Soc.* **284**, 885.

Peacock, J. A. (1999). *Cosmological Physics* (Cambridge: Cambridge University Press).

Peacock, J. A. and Dodds, S. J. (1996). *Mon. Not. Royal Astron. Soc.* 280L, 19.

Peacock, J. A., Cole, S., Norberg, P. *et al.* (2001). *Nature* **410**, 169.

Peebles, P. J. E. (1968). *Astrophys. J. Lett.* **153**, 1.

Peebles, P. J. E. (1971). *Physical Cosmology.* (Princeton: Princeton University Press).

Peebles, P. J. E. (1980). *The Large Scale Structure of the Universe* (Princeton: Princeton University Press).

Peebles, P. J. E. (1981). *Astrophys. J. Lett.* **248**, 885.

Peebles, P. J. E. (1983). *Astrophys. J.* **274**, 1.

Peebles, P. J. E. (1985). *Astrophys. J.* **297**, 350.

Peebles, P. J. E. (1993). *Principles of Physical Cosmology* (Princeton: Princeton University Press).

Peebles, P. J. E. (1999a). In Dekel, A. and Ostriker, J. P., eds, *Formation of Structure in the Universe* (Cambridge: Cambridge University Press), p. 435.

Peebles, P. J. E. (1999b). In Harwitana, M. and Hauser, M. G., eds, *Extragalactic Infrared Background and its Cosmological Implications*, IAU Symposium, vol. 204 (Dordrecht: Kluwer Academic).

Peebles, P. J. E. and Yu, J. T. (1970). *Astrophys. J.* **162**, 815.

Peebles, P. J. E., Seager, S. and Hu, W. (2000). astro-ph/0004389.

Peebles, P. J. E., Seager, S. and Hu, W. (2001). *Astrophys. J. Lett.* **539**, L1.

Penzias, A. (1979). *Rev. Mod. Phys.* **51**, 430.

Penzias, A. A. and Wilson, R. W. (1965). *Astrophys. J.* **142**, 491.

Penzias, A. A. and Wilson, R. W. (1967). *Astron. J.* **72**, 315.

Pequignot, D., Petitjean, P. and Boisson, C. (1991). *Astron. Astrophys.* **251**, 680.

Perlmutter, S., Aldering, G., Goldhaber, G. *et al.* (1999). *Astrophys. J.* **517**, 565.

Perlmutter, S. and Schmidt, B. (2003). astro-ph/0303428.

Pettini, M. and Bowen, D. V. (2001). *Astrophys. J.* **560**, 41.

Pierpaoli, E., Borgani, S., Scott, D. and White, M. (2003). *Mon. Not. Royal Astron. Soc.* **342**, 163.

Pointecouteau, E., Giard, M., Benoit, A. *et al.* (1999). *Astrophys. J. Lett.* **519**, L115.

Polarski, D. and Starobinsky, A. A. (1994). *Phys. Rev. D* **50**, 6123.

Polenta, G., Ade, P. A. R., Bock, J. J. *et al.* (2002). *Astrophys. J. Lett.* **572**, L27.

Polnarev, A. G. (1985). *Astron. Zh.* **29**, 607.

Popa, L., Burigana, C., Finelli, F. and Mandolesi, N. (2000). *Astron. Astrophys.* **363**, 825.

Pozzetti, L., Madau, P., Zamorani, G., Ferguson, H. C. and Bruzual, A. G. (1998). *Mon. Not. Royal Astron. Soc.* **298**, 1133.

Protheroe, R. J. and Biermann, P. L. (1996). *Astropart. Phys.* **6**, 45.

Rao, S. and Briggs, F. (1993). *Astrophys. J.* **419**, 515.

Rauch, M., Miralde-Escude, J., Sargent, W. L. W., *et al.* (1997) *Astrophys. J.* **489**, 7.

Readhead, A. C. S., Myers, S. T., Pearson, T. J. *et al.* (2004). *Science* **306**, 836.

Rees, M. J. (1968). *Astrophys. J. Lett.* **153**, L1.

Rees, M. J. and Sciama, D. W. (1968). *Nature* **217**, 511.

Reese, E. D. (2004). In Freedman, W., ed., *Measuring and Modelling the Universe*, Carnegie Astrophysics Series, vol. 2 (Cambridge: Cambridge University Press), p. 138.

Reiprich, T. H. and Böhringer, H. (2002). *Astrophys. J.* **567**, 716.

Rephaeli, Y. (1995). *Ann. Rev. Astron. Astrophys.* **33**, 541.

Rephaeli, Y. (2001). astro-ph/0110510.

Rice, S. O. (1944). *Bell Syst. Tech. J.* **23**, 282.

Rice, S. O. (1945). *Bell Syst. Tech. J.* **24**, 41.

Ricotti, M., Gnedin, N. and Shull, M. (2001). astro-ph/0110431.

Riess, A. G., Press, W. H. and Kirshner, R. P. (1995). *Astrophys. J.* **438**, L17.

Riess, A. G., Filippenko, A. V., Challis, P. *et al.* (1998). *Astron. J.* **116**, 1009.

Roberts, M. S. and Haynes, M. P. (1994). *Ann. Rev. Astron. Astrophys.* **32**, 115.

Roll, P. G. and Wilkinson, D. T. (1966). *Phys. Rev. Lett.* **16**, 405.

Roth, K. C., Meyer, D. M. and Hawkins, I. (1993). *Astrophys. J.* **413**, L67.

Ryan, S. G., Norris, J. E. and Beers, T. C. (1999). *Astrophys. J.* **523**, 654.

Ryan, S. G., Beers, T. C., Olive, K. A., Fields, B. D. and Norris, J. E. (2000). *Astrophys. J. Lett.* **530**, L57.

Rybicki, G. B. (1984). In Kalkofen, W., ed., *Methods in Radiative Transfer* (Cambridge: Cambridge University Press), chap. 3.

Rybicki, G. B. and Dell'Antonio, I. P. (1994). *Astrophys. J.* **427**, 603.

Saha., A., Sandage, A., Tammann, G. A., Labhardt, L., Macchetto, F. D. and Paragia, N. (1999). *Astrophys. J.* **522**, 803.

Sachs, R. K. and Wolfe, A. M. (1967). *Astrophys. J.* **147**, 73.

Sakai, S., Mould, J. R., Hughes, S. M. G. *et al.* (2000). *Astrophys. J.* **529**, 698.

Sakharov, A. D. (1965). *Zh. Eksp. Teor. Fiz.* **49**, 345.

Sakharov, A. D. (1999). *Nauchnye Trudy* (Moscow: Nauka), p. 213.

Sandage, A. and Tamman, G. A. (1982). *Astrophys. J.* **256**, 339.

Sazin, M. V. (1985). *Mon. Not. Royal Astron. Soc.* **216**, 25.

Schmalzing, J. and Buchert, T. (1997). ApJ 482L.

Schmalzing, J. and Gorski, K. M. (1998). *Mon. Not. Royal Astron. Soc.* **297**, 355.

Schmidt, B. P., Eastman, R. G. and Kirshner, R. (1994). *Astrophys. J.* **432**, 42.

Schmidt, B. P., Suntzeff, N. B., Phillips M. M. (1998). *Astrophys. J.* **506**, 46.

Schmidt, M. (1965). *Astrophys. J.* **141**, 1295.

Scott, D. (1999a) astro-ph/9911325.

Scott, D. (1999b) astro-ph/9912038.

Schuecker, P., Bohringer, H., Collins, C. A. and Guzzo, L. (2003). *Astron. Astrophys.* **398**, 867.

Seager, S., Sasselov, D. and Scott, D. (1999a). astro-ph/9912182.

Seager, S., Sasselov, D. and Scott, D. (1999b). *Astrophys. J. Lett.* **523**, L1.

Seager, S., Sasselov, D. D. and Scott, D. (2000). *Astrophys. J. Suppl.* **128**, 407.

Seljak, U. (1996a) *Astrophys. J.* **463**, 1.

Seljak, U. (1996b) *Astrophys. J.* **482**, 6.

Seljak, U. and Zaldarriaga, M. (1997). *Phys. Rev. Lett.* **78**, 2054.

Seljak, U. and Zaldarriaga, M. (1998). astro-ph/9805010.

Shandarin, S. F., Doroshkevich, A. G. and Zeldovich, Ya. B. (1983). *Sov. Phys. Uspekhi* **139**, 336.

Shklovsky, I. S. (1965). *Astron. Zh.* **42**, 893.

Shklovsky, I. S. (1966). *Astron. Zh.* **43**, 747.

Shmaonov, T. (1957). *Pribory i tekhnika eksperimenta* **1**, 83.

Sigl, G. (2001a). hep-ph/0109202.

Sigl, G. (2001b). astro-ph/0104291.

Silk, J. (1968). *Astrophys. J.* **151**, 459.

Silk, J. and Wilson, M. L. (1981). *Astrophys. J. Lett.* **244**, L37.

Simon, A. C. (1978). *Mon. Not. Royal Astron. Soc.* **180**, 429.

Sironi, G., Limon, M., Marcellino, G. *et al.* (1990). *Astrophys. J.* **357**, 301.

Sironi, G., Bonelli, G., Limon, M. (1991). *Astrophys. J.* **378**, 550.

Skillman, E. and Kennicutt, R. C. (1993). *Astrophys. J.* **411**, 655.

Skillman, E., Terlevich, R. J., Kennicutt, R. C., Garnett, D. R. and Terlevich, E. (1994). *Astrophys. J.* **431**, 172.

Skillman, E., Terlevich, E. and Terlevich, R. (1998). *Space Sci. Rev.* **84**, 105.

Smoot, G. and Davidson, K. (1993). *Wrinkles in Time* (New York: William Morrow).

Smoot, G. and Scott, D. (1997). astro-ph/9711069.

Sommer-Larsen, E. and Dolgov, A. (1999). astro-ph/9912166.

Sommer-Larsen, J., Naselsky, P. D., Novikov, I. D. and Gotz, M. (2004). *Mon. Not. Royal Astron. Soc.* **351**, 125.

Spergel, D., Verde, L., Peiris, H. *et al.* (2003a). *Astrophys. J. Suppl.* **148**, 39.

Spergel, D., Verde, L., Peiris, H. *et al.* (2003b). *Astrophys. J. Suppl.* **148**, 175.

Spitzer, L. and Greenstein, J. L. (1951). *Astrophys. J.* **114**, 407.

Sreekumar, P., Bertsch, D. L., Dingus, B. L. *et al.* (1998). *Astrophys. J.* **494**, 523.

Staggs, S., Jarosik, N. C., Meyer, S. S. *et al.* (1996a). *Astrophys. J.* **458**, 407.

Staggs, S., Jarosik, N. C., Wilkinson, D. T. and Wollack, E. J. (1996b). *Astrophys. J.* **473**, L1.

Stankevich, K. S., Wielebinski, R. and Wilson, W. E. (1970). *Austral, J. Phys.* **23**, 529.

Starobinsky, A. A. (1979). *JETP Lett.* **30**, 682.

Starobinsky, A. A. (1980). *Phys. Lett. B* **91**, 99.

Starobinsky, A. A. (1982). *Phys. Lett. B* **117**, 175.

Starobinsky, A. A. (1983). *Pis'ma Astron. Zh.* **9**, 302.

Starobinsky, A. A. (1985a). *Pis'ma Astron. Zh.* **11**, 133.

Starobinsky, A. A. (1985b). *Pis'ma Astron. Zh.* **11**, 643.

Starobinsky, A. A. (1988). *Sov. Astron. Lett.* **14**, 166.

Starobinsky, A. A. (1992a). *Pis'ma Astron. Zh.* **11**, 133.

Starobinsky, A. A. (1992b). *JETP Lett.* **55**, 489.

Stebbins, A. (1996). astro-ph/9609149.

Stokes, R. A., Partridge, R. B. and Wilkinson, D. T. (1967). *Phys. Rev. Lett.* **19**, 1199.

Strong, A. W., Moskalenko, I. V. and Reimer, O. (2004). *Astrophys. J.* **613**, 962S.

Strukov, I. A. and Skulachev, D. P. (1984). *Pis'ma Astron. Zh.* **10**, 3.

Suntzeff., N. B., Phillips, M. M., Covarrubias, R. *et al.* (1999). *Astron. J.* **117**, 1175.

Sunyaev, R. A. and Zeldovich, Ya. B. (1970a). *Space Sci.* **7**, 3.

Sunyaev, R. A. and Zeldovich, Ya. B. (1970b). *Space Sci.* **9**, 368.

Sunyaev, R. A. and Zeldovich, Ya. B. (1972). *Astron. Astrophys.* **20** 189.

Sunyaev, R. A. and Zeldovich, Ya. B. (1980). Ann. Rev. *Astron. Astrophys.* **18**, 537.

Takeda, M., Hayashida, N., Honda, K. *et al.* (1998). *Phys. Rev. Lett.* **81**, 1163.

Tammann, G. A. (1999). In Klapdor-Kleingrothaus, R. and Baudis, L., eds, *Proceedings of the 2nd International Conference on Dark Matter in Astrophysics and Particle Physics*, Heidelberg, Germany, July 20–25, 1998 (Philadelphia, PA: Institute of Physics), p. 153.

Taytler, D., O'Meara, J., Suzuki, N. and Lubin, D. (2000). *Physics Scripta* **T85**, 12.

Tegmark, M., Silk, J. and Blanchard, A. (1994). *Astrophys. J.* **420**, 484.

Tegmark, M. and Zaldarriaga, M. (2000). *Phys. Rev. Lett.* **85**, 2240.

Tegmark, M., de Oliviera-Costa, A. and Hamilton, A. (2003). *Phys. Rev. D* **68**, 123523.

Tennyson, P. D., Henry, R. C., Feldman, P. D. and Hartig, G. F. (1988). *Astrophys. J.* **330**, 435.

Thaddeus, P. (1972). *Annual Rev. Astron. Astrophys.* **10**, 305.

Thaddeus, P. and Clauser, J. F. (1966). *Phys. Rev. Lett.* **16**, 819.

Thomas, D., Shramm, D., Olive, K. and Fields, B. (1993). *Astrophys. J.* **406**, 669.

Tonry, J. L., Blakesley, J. P., Ajhar, E. A. and Dressler, A. (2000). *Astrophys. J.* **530**, 625.

Tonry, J. L., Schmidt, B. P., Barris, B. *et al.* (2003). *Astrophys. J.* **594**, 1.

Turok, N. (1996). *Astrophys. J.* **473**, L5.

Tytler, D., Fan, X.-M. and Burles, S. (1996). *Nature* **381**, 207.

Udomprasert, P. S., Mason, B. S. and Readhead, A. C. S. (2000). *Bull. Am. Astron. Soc.* **34**, 1142.

Varshalovich, D. A., Ivanchuk, A. V. and Potekhin, A. Yu. (1999). *Zh. Exsp. Teor. Fiz.* **144**, 1001.

Verner, D. A. and Ferland, G. J. (1996). *Astrophys. J. Suppl.* **103**, 467.

Viana, P. T. P. and Liddle, A. R. (1999). astro-ph/9902245.

Vielva, P., Martinez-Gonsales, E., Cayon, L., Diego, J. M., Sanz, J. L. and Toffolatti, L. (2001). *Mon. Not. Royal Astron. Soc.* **326**, 181.

Vishniak, E. T. (1987). *Astrophys. J.* **322**, 597.

Vittorio, N. and Silk, J. (1984). *Astrophys. J. Lett.* **285**, L39.

Wagoner, R. V. (1973). *Astrophys. J.* **179**, 343.

Wang, X., Tegmark, M. and Zaldarriaga, M. (2002). *Phys. Rev. D* **65**, 123001.

Wang, X., Tegmark, M. and Zaldarriaga, M. (2001). *Phys. Rev. D* **65**, 123001.

Weinberg, S. (1972). *Gravitation and Cosmology* (New York: Wiley).

Weinberg, S. (1977). *The First Three Minutes: A Modern View of the Origin of the Universe* (New York: Basic Books).

Weinberg, S. (2001a). *Phys. Rev. D* **64**, 123511.

Weinberg, S. (2001b). *Phys. Rev. D* **64**, 123512.

Welch, W. J., Keachie, S., Thornton, D. D. and Wrixon, G. (1967). *Phys. Rev. Lett.* **18**, 1068.

Weller, C. S. (1983). *Astrophys. J.* **268**, 899.

White, S. D. M. and Fabian, A. C. (1995). *Mon. Not. Royal Astron. Soc.* **273**, 72.

White, S. D. M., Efstathiou, G. and Frenk, C. S. (1993). *Mon. Not. Royal Astron. Soc.* **262**, 1023.

White, M. and Hu, W. (1996). astro-ph/9606138.

Willis, T. D. (2002). astro-ph/0201515.

Wilson, M. L. (1983). *Astrophys. J.* **273**, 2.

Wilson, M. L. and Silk, J. (1981). *Astrophys. J.* **243**, 14.

Winitzki, S. and Kosowsky, A. (1997). *New Astron.* **3**, 75.

Woosley, S. and Weaver, T. (1986). In Mihalas, D. and Winkler, K. H., eds, *Radiation Hydrodynamics*, IAU Colloquium 89 (Dordrecht: Reidel), p. 91.

Wright, E. L. (1979). *Astrophys. J.* **232**, 348.

Wu, J. H. P., Balbi, A., Borrill, J. *et al.* (2001a). *Astrophys. J. Suppl.* **132**, 1.

Wu, J. H. P., Balbi, A., Borrill, J. *et al.* (2001b). *Phys. Rev. Lett.* **87**, 251303.

Yahil, A., Tammann, G. A. and Sandage, A. (1977). *Astrophys. J.* **217**, 903.

Yoshida, S. and Dai, H. (1998). *J. Phys. G.* **24**, 905.

Yoshida, S., Hayashida, N., Honda, K. *et al.* (1995). *Astropart. Phys.* **3**, 105.

Zabotin, N. A. and Naselsky, P. D. (1982a). *Astron. Zh.* **42**, 893.

Zabotin, N. A. and Naselsky, P. D. (1982b). *Pis'ma Astron. Zh.* **8**, 67.

Zabotin, N. A. and Naselsky, P. D. (1983). *Astron. Zh.* **9**, 335.

Zabotin, N. A. and Naselsky, P. D. (1985). *Astron. Zh.* **29**, 614.

Zaldarriaga, M. (1996). astro-ph/960805.

Zaldarriaga, M. (1997). *Phys. Rev. D* **55**, 1822.

Zaldarriaga, M. (2004). In Freedman, W., ed. *Measuring and Modelling the Universe*, Carnegie Astrophysics Series, vol. 2 (Cambridge: Cambridge University Press), p. 309.

Zaldarriaga, M. and Harari, D. (1995). *Phys. Rev. D* **52**, 3276.

Zaldarriaga, M. and Seljak, U. (1997). *Phys. Rev. D* **55**, 1830.

Zaldarriaga, M. and Seljak, U. (1998). *Phys. Rev. D* **58**, 023003.

Zaldarriaga, M., Seljak, U. and Bertshinger, E. (1998). *Astrophys. J.* **494**, 491.

Zamorani, G. (1993). In Calzetti, D., Livio, M. and Madau, P., eds, *Extragalactic Background Radiation*. Proceedings of the Extragalactic Background Radiation Meeting, Baltimore 1993 (Cambridge: Cambridge University Press), p. 37.

Zaritsky, D., Smoth, R., Frenk, C. and White, S. D. M. (1997). *Astrophys. J.* **478**, 39.

Zatsepin, G. T. and Kuzmin, V. A. (1966). *Pis'ma Zh. Eksp. Teor. Fiz.* **4**, 114.

Zeldovich, Ya. B. (1970). *Astron. Astrophys.* **5**, 84.

Zeldovich Ya. B. and Novikov, I. D. (1966). *Astron. Zh.* **43**, 758.

Zeldovich, Ya. B. and Novikov, I. D. (1983). *Relativistic Astro-physics*, vol. II (Chicago: University of Chicago Press).

Zeldovich, Ya. B. and Sunyaev, R. A. (1969). *Astrophys. & Space Sci.* **4**, 301.

Zeldovich, Ya. B. and Sunyaev, R. A. (1970). *Astrophys. & Space Sci.* **7**, 20.

Zeldovich, Ya. B., Kurt, V. G. and Sunyaev, R. A. (1969). *JETF* **28**, 146.

Zwicky, F. (1957). Morphological Astronomy (Berlin: Springer).

Index

Printed in the United States
By Bookmasters